MONOGRAPHS AND RESEARCH NOTES IN MATHEMATICS

Nonlinear Reaction-Diffusion-Convection Equations

Lie and Conditional Symmetry, Exact Solutions, and Their Applications

Roman Cherniha
Mykola Serov
Oleksii Pliukhin

CRC Press
Taylor & Francis Group
Boca Raton London New York

CRC Press is an imprint of the
Taylor & Francis Group, an **informa** business

A CHAPMAN & HALL BOOK

MONOGRAPHS AND RESEARCH NOTES IN MATHEMATICS

Series Editors

John A. Burns
Thomas J. Tucker
Miklos Bona
Michael Ruzhansky

Published Titles

Actions and Invariants of Algebraic Groups, Second Edition, Walter Ferrer Santos and Alvaro Rittatore

Analytical Methods for Kolmogorov Equations, Second Edition, Luca Lorenzi

Application of Fuzzy Logic to Social Choice Theory, John N. Mordeson, Davender S. Malik and Terry D. Clark

Blow-up Patterns for Higher-Order: Nonlinear Parabolic, Hyperbolic Dispersion and Schrödinger Equations, Victor A. Galaktionov, Enzo L. Mitidieri, and Stanislav Pohozaev

Bounds for Determinants of Linear Operators and Their Applications, Michael Gil'

Complex Analysis: Conformal Inequalities and the Bieberbach Conjecture, Prem K. Kythe

Computation with Linear Algebraic Groups, Willem Adriaan de Graaf

Computational Aspects of Polynomial Identities: Volume I, Kemer's Theorems, 2nd Edition Alexei Kanel-Belov, Yakov Karasik, and Louis Halle Rowen

A Concise Introduction to Geometric Numerical Integration, Fernando Casas and Sergio Blanes

Cremona Groups and Icosahedron, Ivan Cheltsov and Constantin Shramov

Delay Differential Evolutions Subjected to Nonlocal Initial Conditions Monica-Dana Burlică, Mihai Necula, Daniela Roşu, and Ioan I. Vrabie

Diagram Genus, Generators, and Applications, Alexander Stoimenow

Difference Equations: Theory, Applications and Advanced Topics, Third Edition Ronald E. Mickens

Dictionary of Inequalities, Second Edition, Peter Bullen

Elements of Quasigroup Theory and Applications, Victor Shcherbacov

Finite Element Methods for Eigenvalue Problems, Jiguang Sun and Aihui Zhou

Integration and Cubature Methods: A Geomathematically Oriented Course, Willi Freeden and Martin Gutting

Introduction to Abelian Model Structures and Gorenstein Homological Dimensions Marco A. Pérez

Iterative Methods without Inversion, Anatoly Galperin

Iterative Optimization in Inverse Problems, Charles L. Byrne

Line Integral Methods for Conservative Problems, Luigi Brugnano and Felice Iavernaro

Lineability: The Search for Linearity in Mathematics, Richard M. Aron, Luis Bernal González, Daniel M. Pellegrino, and Juan B. Seoane Sepúlveda

Published Titles Continued

Modeling and Inverse Problems in the Presence of Uncertainty, H. T. Banks, Shuhua Hu, and W. Clayton Thompson

Monomial Algebras, Second Edition, Rafael H. Villarreal

Noncommutative Deformation Theory, Eivind Eriksen, Olav Arnfinn Laudal, and Arvid Siqveland

Nonlinear Functional Analysis in Banach Spaces and Banach Algebras: Fixed Point Theory Under Weak Topology for Nonlinear Operators and Block Operator Matrices with Applications, Aref Jeribi and Bilel Krichen

Nonlinear Reaction-Diffusion-Convection Equations: Lie and Conditional Symmetry, Exact Solutions, and Their Applications, Roman Cherniha, Mykola Serov, and Oleksii Pliukhin

Optimization and Differentiation, Simon Serovajsky

Partial Differential Equations with Variable Exponents: Variational Methods and Qualitative Analysis, Vicenţiu D. Rădulescu and Dušan D. Repovš

A Practical Guide to Geometric Regulation for Distributed Parameter Systems Eugenio Aulisa and David Gilliam

Reconstruction from Integral Data, Victor Palamodov

Signal Processing: A Mathematical Approach, *Second Edition*, Charles L. Byrne

Sinusoids: Theory and Technological Applications, Prem K. Kythe

Special Integrals of Gradshteyn and Ryzhik: the Proofs – Volume I, Victor H. Moll

Special Integrals of Gradshteyn and Ryzhik: the Proofs – Volume II, Victor H. Moll

Spectral and Scattering Theory for Second-Order Partial Differential Operators, Kiyoshi Mochizuki

Stochastic Cauchy Problems in Infinite Dimensions: Generalized and Regularized Solutions, Irina V. Melnikova

Submanifolds and Holonomy, Second Edition, Jürgen Berndt, Sergio Console, and Carlos Enrique Olmos

Symmetry and Quantum Mechanics, Scott Corry

The Truth Value Algebra of Type-2 Fuzzy Sets: Order Convolutions of Functions on the Unit Interval, John Harding, Carol Walker, and Elbert Walker

Variational-Hemivariational Inequalities with Applications, Mircea Sofonea and Stanislaw Migórski

Willmore Energy and Willmore Conjecture, Magdalena D. Toda

Forthcoming Titles

Groups, Designs, and Linear Algebra, Donald L. Kreher

Handbook of the Tutte Polynomial, Joanna Anthony Ellis-Monaghan and Iain Moffat

Microlocal Analysis on R^n and on NonCompact Manifolds, Sandro Coriasco

Practical Guide to Geometric Regulation for Distributed Parameter Systems, Eugenio Aulisa and David S. Gilliam

CRC Press
Taylor & Francis Group
6000 Broken Sound Parkway NW, Suite 300
Boca Raton, FL 33487-2742

© 2018 by Taylor & Francis Group, LLC
CRC Press is an imprint of Taylor & Francis Group, an Informa business

No claim to original U.S. Government works

Printed on acid-free paper
Version Date: 20170921

International Standard Book Number-13: 978-1-4987-7617-2 (Hardback)

This book contains information obtained from authentic and highly regarded sources. Reasonable efforts have been made to publish reliable data and information, but the author and publisher cannot assume responsibility for the validity of all materials or the consequences of their use. The authors and publishers have attempted to trace the copyright holders of all material reproduced in this publication and apologize to copyright holders if permission to publish in this form has not been obtained. If any copyright material has not been acknowledged please write and let us know so we may rectify in any future reprint.

Except as permitted under U.S. Copyright Law, no part of this book may be reprinted, reproduced, transmitted, or utilized in any form by any electronic, mechanical, or other means, now known or hereafter invented, including photocopying, microfilming, and recording, or in any information storage or retrieval system, without written permission from the publishers.

For permission to photocopy or use material electronically from this work, please access www.copyright.com (http://www.copyright.com/) or contact the Copyright Clearance Center, Inc. (CCC), 222 Rosewood Drive, Danvers, MA 01923, 978-750-8400. CCC is a not-for-profit organization that provides licenses and registration for a variety of users. For organizations that have been granted a photocopy license by the CCC, a separate system of payment has been arranged.

Trademark Notice: Product or corporate names may be trademarks or registered trademarks, and are used only for identification and explanation without intent to infringe.

Library of Congress Cataloging-in-Publication Data

Names: Cherniha, Roman, author. | Serov, Mykola, author. | Pliukhin, Oleksii, author.
Title: Nonlinear reaction-diffusion-convection equations : Lie and conditional symmetry, exact solutions, and their applications / by Roman Cherniha, Mykola Serov, Oleksii Pliukhin.
Description: Boca Raton, Florida : CRC Press, [2018] | Includes bibliographical references and index.
Identifiers: LCCN 2017028035| ISBN 9781498776172 (hardback) | ISBN 9781315154848 (e-book) | ISBN 9781498776196 (e-book) | ISBN 9781351650878 (e-book) | ISBN 9781351641364 (e-book)
Subjects: LCSH: Reaction-diffusion equations. | Reaction-diffusion equations--Numerical solutions. | Differential equations, Nonlinear | Differential equations, Partial. | Symmetry (Mathematics)
Classification: LCC QA377 .C444 2018 | DDC 515/.3534--dc23
LC record available at https://lccn.loc.gov/2017028035

Visit the Taylor & Francis Web site at
http://www.taylorandfrancis.com

and the CRC Press Web site at
http://www.crcpress.com

Contents

Preface	ix
List of Figures	xv
List of Tables	xvii

1 Introduction — 1

- 1.1 Nonlinear reaction-diffusion-convection equations in mathematical modeling — 1
- 1.2 Main methods for exact solving nonlinear reaction-diffusion-convection equations — 3
- 1.3 Lie symmetry of differential equations: historical review, definitions and properties — 8

2 Lie symmetries of reaction-diffusion-convection equations — 19

- 2.1 Symmetry of the linear diffusion equation — 19
- 2.2 Symmetry of the nonlinear diffusion equation — 21
- 2.3 Equivalence transformations and form-preserving transformations — 29
  - 2.3.1 The group of equivalence transformations — 30
  - 2.3.2 Form-preserving transformations — 32
- 2.4 Determining equations for reaction-diffusion-convection equations — 36
- 2.5 Complete description of Lie symmetries of reaction-diffusion-convection equations — 39
  - 2.5.1 Principal algebra of invariance — 39
  - 2.5.2 Necessary conditions for nontrivial Lie symmetry — 39
  - 2.5.3 Lie symmetry classification via the Lie–Ovsiannikov algorithm — 49
  - 2.5.4 Application of form-preserving transformation — 60
- 2.6 Nonlinear equations arising in applications and their Lie symmetry — 69
  - 2.6.1 Heat (diffusion) equations with power-law nonlinearity — 69
  - 2.6.2 Diffusion equations with a convective term — 71

Contents

    2.6.3 Nonlinear equations describing three types of transport mechanisms 73

3 Conditional symmetries of reaction-diffusion-convection equations   77

3.1 Conditional symmetry of differential equations: historical review, definitions and properties 77

3.2 Q-conditional symmetry of the nonlinear heat equation . . . 83

3.3 Determining equations for finding Q-conditional symmetry of reaction-diffusion-convection equations 87

3.4 Q-conditional symmetry of reaction-diffusion-convection equations with constant diffusivity 92

3.5 Q-conditional symmetry of reaction-diffusion-convection equations with power-law diffusivity 100

    3.5.1 The case of proportional diffusion and convection coefficients 101

    3.5.2 The case of different diffusion and convection coefficients 107

3.6 Q-conditional symmetry of reaction-diffusion-convection equations with exponential diffusivity 111

    3.6.1 Solving the nonlinear system (3.166) 118

    3.6.2 Solving the nonlinear system (3.169) 125

3.7 Nonlinear equations arising in applications and their conditional symmetry 129

4 Exact solutions of reaction-diffusion-convection equations and their applications   135

4.1 Classification of exact solutions from the symmetry point of view 135

4.2 Examples of exact solutions for some well-known nonlinear equations 138

4.3 Solutions of some reaction-diffusion-convection equations arising in biomedical applications 143

    4.3.1 The Fisher and Murray equations 143

    4.3.2 The Fitzhugh–Nagumo equation and its generalizations 146

4.4 Solutions of reaction-diffusion-convection equations with power-law diffusivity 156

    4.4.1 Lie's solutions of an equation with power-law diffusion and convection 156

    4.4.2 Non-Lie solutions of some equations with power-law diffusion and convection 159

4.5 Solutions of reaction-diffusion-convection equations with exponential diffusivity 176

	4.5.1	Lie's solutions of an equation with exponential diffusion and convection	176
	4.5.2	Non-Lie solutions of an equation with exponential diffusion and convection	180
	4.5.3	Application of the solutions obtained for population dynamics	188

5 The method of additional generating conditions for constructing exact solutions   191

5.1	Description of the method and the general scheme of implementation	191
5.2	Application of the method for solving nonlinear reaction-diffusion-convection equations	195
	5.2.1 Reduction of the nonlinear equations (5.10) and (5.11) to ODE systems	196
	5.2.2 Exact solutions of the nonlinear equations (5.10) and (5.11)	201
	5.2.3 Application of the solutions obtained for solving boundary-value problems	211
5.3	Analysis of the solutions obtained and comparison with the known results	216

References   219

Index   239

Preface

Second-order partial differential equations (PDEs) have played a crucial role in mathematical modeling a wide range of processes in natural and life sciences since the 18th century. Typically linear PDEs are used in order to describe various processes, while *nonlinear partial differential equations* have been widely involved for such purposes only since the beginning of the 20th century. Nowadays it is generally accepted that a huge number of real world processes arising in physics, biology, chemistry, material sciences, engineering, ecology, economics etc. can be adequately described only by nonlinear PDEs.

At the present time, there is no existing general theory for integration of nonlinear PDEs, hence construction of particular exact solutions for these equations remains an important mathematical problem. Finding exact solutions that have a clear interpretation for the given process is of fundamental importance. In contrast to linear PDEs, the well-known principle of linear superposition cannot be applied to generate new exact solutions for nonlinear PDEs. Thus, the classical methods for solving linear PDEs are not applicable to nonlinear PDEs. Of course, a change of variables can sometimes be found that transforms a nonlinear PDE into a linear equation (the classical example is the Cole–Hopf substitution for the Burgers equation). It was stated by W.F. Ames in 1965 that "transformations are perhaps the most powerful general analytic tool currently available in this area". However, finding exact solutions of a large majority of nonlinear PDEs requires new methods. Nowadays, 50 years later, the most powerful methods for construction of exact solutions to nonlinear PDEs are the symmetry-based methods, in particular the Lie method and the method of nonclassical (i.e., non-Lie) symmetries.

The Lie method (the terminology "the Lie symmetry analysis" and "the group analysis" are also used) is based on finding Lie's symmetries of a given PDE and using the symmetries obtained for the construction of exact solutions. The method was created by the prominent Norwegian mathematician Sophus Lie in the 1880s. It should be pointed out that Lie's works on application Lie groups for solving PDEs were almost forgotten during the first half of the 20th century. In the end of the 1950s, L.V. Ovsiannikov inspired by Birkhoff's works devoted to application of Lie groups in hydrodynamics, rewrote Lie's theory using modern mathematical language and published a monograph in 1962, which was the first book (after Lie's works) devoted fully to this subject. The Lie method was essentially developed by L.V. Ovsiannikov, W.F. Ames, G. Bluman, W.I. Fushchych, N. Ibragimov, P. Olver, and other researchers in the 1960s–1980s. Several excellent textbooks devoted to the Lie method were published during the last 30 years, therefore one may claim that it is the well-established theory at the present time. Notwithstanding the method still attracts the attention of many researchers and new results are published on a regular basis. In particular, solving the so-called problem of group classification (Lie symmetry classification) still remains a highly nontrivial task and such problems are not solved for several classes of PDEs arising in real world applications.

On the other hand, it is well-known that some nonlinear RDC equations arising in applications have a "poor" Lie symmetry. For example, the Fisher and Fitzhugh–Nagumo equations, which are widely used in mathematical biology, are invariant only under the time and space translations. The Lie method is not efficient for such equations since it enables one to construct only those exact solutions, which can be obtained without using this cumbersome algorithm. Taking into account this fact, one needs to apply other approaches for solving such equations. The best known among them is the method of nonclassical symmetries proposed by G. Bluman and J. Cole in 1969. Although this approach was suggested almost 50 years ago its successful applications for solving nonlinear equations were accomplished only in the 1990s owing to D.J. Arrigo, P. Broadbridge, P. Clarkson, J.M. Hill, E.L. Mansfield, M.C. Nucci, P. Olver, E. Pucci, G. Saccomandi, E.M. Vorob'ev, P. Winternitz and others. A prominent role in applications and further development of the nonclassical symmetry method belongs to the Ukrainian school of symmetry analysis, which was created in the early 1980s and led by W.I. Fushchych (V.I. Fyshchich) until 1997 when he passed away. In particular, a concept of conditional symmetry was worked out and its applications to a wide range of nonlinear PDEs were realized by M. Serov, I. Tsyfra, R. Zhdanov, R. Popovych, R. Cherniha and others. Notably, following Fushchysh's proposal dating back to 1988, we continuously use the terminology "Q-conditional symmetry" instead of "nonclassical symmetry".

We also note that several other approaches for solving nonlinear PDEs (in particular, evolution equations) were independently suggested in the 1990s–2000s. Not pretending to completeness and precise statement, the following of them should be mentioned: the method of linear invariant subspaces, the method of generalized conditional symmetries, the method of heir-equations, the method of linear determining equations, the method of additional generating conditions etc. Notwithstanding some of these methods formally do not use any symmetries, a deep analysis shows that they are related to symmetry-based methods.

The main mathematical object of this book is the class of nonlinear reaction-diffusion-convection equations (RDC). In our opinion, nonlinear RDC equations possess the most important role among other nonlinear equations. One cannot imagine a correct mathematical model describing heat and mass transfer, filtration of liquid, solute transport in tissue, diffusion in chemical reactions, tumor growth and many other processes without RDC equations. The importance of RDC equations in real world applications follows from the fact that they model three main transport mechanisms: diffusion (heat transfer), reaction (source/sink), and convection (advection). Thus, they have been extensively studied by means of different mathematical methods and techniques, including symmetry-based methods.

This book is devoted to (i) search Lie and Q-conditional (non-classical) symmetries of nonlinear RDC equations; (ii) constructing exact solutions using the symmetries obtained and using the method of additional generating conditions; (iii) applications of the solutions derived for solving some biologically and physically motivated problems.

The monograph summarizes in a unique way the results derived by the authors during the last 20 years. Notably, the first joint paper was written by R. Cherniha and M. Serov about 20 years ago during our stay at the Mathematisches Forschungsinstitut Oberwolfach (Germany), while the last joint paper of M. Serov and O. Pliukhin was published in 2015. It should be pointed out that a number of misprints, inexactness and even mistakes arising in our papers were corrected during the book preparation, and new unpublished results were included in Chapters 3, 4 and 5. Moreover, our results are supplemented by those obtained by other authors. As a result, the reader will realize a huge progress, which has been done in study of nonlinear RDC equations by means of symmetry-based methods since the 1990s.

The book presents a most complete (at the present time) description of Lie and conditional symmetries for nonlinear RDC equations, which are very common in real world applications. The most interesting subclasses from this class (like equations with power-law and exponential nonlinearities) are extensively studied. In particular, an essential stress is made on finding symmetries and exact solutions for the widely used equations in bio-medical applications, including Fisher, Murray, Fitzhugh-Nagumo and Kolmogorov-Petrovskii-Piskunov type equations. Concerning the equations listed above and their generalization, a number of examples are presented, in which the relevant real world models are analytically solved, and a biological/physical interpretation of the solutions obtained is provided.

In Introduction (Chapter 1), some mathematical models based on nonlinear RDC equations are discussed, methods for constructing exact solutions of nonlinear PDEs are briefly presented together with a short historical review. The remaining part of this chapter is devoted to the main notions, definitions, and theorems, which form theoretical background of the Lie method and other symmetry-based methods.

Chapter 2 is partly devoted to the linear and nonlinear diffusion (heat) equations, including the multi-dimensional case. Here we present the well-known results of the Lie symmetry classification (the group classification) together with some applications for constructing exact solutions. The main part of Chapter 2 is devoted to the complete Lie symmetry classifications (LSC) of the general class of RDC equations

$$u_t = [A(u)u_x]_x + B(u)u_x + C(u),$$

where $u = u(t,x)$ is an unknown function, $A(u), B(u), C(u)$ are arbitrary smooth functions and the subscripts t and x denote differentiation with respect to these variables. First, LSC is derived using the well-known Lie-Ovsiannikov approach based on the equivalence transformations. Afterwards the second LSC is obtained via so-called form-preserving transformations. An extensive discussion including nontrivial examples is presented to show advantages of application of the form-preserving transformations in order to solve the LSC problem.

Chapter 3 is devoted non-Lie symmetries of nonlinear PDEs. First, we present a historical review concerning conditional symmetry of PDEs, introduce notion of Q-conditional symmetry (nonclassical symmetry) and repeat some well-known results about non-Lie symmetry of nonlinear diffusion equations. The main part of Chapter 3 is devoted to the Q-conditional symmetry classifications of the general class of RDC equations. In contrast to the LSC problem, the result is incomplete, however, a complete classification is derived for several important (from applicability point of view) subclasses of the general class. Probably the most important among them is the Burgers type equations of the form

$$u_t = u_{xx} + \lambda u u_x + C(u), \quad \lambda \in \mathbb{R}.$$

In fact, the above class of equations contains as particular cases the Fisher, Murray, and Fitzhugh–Nagumo equations and their natural generalizations used widely in modeling of biomedical and ecological processes.

Chapter 4 is fully devoted to construction of exact solutions. A wide range of RDC equations are examined in order to search for both Lie and non-Lie solutions. Several examples are presented, which show how nontrivial exact solutions can be constructed for some well-known nonlinear equations. In particular, our attention is addressed to the nonlinear RDC equations with constant and power-law diffusivities, arising in bio-medical and ecological applications. For such equations, we construct exact solutions, examine their properties and (in some cases) provide their biological interpretation. The RDC equations with exponential nonlinearities are also under study in this section.

Chapter 5 is devoted to the method of additional generating conditions and its application for solving some nonlinear RDC equations. This method can be treated as a particular case of the method of differential constraints. Basic ideas of the method of differential constraints have roots in Darboux's works. In the 1960s, N.N. Yanenko formalized the method using the modern mathematical language. However, he has not provided any constructive algorithm for finding the compatible differential constraints for a PDE in question. The method of additional generating conditions solves this problem for PDEs, which can be reduced to those with quadratic nonlinearities. The chapter contains a detailed description of the method, examples demonstrating its efficiency in the case of the nonlinear RDC equations with power-law and exponential nonlinearities. An extensive comparison of the solutions obtained with those derived via other techniques is also presented.

Chapters 2, 3 and 4, which form the main part of this monograph and are essentially connected each with another, were written by all the authors and they contributed equally to these chapters.

Chapters 1 and 5 were written by R. Cherniha.

The book is a monograph. Its academic level suits graduate students and higher. Some parts of the book may be used in "Mathematical Biology" and "Nonlinear Partial Differential Equations" courses for master students and in the final year of undergraduate studies. Nowadays such courses are common in all leading universities over the world.

The book was typeset in LaTeX using the CRC Press templates, the figures were drawn using the computer algebra package `Maple` and some calculations were done using `Mathematica`.

Last but not the least, we are grateful to our colleagues and our teacher Wilhelm Fushchych (1936–1997), who was the supervisor for R. Cherniha and M. Serov in the 1980s. This book could not have been written without his innovative scientific ideas and many years of his support. The authors thank their Ukrainian colleagues for fruitful discussions, valuable critique, and helpful suggestions, which helped us to write this modest work. Especially, we are grateful to Vasyl' Davydovych, Sergii Kovalenko, Inna Rassokha, and Valentyn Tychynin.

R. Cherniha is indebted to John R. King for valuable discussions in 2013–2015, when he was Marie Curie Fellow at the University of Nottingham. R. Cherniha is also grateful to Malte Henkel, Phil Broadbridge, Changzheng Qu, and Jacek Waniewski for fruitful discussions and valuable comments, which inspired us for further research and generated new ideas.

Finally, we would like to thank our families for their incredible patience and support for this book project. Especially, each of us wants very much to reaffirm his love and thankfulness to his wife Nataliya, Mariya, and Antonina, respectively.

Kyiv, Ukraine
Poltava, Ukraine

Roman Cherniha
Mykola Serov, Oleksii Pliukhin

List of Figures

4.1	Exact solution (4.33)	146
4.2	Exact solution (4.72)	153
4.3	Exact solution (4.100)	162
4.4	Exact solution (4.102)	163
4.5	Exact solution (4.107)	164
4.6	Exact solution (4.112)	165
4.7	Exact solution (4.112)	166
4.8	Exact solution (4.112)	167
4.9	Exact solution (4.112)	168
4.10	Exact solution (4.140)	173
4.11	Exact solution (4.142)	174
4.12	Exact solution (4.142)	175
4.13	Exact solution (4.202)	190
5.1	Exact solution from Table 5.1 case 3	208
5.2	Exact solution from Table 5.1 case 3	211
5.3	Exact solution from Table 5.2 case 1	212
5.4	Exact solution from Table 5.3 case 1	213
5.5	Exact solution (5.98)	214

List of Tables

2.1	All possible extensions of the algebra A^{pr} of equations from the PDE class (2.9).	25
2.2	All possible extensions of the algebra A^{pr} of equations from the PDE class (2.29).	27
2.3	All possible extensions of the algebra A^{pr} of equations from the PDE class (2.27).	28
2.4	Simplification of the RDC equations from Theorem 2.9 using ETs (2.36)	48
2.5	The complete LSC of equations of the form (2.35) using the group of ETs $\mathcal{E}$	50
2.6	Simplification of the RDC equations from Table 2.5 by means of FPTs	61
2.7	The complete LSC of equations of the form (2.35) using FPTs	63
3.1	A complete list of exact solutions of the nonlinear system (3.166)	124
3.2	A complete list of the solutions $a = -\partial_x \ln|\gamma|$, $\alpha = -\partial_t \ln|\gamma|$ of system (3.169)	130
4.1	Lie's ansätze and reduction equations for Eq. (4.9).	140
4.2	Lie's ansätze and reduction equations for Eq. (4.87).	157
4.3	Lie's ansätze and reduction equations for Eq. (4.146).	177
5.1	Exact solutions of the nonlinear RDC equation (5.10) with $\alpha \neq -2$ and $\lambda = 0$.	209
5.2	Exact solutions of the nonlinear RDC equation (5.10) with $\alpha = -2$ and $\lambda = 0$.	210
5.3	Exact solutions of the nonlinear RDC equation (5.11).	210

Acronyms

DE(s) – determining equation(s)

Eq(s). – equation(s)

ET(s) – equivalence transformation(s)

FN – Fitzhugh–Nagumo

FPT(s) – form-preserving transformation(s)

KPP – Kolmogorov–Petrovskii–Piskunov

LSC – Lie symmetry classification

MAGC – method of additional generating conditions

MAI(s) – maximal algebra(s) of invariance

MGI – maximal group of invariance

ODE – ordinary differential equation

QSC – Q-conditional symmetry classification

PDE – partial differential equation

RD – reaction-diffusion

RDC – reaction-diffusion-convection

w.r.t. – with respect to

Chapter 1

Introduction

1.1	Nonlinear reaction-diffusion-convection equations in mathematical modeling	1
1.2	Main methods for exact solving nonlinear reaction-diffusion-convection equations	3
1.3	Lie symmetry of differential equations: historical review, definitions and properties	8

1.1 Nonlinear reaction-diffusion-convection equations in mathematical modeling

Since the 17th century when G. Leibnitz and I. Newton discovered differential and integral calculus, differential equations are the most powerful tools for mathematical modeling various processes in physics, chemistry, biology, medicine, ecology, economics etc. Of course, pioneering models were created in order to express some classical laws (like the second Newton law) in physics and astronomy, later differential equations came to be used for describing a wide range of processes not only in physics but also in other natural and life sciences. It can be noted that almost all mathematical models created before the end of the 19th century were based on ordinary differential equations (ODEs) and linear partial differential equations (PDEs). *Nonlinear PDEs* are widely used in mathematical modeling real world processes since the beginning of the 20th century only. Probably, one of the first attempts in applying and solving a nonlinear PDE of the parabolic type was made by J. Boussinesq who studied the porous diffusion equation describing the water filtration in soil [32]. One of the first applications of nonlinear reaction-diffusion (RD) equations in biology was proposed by R.A. Fisher [98, 99].

Nowadays, i.e., 100 years later, it is generally accepted that a huge number of real processes arising in physics, biology, chemistry, material sciences, engineering, ecology, economics etc. can be adequately described only by *nonlinear PDEs* (or systems of such equations). The most widely used type of equations for modeling such processes are the nonlinear reaction-diffusion-convection (RDC) (advection) equations. Since 1952 when A.C. Turing published the remarkable paper [240], in which he proposed a revolutionary idea about mechanism of morphogenesis (the development of structures in an organism during the life), nonlinear RD equations (including those with convective terms) play a crucial role in real world applications and have been extensively studied by means of different mathematical methods/techniques. As a result, in the 1970s several monographs were published, which are devoted to study and application of the nonlinear reaction-diffusion-convection (RDC) equations in physics [5, 6, 158], biology [97, 179] and chemistry [10, 11]. In our opinion, these books had a great impact attracting many scholars to study the nonlinear RDC equations and use them for modeling real world processes. Since that time many other excellent monographs and textbooks appeared, especially for models related to life sciences (see, e.g., [35, 96, 160, 181, 182, 194, 231, 245]). We concentrate ourselves mostly on biologically motivated models in what follows.

Typically the RDC equation describing a process in the 1D space approximation has the form

$$u_t = [A(u)u_x]_x + B(u)u_x + C(u),$$

where A, B and C are some given functions, while $u(t, x)$ is an unknown function (hereafter the lower indices t and x denote differentiation with respect to (w.r.t.) these variables). In the models related to biomedical applications, the function $u(t, x)$ means the concentration of cells (population, drugs, molecules). The functions A, B and C are related to the three most common types of transport mechanisms occurring in real world processes. The diffusivity $A > 0$ (typically it is a constant) is the main characteristic of the diffusion process, the term $B(u)u_x$ (B typically means velocity, which can be positive and/or negative) describes the convective transport (in contrast to diffusion, one is not random) and the reaction term $C(u)$ describes the process kinetics (for example, this function presents interaction of the population u with the environment). A natural multidimensional analog of the above equation reads as

$$u_t = \nabla \cdot (A(u)\nabla u) + V(u) \cdot \nabla u + C(u). \tag{1.1}$$

Here u is the function of t and $x_1, \ldots, x_n$, $\nabla = \left(\frac{\partial}{\partial x_1}, \ldots, \frac{\partial}{\partial x_n}\right)$, the velocity $B(u)$ is replaced by the velocity vector $V(u)$ and $\cdot$ means the scalar product.

In (1.1) the variable diffusivity $D(u)$ (typically it is a power-law function but can be the function with a more complicated structure [40, 160]) arises in more and more modeling situations of biomedical importance from diffusion of genetically engineered organisms in heterogeneous environments to the effect of white and grey matter in the growth and spread of brain tumors [160, 182]. For example, the power-law diffusivity occurs as an extension of the classical diffusion model, when there is an increase in diffusion due to population pressure (see Section 11.3 in [182] and references therein).

The velocity vector V can be a vector function depending on the concentration u (if $V = const$ then the term $V \cdot \nabla u$ is removable from (1.1) by the Galilei transformations) and the simplest case when V is linear w.r.t. u was firstly studied in [179]. Notably, the velocity vector typically has the structure

$V = (B(u), 0 \ldots, 0)$ in real world models (for example, the evolution of fish population in a river can be adequately described if one takes into account the stream velocity only in the direction x_1 and neglects in other directions).

Although the reaction term C can possess a great variety of forms depending on the model in question (see examples in the books cited above), the most typical form of the reaction term is $C(u) = \lambda_1 u^p - \lambda_2 u^q$ with the positive exponents p and q (see Sections 11.3 and 13.4 in [182]). Depending on values of p and q several well-known equations arising in biomedical applications can be identified. For example, setting $p = 1$, $q = 2$, $A(u) = d = const$ and $V = 0$, the famous Fisher equation

$$u_t = d\Delta u + u(1 - u),$$

(here Δ is the Laplace operator) is obtained describing the spread in space of a favored gene in a population. Setting $p = 2$ and $q = 3$, the Huxley equation [43, 88]

$$u_t = d\Delta u + u^2(1 - u)$$

is derived, which can be thought as a limiting case of the famous Fitzhugh–Nagumo (FN) equation [100, 185]

$$u_t = d\Delta u + u(u - \delta)(1 - u), \ 0 < \delta < 1.$$

The latter is a simplification of the celebrated Hodgkin–Huxley model [133] describing the ionic current flows for axonal membranes. The FN equation reduces the Hodgkin–Huxley model, which has a very complex structure, and describes the nerve impulse propagation. The function $u(t, x)$ means the electric potential across the cell membrane.

In conclusion, we note that there are real world processes, which are described by the RDC equations involving coefficients A, B and C depending on derivatives of the function u (see, e.g., the recent paper [69] and works cited therein). Examination of such equations lies beyond the scope of this monograph.

1.2 Main methods for exact solving nonlinear reaction-diffusion-convection equations

As it was already pointed out, it is a generally accepted fact at the present time that a huge number of real processes arising in physics, biology, chemistry etc. can be adequately described by *nonlinear* partial differential equations (PDEs) only. On the other hand, the well-known principle of linear superposition cannot be applied to generate new exact solutions to nonlinear PDEs. Thus, the classical methods (the Fourier method, the Green function method, the method of the Laplace transformations, and so forth) are not applicable for solving nonlinear equations.

At the present time, there are many methods/techniques, which allow us to construct particular solutions of some nonlinear PDEs, however those are applicable to correctly-specified classes of PDEs only and any general integration theory is unknown. While there is no existing general theory for integrating nonlinear PDEs, construction of particular exact solutions for these equations is *a nontrivial and important problem*. Finding exact solutions that have a physical, chemical or biological interpretation is of *fundamental importance*.

One may say that the oldest technique for solving nonlinear differential equation is finding an appropriate transformation for a given PDE. In fact, a change of variables can sometimes be found that transforms the given nonlinear PDE into a linear equation. Transformation of the Burgers equation into the linear heat equations via the Cole–Hopf substitution is the classical example in this direction. However, finding exact solutions of most nonlinear PDEs generally requires other methods/thechniques than those for linear equations.

Nowadays the most powerful methods for construction of exact solutions for a wide range of classes of nonlinear PDEs are symmetry-based methods. All these methods have the common idea stating that exact solutions (at least particular ones) can be found for a given PDE provided its symmetry (a set of symmetries) is known. These methods originated from the Lie method, which was created by the prominent Norwegian mathematician Sophus Lie in the end of 19th century [166, 167] (see also the reprints in [168, 169]). The method was essentially developed using modern mathematical language by L.V. Ovsiannikov, G. Bluman, N. Ibragimov, W.F. Ames and some other researchers in the 1960s–1970s. Although the technique of the Lie method is well-known, the method still attracts attention of researchers and new results are published on a regular basis. In the next section a short historical review, basic notions, examples and theorems of the Lie method are presented.

In 1969, G. Bluman and J. Cole introduced an essential generalization of the Lie symmetry notion [26], which later was called nonclassical symmetry (in order to distinguish the new kind of symmetry from the classical Lie symmetry). Although nonclassical symmetries were not used for examination of nonlinear PDEs almost for 20 years (until the late 1980s), nowadays it is a powerful tool for constructing exact solutions of nonlinear equations. In particular, many important results for evolution PDEs were obtained during the last two decades. In the first section of Chapter 3, a short historical review, basic notions and theorems of the nonclassical method are presented. Notably, the terminology "nonclassical symmetry" is not generally accepted because notions "Q-conditional symmetry" and "reduction operator" are widely used too (see discussion on this matter in Section 3.1).

In the 1980s–1990s, a few new types of symmetries were introduced, which also allow us to construct exact solutions of nonlinear PDEs. The notion of conditional symmetry was suggested by Fushchych and his collaborators [112], [114, Section 5.7]. Note that the notion of nonclassical symmetry can be derived as a particular case from conditional symmetry but not vice versa (see highly nontrivial examples in Section 3.1 and in paper [63, 64]). Weak symmetry was suggested in [198, 199], potential symmetry was introduced in [29, 30, 159], while generalized conditional symmetry was independently formulated in [101, 173] and [251] (the terminology "conditional Lie–Bäcklund symmetry" was used in the latter).

The crucial idea used for introducing new types of symmetries can be formulated as follows. Let us consider an arbitrary PDE. For simplicity we restrict ourselves to the second-order two-dimensional equation

$$L(t, x, u, u_t, u_x, u_{tt}, u_{tx}, u_{xx}) = 0. \tag{1.2}$$

Hereafter $u = u(t, x)$ is an unknown smooth function[1], while L is a given smooth function. Almost all known in literature symmetries of Eq. (1.2) can be written as the differential operator

$$X = \xi^0 \partial_t + \xi^1 \partial_x + \eta \partial_u, \tag{1.3}$$

where the coefficients ξ^0, ξ^1 and η should be found according to the definition (criteria) of symmetry in question. In the case of the Lie, nonclassical and conditional symmetries, the coefficients depend on dependent and independent variables at maximum (those can be simply constants) and the corresponding criteria for their finding are presented in Sections 1.3 and 3.1.

However, the operator X has a more complicated structure if one is looking for other types of symmetries. For example, the coefficient η depends on derivatives of the function u in the case of generalized conditional symmetries [101, 173, 220].

In the case of potential symmetries, the coefficients depend on integrals of u, so that nonlocal operators are obtained. Thus, the terminology "nonlocal symmetry" is also used (see, e.g., [241] and references therein). A substantial number of examples involving potential symmetries for examination of nonlinear PDEs is presented in [25] (see also references therein).

A vast literature (see, e.g., [196] and citations therein) is devoted to higher-order symmetries (the terminology "generalized" and "Lie–Bäcklund" symmetry is also used) of the form

$$Z = X + \varsigma^0 \partial_{u_t} + \varsigma^1 \partial_{u_x} + \varsigma^{11} \partial_{u_{xx}} + \varsigma^{01} \partial_{u_{tx}} + \ldots, \tag{1.4}$$

where coefficients depend on derivatives of the function u. Such symmetries were introduced by Noether in her remarkable work [190]. Here we want only to stress that integrability of nonlinear PDEs via the method of inverse scattering problem [1, 93] is related with higher-order symmetries and conservation laws [177].

---

[1] Throughout the book the notion "smooth function" means that one is differentiable with respect to (w.r.t.) its variables up to the equation order, i.e., in the case of Eq. (1.2), u is the twice differentiable function w.r.t. t and x (at least in an open domain)

At the present time, it is a widely accepted hypothesis that each known exact solution of a given nonlinear PDE can be derived using an appropriate symmetry. On the other hand, there are some efficient methods/techniques allowing to construct exact solutions without knowledge of any symmetry. Here we are not going to present all of them because nowadays there are too many techniques proposed by a huge number of authors and it is very difficult to classify them (notably some new methods are particular cases of those proposed in the 1990s and earlier). In our opinion, the most general approach called the method of differential constraints was formulated in the 1960s [249] and was further developed in monograph [232]. Actually, basic ideas of the method of differential constraints have roots in Darbouxs works [103](see also excellent historical reviews on this matter in [214] and [119]). The main idea of the method of differential constraints is very simple: to define suitable constraint(s) for a given PDE in such a way that the overdetermined system obtained will be compatible and can be (partly) solved using the existing methods. Methods for solving overdetermined systems of differential equations were known since the first half of the 20th century [42, 139] and can be successfully applied in many cases (for instance, see examples in [232]). However, the main problem of the method is how to define the suitable constraint(s). At the present time, the corresponding algorithm does not exist in general case and one may claim that the method of differential constraints is rather a fruitful idea without a constructive algorithm. Interestingly, the symmetry-based methods implicitly use this idea in a constructive way. In fact, in order to find exact solutions, one solves the given nonlinear PDE (system of PDEs) together with the differential constraint(s) generated by a symmetry operator.

There are several techniques, which propose to use the correctly-specified differential constraints in order to find exact solutions for some correctly-specified classes of PDEs. In particular, the method of additional generating conditions [45, 46, 48] and the method of determining equations [143, 144] were independently worked out in the 1990s. Both methods are very similar and Chapter 5 is devoted to the first of them.

Several approaches based on substitutions of the special form, which are often called *ansatz*[2], should be mentioned. Such substitution reduces the given PDE to a simpler equation (e.g., ODE) or a system of simpler equations, which can be integrated (at least partly). In order to construct the relevant ansatz, either some physical (biological, chemical etc.) motivation or an ad hoc approach are used. The most typical is the plane wave ansatz

$$u = \phi(\omega), \ \omega = x - vt, \ v \in \mathbb{R}, \tag{1.5}$$

which reduces Eq. (1.2) to an ODE provided the function L does not depend on t and x. Although solving the nonlinear ODE obtained can be also a nontrivial problem, the solution (at least partial) can be usually found by using the classical methods or handbooks like [142, 212]. Notably, there exist some

---

[2]Ansatz (pl. ansätze) is a German word

recent techniques (e.g., the tanh- and exp-function methods) allowing to construct such solutions of the nonlinear ODE obtained, which lead to traveling waves (fronts) of the given PDE (see, e.g., [174, 175, 246]), however their wide applicability is often questionable [161].

The second classical ansatz follows from the Furrier method (the method of separation of variables) and reads as

$$u = T(t)X(x), \tag{1.6}$$

where $T(t)$ and $X(x)$ are to-be-determined functions. Depending on the form of L, PDE (1.2) can be reduced to a system of two ODEs. Both ansätze are well-known since the 19th century and related with Lie symmetry of the equation in question. The first one reflects invariance under the time and space translations, the second is related to scale invariance of the given equation. Notably, J. Boussinesq was the first to construct an exact solution of the nonlinear physically motivated problem using ansatz (1.6) [32] (see Section 4.2 for details).

Several new ansätze were suggested during the recent decades in order to construct exact solutions for nonlinear PDEs, especially evolution equations. Many of them can be united and written in the form of the ansatz with separated variables

$$A(u) = \varphi_0(t)g_0(x) + ... + \varphi_{m-1}(t)g_{m-1}(x). \tag{1.7}$$

Of course, (1.7) is the straightforward generalization of ansatz (1.6). Here some functions are known, others should be found. Typically, the functions $\varphi_0, \ldots, \varphi_{m-1}(t)$ are unknown in the case of evolution equations, the functions $g_0(x), \ldots, g_{m-1}(x)$ are specified by either physically motivated reasons or an ad hoc approach, while $A(u)$ reflects usually nonlinearities in the given PDE. As a result, a m-dimensional system of ODEs should be obtained for finding $\varphi_0, \ldots, \varphi_{m-1}(t)$ by using ansatz (1.7). For example, ansatz (1.7) with the power-law functions $g_k = x^k$, i.e.,

$$A_0(U) = \varphi_0(t) + \varphi_1(t)x + ... + \varphi_{m-1}(t)x^{m-1}, \tag{1.8}$$

was extensively used for constructing exact solutions of reaction-diffusion (RD) equations with power and exponential nonlinearities in the 1990s [117, 118, 149, 151, 148, 237]. It should be mentioned that ansatz (1.8) was also applied in the earlier papers [200, 238], while later, in the 2000s, it was shown that the ansatz is applicable for reduction of some nonlinear PDEs when the functions $g_k = x^k$ contain rational (and even irrational) exponents k (see, e.g., examples in [84, 119]). Finally, a generalization of ansatz (1.8) for solving the multi-dimensional PDEs was suggested in [150] (see also examples in [119]).

Several cases of ansatz (1.7) with the functions $g_k \neq x^k$ were used for construction of new solutions to nonlinear evolution PDEs (see, e.g., [48, 52, 118, 237]). Paper [22, 200] should be quoted as one of the first in this direction.

In the 1990s, the method of linear invariant subspaces was developed [118, 237] and was extensively applied for constructing exact solutions of a wide range of nonlinear PDEs (almost all the results are summarized in monograph [119]). One may note that the method is reduced to finding exact solutions in the form (1.7) provided the functions $g_0(x), \ldots, g_{m-1}(x)$ generate the basis of an invariant subspace for a given PDE.

Finally, the direct method [87], which is also called the Clarkson–Kruskal method, should be mentioned. The basic idea of the method is to reduce PDE (1.2) to ODE (a system of ODEs) using the ansatz

$$u = \varphi(t, x, \omega), \qquad (1.9)$$

where the function $\omega(z)$ is a solution of some ODE, while the variable $z(t,x)$ should be additionally defined (e.g., it can be an invariant variable corresponding to a Lie symmetry). The function φ is usually assumed to be linear with respect to $\omega(z)$. It turns out that such ansatz leads to new exact solutions for some PDEs arising in real world applications [86, 87]. It was shown later that a consistency criteria can be formulated when the the direct method and the nonclassical method [26] lead to the same solutions [13]. On the other hand, there are examples of the exact solutions found by the nonclassical method, which are not obtainable by the direct method [13, 193], hence the nonclassical method seams to be a more general tool for finding exact solution of nonlinear PDEs.

1.3 Lie symmetry of differential equations: historical review, definitions and properties

The Lie method was created by the famous Norwegian mathematician Sophus Lie. In the 1880s–1890s, he published several papers and books, which had an essential impact on development of pure and applied mathematics in the 20th century. It should be noted that his works were published in German and were not translated in English for a long time (to the best of our knowledge, some of them still exist in German only). The most important among Lie's works from the applicability point of view are papers [166] and [167].

Research of S. Lie was inspired by the Galois group theory for solving algebraic equations. His aim was to create a similar theory for solving differential equations. In particular, he created *the theory of continuous groups, i.e., Lie groups* [170, 171, 172], which plays a key role in modern mathematics and its applications.

In the end of the 1950s, L.V. Ovsiannikov inspired by Birkhoff's works on hydrodynamics [23], rewrote the basic notions of the Lie method using modern terminology and published the first monograph (after S. Lie) devoted to the

Lie method and its development [202]. His book accelerated a great interest to the subject during the 1960s–1980s. As a result, a huge number of papers and several books were published. The best known books (one may refer to them as classical) were published in that time [6, 27, 29, 114, 196, 203] (notably the Russian version of [114] was published in 1989). Many results derived during this period (and till the middle of the 1990s) are summarized in the collective monographs [7, 8, 136].

At the present time, the Lie method and related symmetry-based methods are still the most popular methods for construction of exact solutions of ODEs, PDEs (including multidimensional PDEs), integro-differential equations, difference equations and some others. There is a huge number of papers and several excellent books devoted to theoretical foundations of the Lie method and its application published in this millennium. In particular, the books [12, 24, 25, 135] are devoted to presentation of the Lie method and related symmetry-based methods. They are partly based on the relatively recent results and the modern terminology is used therein.

Taking into account that the Lie method is a well-established method with rigorous theoretical foundations, here we briefly present the basic notions and theorems only. The reader may find much more details in the monographs and textbooks cited above. Because this monograph is devoted mostly to two-dimensional PDEs, we restrict ourselves to the two-dimensional case from the very beginning. The relevant generalization on the multi-dimensional case can be done in a straightforward way (or found elsewhere).

Let us consider the k-order PDE

$$L\left(t, x, u, \underset{1}{u}, \ldots, \underset{k}{u}\right) = 0, \quad k \geq 1, \quad (1.10)$$

where $u = u(t, x)$ is an unknown function, $\underset{s}{u}$ means a totality of s-order derivatives of $u(t, x)$ ($s = 1, 2, \ldots, k$) and L is a given smooth function of their arguments. On the other hand, we suppose that the transformations

$$\bar{t} = f^0(t, x, u, v), \quad \bar{x} = f^1(t, x, u, v), \quad \bar{u} = g(t, x, u, v) \quad (1.11)$$

(here $v = (v_1, \ldots, v_r)$ are group parameters and f^0, f^1 and g are the given smooth function) generate the r-parameter Lie group G^r. According to the known algorithm, the k-th order extension (prolongation) of G^r can be constructed (see for details, e.g., [24, Chapter 2]) acting in the prolonged space of the variables

$$t, \ x, \ u, \ \underset{1}{u}, \ \ldots, \ \underset{k}{u}. \quad (1.12)$$

In the prolonged space (1.12), each k-th order PDE can be considered as a manifold (surface), i.e., it is a geometrical object.

Definition 1.1 *The r-parameter Lie group G^r is called the invariance group (the group of point symmetries) of PDE (1.10), if its k-th order extension leaves invariant the surface defined by (1.10) in the prolonged space (1.12).*

It follows immediately from Definition 1.1 that an arbitrary solution of PDE (1.10) is mapped to another (or the same) solution of this equation under action of each transformation from G^r. So, the set of all solutions of PDE (1.10) is invariant under G^r.

It is well-known that there is a bijective correspondence between Lie groups and Lie algebras of the same dimensionality. Thus, having the invariance group G^r of PDE (1.10), one can easily construct the corresponding algebra of invariance and vice versa. In fact, the following statements take place.

Theorem 1.1 *Let us consider the continuous transformations, which correspond to the r-parameter group G^r (1.11). Then this group generates the r-dimensional Lie algebra AG^r with the basic operators*

$$X_j = \xi_j^0(t,x,u)\partial_t + \xi_j^1(t,x,u)\partial_x + \eta_j(t,x,u)\partial_u, \quad j=1,\ldots,r \qquad (1.13)$$

where the functions ξ_j^0, ξ_j^1 and η_j are determined as follows

$$\xi_j^0 = \left.\frac{\partial f^0(t,x,u)}{\partial v_j}\right|_{v=0}, \quad \xi_j^1(t,x,u) = \left.\frac{\partial f^1}{\partial v_j}\right|_{v=0}, \quad \eta_j(t,x,u) = \left.\frac{\partial g}{\partial v_j}\right|_{v=0}. \qquad (1.14)$$

Theorem 1.2 *Let us consider the r-dimensional Lie algebra AG^r with the basic operators (1.13). Then this algebra generates the r-parameter Lie group G^r (1.11), where the functions $\bar{t} = f^0(x,u,v)$, $\bar{x} = f^1(x,u,v)$ and $\bar{u} = g(x,u,v)$ form a unique solution of the initial problem*

$$\frac{\partial \bar{t}}{\partial v_j} = \xi_j^0(\bar{t},\bar{x},\bar{u}), \quad \frac{\partial \bar{x}}{\partial v_j} = \xi_j^1(\bar{t},\bar{x},\bar{u}), \quad \frac{\partial \bar{u}}{\partial v_j} = \eta_j(\bar{t},\bar{x},\bar{u})$$
$$f^0(x,u,0) = t, \ f^1(x,u,0) = x, \ g(x,u,0) = u. \qquad (1.15)$$

Remark 1.1 *To guarantee a unique solution of the initial problem (1.15) (at least in a vicinity of the point $v = (v_1,\ldots,v_r) = 0$), one should assume that the coefficient of the Lie operators (1.13) are sufficiently smooth.*

Theorems 1.1 and 1.2 lead to the following consequence: each invariance group of the given PDE is known provided the corresponding Lie algebra of invariance is found and vice versa. Because the operators of any Lie algebras create a linear vector space, usually it is more convenient to deal with Lie algebras if differential equations are under study. Let us present an example in order to illustrate the above definition and theorems.

Example 1.1 *Consider the nonlinear equation*

$$u_t = (uu_x)_x, \qquad (1.16)$$

where $u(t,x)$ is an unknown function. One may say that Eq. (1.16) is a classical example among nonlinear parabolic equations because one is used in a wide range of applications. This equation was introduced by J. Boussinesq [32] and is often named after him (the name "porous diffusion equation" is also used).

Let us consider the set of transformations

$$\bar{t} = e^{2v_2}t + v_0, \quad \bar{x} = e^{v_2}x + v_1, \quad \bar{u} = u, \tag{1.17}$$

where v_0, v_1 and v_2 are arbitrary real parameters. One easily checks that it is the three-parameter Lie group G^3. Assume that $u(t,x)$ is an arbitrary solution of the Boussinesq equation and examine the function $\bar{u}(\bar{t},\bar{x})$. Taking into account (1.17), one obtains

$$\bar{u}_{\bar{t}} = e^{-2v_2}u_t, \quad \bar{u}_{\bar{x}\bar{x}} = e^{-2v_2}u_{xx}$$

so that $\bar{u}(\bar{t},\bar{x}) = u(e^{2v_2}t+v_0, e^{v_2}t+v_1)$ is also the exact solution of Eq. (1.16). Thus, the Lie group G^3 is the invariance group of this equation.

Let us apply Theorem 1.1 for finding the corresponding Lie algebra. Substituting (1.17) into formulae (1.14), we easily calculate the coefficients of the basic operators:

$$\xi_0^0 = \left.\frac{\partial(e^{2v_2}t+v_0)}{\partial v_0}\right|_{v=0} = 1, \quad \xi_0^1 = \left.\frac{\partial(e^{v_2}x+v_1)}{\partial v_0}\right|_{v=0} = 0,$$

$$\xi_1^0 = \left.\frac{\partial(e^{2v_2}t+v_0)}{\partial v_1}\right|_{v=0} = 0, \quad \xi_1^1 = \left.\frac{\partial(e^{v_2}x+v_1)}{\partial v_1}\right|_{v=0} = 1,$$

$$\xi_1^0 = \left.\frac{\partial(e^{2v_2}t+v_0)}{\partial v_2}\right|_{v=0} = 2t, \quad \xi_2^1 = \left.\frac{\partial(e^{v_2}x+v_1)}{\partial v_2}\right|_{v=0} = x,$$

$$\eta_j = \left.\frac{\partial u}{\partial v_j}\right|_{v=0} = 0, \; j = 0, 1, 2. \tag{1.18}$$

Finally, substituting these coefficients into (1.13), we arrive at the three-dimensional Lie algebra with the basic operators

$$X_0 = \partial_t, \; X_1 = \partial_1, \; X_2 = 2t\partial_t + x\partial_x.$$

An operator of invariance (nowadays the terminology "Lie (point) symmetry" is widely used) has the form of a first-order quasilinear differential operator

$$X = \xi^0(t,x,u)\partial_t + \xi^1(t,x,u)\partial_x + \eta(t,x,u)\partial_u. \tag{1.19}$$

If (1.19) is a Lie symmetry of PDE (1.10) then the coefficients $\xi^0(t,x,u), \xi^1(t,x,u)$ and $\eta(t,x,u)$ can be found using the definition (the invariance criteria).

Definition 1.2 *Operator (1.19) is called Lie symmetry of PDE (1.10) if the following invariance criteria is satisfied:*

$$\left.X(L)\right|_{\substack{k \\ \mathcal{M}}} = 0, \tag{1.20}$$

where the differential operator $\underset{k}{X}$ is the k-order prolongation of operator (1.19) and the manifold $\mathcal{M}$ is defined by the equation in question, i.e., it is a smooth manifold $L\left(t, x, u, \underset{1}{u}, \ldots, \underset{k}{u}\right) = 0$ in the prolonged space of the variables (1.12)

The operator

$$\underset{k}{X} = \underset{k-1}{X} + \underset{k}{\eta} \cdot \partial_{\underset{k}{u}}$$

(here $\partial_{\underset{k}{u}}$ is the vector consisting a totality of the operators $\frac{\partial}{\partial u_k}$ with u_k belonging to the set $\underset{k}{u}$, while "$\cdot$" means the scalar product) is the kth-order prolongation of the operator $X \equiv \underset{0}{X}$. Its coefficients are expressed via the functions ξ^0, ξ^1 and η by the well-known formulae, which can be written in the form (see, e.g., [203])

$$\underset{k}{\eta} = D \underset{k-1}{\eta} - uD\underset{k}{\xi}. \tag{1.21}$$

In the case $k = 2$, one obtains

$$\underset{1}{\eta} = (\zeta^0, \zeta^1), \quad \underset{2}{\eta} = (\sigma^{00}, \sigma^{01}, \sigma^{11}), \quad D = (D_t, D_x),$$

$$D_t = \partial_t + u\partial_{u_t} + u_t\partial_{u_{tt}} + u_x\partial_{u_{tx}}, \quad D_x = \partial_x + u\partial_{u_x} + u_t\partial_{u_{xt}} + u_x\partial_{u_{xx}}.$$

Thus, operator $\underset{2}{X}$, which is the second prolongation of the operator X, has the form

$$\underset{2}{X} = X + \zeta^0\partial_{u_t} + \zeta^1\partial_{u_x} + \sigma^{00}\partial_{u_{tt}} + \sigma^{01}\partial_{u_{tx}} + \sigma^{11}\partial_{u_{xx}}, \tag{1.22}$$

where

$$\begin{aligned}
\zeta^0 &= D_t\eta - u_tD_t\xi^0 - u_xD_t\xi^1, \\
\zeta^1 &= D_x\eta - u_tD_x\xi^0 - u_xD_x\xi^1, \\
\sigma^{00} &= D_t\zeta^0 - u_{tt}D_t\xi^0 - u_{tx}D_t\xi^1, \\
\sigma^{01} &= D_x\zeta^0 - u_{tt}D_x\xi^0 - u_{tx}D_x\xi^1, \\
\sigma^{11} &= D_x\zeta^1 - u_{xt}D_x\xi^0 - u_{xx}D_x\xi^1.
\end{aligned} \tag{1.23}$$

It turns out that a set of all Lie symmetry operators of the given PDE always form a Lie algebra. Because a subset of this set may form another Lie algebra, the following notion is important.

Definition 1.3 *A Lie algebra of invariance of the given PDE is called the maximal algebra of invariance (MAI) if one contains each Lie algebra of invariance of this equation as a subalgebra. The Lie group, which corresponds to MAI, is called the maximal group of invariance (MGI).*

The problem of finding MAI for the given PDE is reduced to solving an overdetermined system of linear PDEs (although the equation in question can be nonlinear!). Nowadays it is a straightforward routine, which can be realized using computer algebra packages like Maple, Mathematica, Reduce etc (see, e.g., [85, 130, 248]).

However, the problem becomes nontrivial if PDE contains arbitrary functions as coefficients. In this case the *Lie symmetry classification (LSC) problem* (terminology "group classification problem" is also used) arises. Let us consider the class of PDEs

$$L\left(t, x, u, \underset{1}{u}, \ldots, \underset{k}{u}, \theta\right) = 0, \tag{1.24}$$

where θ is an arbitrary smooth function (can be two or more functions), which depends on the variables t, x, u (dependence on derivatives of u can be as well). In the models describing real world processes, the function θ (with appropriate restrictions) has a clear physical (biological, chemical etc.) interpretation. Obviously, Lie symmetry of PDEs belonging to class (1.24) depends essentially on the form of θ.

Remark 1.2 *It may happen that MAI of PDEs involving θ as a parameter (not a function !) depends on the value of θ. The classical example is the diffusion equation with power-law diffusivity $d(u) = u^k$ (see Section 2.2).*

Two types of the LSC problem are usually distinguished.

1. To find all the correctly-specified forms of θ when PDE (1.24) admits a fixed Lie algebra.

2. To describe all possible Lie algebras (Lie groups), which can be admitted by PDEs from the given class depending on the form of θ.

The first type is very important for examination of mathematical models in theoretical physics because all classical equations (the heat and wave equations, the Laplace and Schrödinger equations, the Maxwell and Dirac systems, etc.) admit Lie symmetries, which reflect the fundamental laws of physics. If one is looking for some nonlinear analogs of the classical equations then the generalizations to be constructed should preserve the relevant Lie symmetry (the reader can find references to an extensive discussion on this topic in [64, 110, 114, 222]).

The second type is more popular among mathematicians who are dealing with PDEs arising in different branches of applied mathematics. In this case, the well-established class of PDEs is usually taken from literature and examined using the Lie method. Chapter 2 is devoted to such type of the LSC problems.

Now we turn to the operator X (1.19). Throughout the book we use the notation $X(u) = 0$ for the first-order PDE

$$\xi^0 u_t + \xi^1 u_x - \eta = 0, \tag{1.25}$$

which is generated by the Lie symmetry (1.19) and often is called the *invariance surface condition* or the *characteristic equation* of X. Rigorously speaking, one should write the invariance surface condition in the form (see, e.g., [24] for details)

$$X(u - u_0(t,x))\Big|_{u=u_0(t,x)} = 0,$$

which leads exactly to the above first-order PDE for an arbitrary solution $u = u_0(t,x)$.

S. Lie established a remarkable property of Eq. (1.25), which allow us to apply Lie symmetries for the exact reduction of the given PDE to ODEs or algebraic equations (in the case of multidimensional PDEs such reduction is more complicated and usually leads to another PDE of a lower dimensionality). In fact, because (1.25) is the quasi-linear first-order PDE, its general solution can be always constructed in explicit (or implicit) form and one depends on a function of a single variable (two variables if the general solution is found in an implicit form). In the case of evolution PDEs (such equations are the main subject of this book), a typical form of the general solution of (1.25) is

$$u = g(t,x) + \varphi\big(\omega(t,x)\big)G(t,x), \qquad (1.26)$$

where φ is an arbitrary function of the variable $\omega(t,x)$, while $g(t,x)$ and $G(t,x)$ are the known functions. The above property guarantees that substitution (1.26) reduces the given PDE (1.10) to the equation

$$L_1\left(\omega, \varphi, \varphi_\omega^{(1)}, \varphi_\omega^{(2)}, ..., \varphi_\omega^{(k_1)}\right) = 0, \qquad (1.27)$$

which is ODE of the order $k_1 \leq k$ for the unknown function φ. Eq. (1.27) is called the *reduction equation*. Thus, each exact solution of the reduction equation (1.27) can be easily transformed into a particular solution of PDE (1.10) using formula (1.26).

Substitution (1.26) is called *ansatz* (Lie's ansatz) as it was proposed by S. Lie in his works. Taking into account that there are also ansätze, which are not related with Lie symmetry, the following definition is useful.

Definition 1.4 *A local substitution reducing a given PDE into another PDE (ODE or algebraic equation) with a smaller number of independent variables is called ansatz.*

If an ansatz is related with a Lie symmetry from MAI of the given equation then it is Lie's ansatz, otherwise the notion non-Lie ansatz will be used. Because each linear PDE admits infinite-dimensional MAI, which reflects the principle of linear superposition of solutions, an infinite number of Lie's ansätze can be constructed. On the other hand, there are several classical methods for solving linear PDEs, therefore we concentrate ourselves on nonlinear PDEs in what follows.

Lie symmetry of differential equations: historical review, definitions 15

Definition 1.5 *An exact solution of PDE in question that can be found using a Lie symmetry operator (i.e., Lie's ansatz) and (possible) subsequent application of transformations from its invariance group is called invariant solution (group-invariant solution, similarity solution, Lie solution).*

Now we present the next example, which is a natural continuation of Example 1.1 and illustrate the above definitions.

Example 1.2 *Here we continue examination of Eq. (1.16). The Lie algebra of invariance derived in Example 1.1 is not maximal because MAI of Eq. (1.16) is the four-dimensional Lie algebra* [203]

$$AG^4 = \langle \partial_t, \partial_x, 2t\partial_t + x\partial_x, x\partial_x + 2u\partial_u \rangle. \tag{1.28}$$

Thus, taking any linear combination of the above operators, one can construct a Lie's ansatz. Notably, several Lie's ansätze were derived by ad hoc approaches (i.e., without application of the Lie method). For example, taking the linear combination $\partial_t + v\partial_x$ (here v is an arbitrary parameter, which usually means velocity) and using the invariance surface condition (1.25), the linear first-order PDE $u_t + vu_x = 0$ is obtained. So, the corresponding ansatz has the form

$$u = \varphi(\omega), \quad \omega = x - vt, \tag{1.29}$$

where $\varphi(\omega)$ is to be determined function. On the other hand, (1.29) is the well-known substitution for finding the plane wave solutions (particularly traveling fronts, solitary waves, kinks etc.) and is applicable for a wide range of PDEs.

Let us construct a nontrivial ansatz using the Lie symmetry $x\partial_x + 2u\partial_u$. In this case, the invariance surface condition takes the form $xu_x - 2u = 0$, hence one obtains the ansatz

$$u = x^2 \varphi(\omega), \quad \omega = t. \tag{1.30}$$

Substituting ansätze (1.29) and (1.30) into Eq. (1.16), one derives two reduction equations

$$(\varphi \varphi_\omega)_\omega + v\varphi_\omega = 0, \quad \varphi_\omega = 4\varphi^2. \tag{1.31}$$

Obviously, both ODEs can be easily integrated, hence the corresponding particular solutions of the Boussinesq equation are obtained using formulae (1.29) and (1.30).

In contrast to the above cases, the problem of finding exact solutions for given nonlinear PDEs via application of Lie symmetries can be a nontrivial task. For example, if one takes the linear combination of the last two operators from AG^4, i.e., $2t\partial_t + (1+v)x\partial_x + 2vu\partial_u$ then the corresponding ansatz and reduction equation are much more complicated (its analog will be examined in Section 4.2). As a result, invariant solutions of Eq. (1.16) cannot be constructed for an arbitrary parameter v. Notably, a solution corresponding to the parameter $v = -1$ firstly was derived in [32].

Because construction of exact solutions is the most important feature of the Lie method (we remind the reader that S. Lie initially was going to create a general theory of integrability of ODEs and PDEs), one needs to develop an algorithm how to derive (at least to describe) all invariant solutions of the equation in question.

The algorithm for constructing invariant solutions of nonlinear PDEs consists of the following main steps.

1. To find MAI of the equation in question using the invariance criteria (1.20).

2. To construct all inequivalent Lie's ansätze by solving the characteristic equations of the form (1.25) for the corresponding operators from MAI.

3. To construct the list of reduction equations using the ansätze derived.

4. To solve the reduction equations using the existing techniques and to construct exact solutions of the equation in question.

5. To construct formulae of multiplication of the solutions obtained using Definition 1.1 and group transformations from MGI.

It should be stressed that all the steps of the algorithm often cannot be realized because the fourth step essentially depends on the form of the reduction equations. So, some particular solutions of those can be constructed only. However, one may provide a qualitative analysis of nonintegrable reduction equations in order to establish properties of the solutions of PDE in question (see, e.g., [68]).

Other steps of the algorithm can be realized for a wide range of PDEs (including multidimensional ones) arising in real world applications. We briefly present some remarks to each step only. As it was already noted, MAI can be found using computer algebra packages (we assume that the given equation does not contain arbitrary functions as coefficients).

The second step is completed in a straightforward way provided MAI is of a low dimensionality (for example, the MAI dimensionality is $r = 2, 3$ or 4). In such cases, one takes a linear combination of the basic operators (with arbitrary coefficients, say, λ_i, $i = 1, \ldots, r$) and obtains Eq. (1.25) involving r parameters. Solving such equation with a small number of arbitrary parameters usually is a simple task, so that all possible ansätze are derived. However, the task becomes a nontrivial problem if MAI has the high dimensionality (or one is infinite-dimensional). Nonlinear multidimensional PDE arising in real world applications usually are invariant under such algebras. For example, the well-known Schrödinger equation with cubic nonlinearity in 1D space (i.e., the two-dimensional PDE) is invariant under 5-dimensional MAI, while one in 3D space admits 12-dimensional MAI (see, e.g., [108, 109, 114]). In the case of MAI of a high dimensionality, the additional problem occurs, which was formulated by Ovsiannikov in the 1960s [202, 203]. One needs to construct systems of inequivalent (nonconjugated) subalgebras of MAI of the dimensionalities $1, \ldots, r-1$. Each such system is called the optimal system of n-dimensional subalgebras. It turns out that this problem, which is a pure algebraic one, can be completely solved for all possible Lie algebras of the low dimensionalities [204, 205, 206] but the problem is still unsolved in the general case. To the best of our knowledge, the problem is not solved for six- and higher-dimensional Lie algebras at the present time (see also the extensive discussion in [163]). In the case of two-dimensional PDEs, the problem simplifies because one needs to construct only the optimal system of one-dimensional subalgebras for the given Lie algebra.

The third step of the algorithm consists of simple calculations provided all inequivalent ansätze are derived.

The last step of the algorithm is not important from the rigorous mathematical point of view. In fact, one can claim that all solutions of the given PDE are equivalent (although they may have very different properties!) if they can be mapped to a fixed invariant solution by a transformation from MGI. Thus, it is enough to construct the so-called optimal system of invariant solutions corresponding to subalgebras belonging the given optimal system (see, e.g., [196]). However, the fifth step is important from applicability point of view. Let us take an exact solution, which is trivial and without any physical (biological) interpretation. The solution can be multiplied to a set of solutions by the transformations belonging to MGI. As a result, some of the solutions obtained may admit an interpretation of the process described by the equation in question. For example, the constant solution of the classical heat equation $u_t = u_{xx}$ can be mapped to its fundamental solution $u = \frac{1}{2\sqrt{t}} \exp(-\frac{x^2}{4t})$ by the invariance transformations (and taking a limiting case) of the equation. Notably, the relevant formula of the multiplication of exact solutions of the heat equation was derived by P. Appell [9] without knowledge of the Lie method. On the other hand, this formula was generalized on some nonlinear PDEs and nonlinear systems of PDEs using the Lie method (see examples in [44, 50, 108, 109, 110]).

Chapter 2

Lie symmetries of reaction-diffusion-convection equations

2.1	Symmetry of the linear diffusion equation	19
2.2	Symmetry of the nonlinear diffusion equation	21
2.3	Equivalence transformations and form-preserving transformations	29
	2.3.1 The group of equivalence transformations	30
	2.3.2 Form-preserving transformations	32
2.4	Determining equations for reaction-diffusion-convection equations	36
2.5	Complete description of Lie symmetries of reaction-diffusion-convection equations	39
	2.5.1 Principal algebra of invariance	39
	2.5.2 Necessary conditions for nontrivial Lie symmetry	39
	2.5.3 Lie symmetry classification via the Lie–Ovsiannikov algorithm	49
	2.5.4 Application of form-preserving transformation	60
2.6	Nonlinear equations arising in applications and their Lie symmetry	69
	2.6.1 Heat (diffusion) equations with power-law nonlinearity	69
	2.6.2 Diffusion equations with a convective term	71
	2.6.3 Nonlinear equations describing three types of transport mechanisms	73

2.1 Symmetry of the linear diffusion equation

The linear heat (diffusion) equation

$$u_t = u_{xx} \tag{2.1}$$

(here the function $u(t,x)$ means temperature or concentration) is the most common particular case among reaction-diffusion-convection (RDC) equations. Sophus Lie was the first to calculate the maximal invariance algebra

(MAI) of this equation in his seminal work [166]. It is the generalized Galilei algebra with the following basic operators:

$$\partial_t \equiv \tfrac{\partial}{\partial t}, \quad \partial_x \equiv \tfrac{\partial}{\partial x},$$
$$I = u\partial_u \equiv u\tfrac{\partial}{\partial u}, \quad G = t\partial_x - \tfrac{1}{2}xI,$$
$$D = 2t\partial_t + x\partial_x - \tfrac{1}{2}I \qquad (2.2)$$
$$\Pi = t^2\partial_t + tx\partial_x - \tfrac{1}{2}(\tfrac{1}{2}x^2 + t)I,$$

and an infinite-dimensional algebra generated by the operators of the form

$$X_\infty = v(t,x)\partial_u, \qquad (2.3)$$

where the function $v(t,x)$ is an arbitrary solution of the equation $v_t = v_{xx}$. It is well-known that the operators ∂_t and ∂_x generate the group of time and space translations. The operators G, D and Π generate the groups of the Galilei, scale and projective transformations, respectively. In explicit form they are presented, e.g., in [196] (see also [107, 114] for the multidimensional heat equation). Hereafter the operator I will be called the unit operator in order to underline that the commutator $[G, \partial_x] = \tfrac{1}{2}I$ in contrast to the classical Newton equation $\ddot{x}(t) = 0$, for which such commutator vanishes because $G = t\partial_x$. Note that I is often called the mass operator in theoretical physics in context of the free Schrödinger equation, which has the same Lie symmetry. Finally, the operator X_∞ is a common operator for any linear equation and one reflects the well-known principle of linear superposition of solutions for linear PDEs. We do not take into account such operators in what follows because this monograph is devoted to *nonlinear PDEs*.

The above result about symmetry of the heat equation (2.1) can be generalized on the multidimensional case. In fact, the $(1+n)$-dimensional heat equation

$$u_t = \Delta u, \qquad (2.4)$$

(here Δ is the Laplacian and u is the function of the time t and the space variables $x_1, \ldots, x_n$) is invariant under $\tfrac{n^2+3n+8}{2}$-dimensional algebra with the basic operators

$$\partial_t, \quad \partial_a \equiv \tfrac{\partial}{\partial x_a} \quad (a = 1, \ldots, n), \qquad (2.5)$$

$$I = u\partial_u, \quad G_a = t\partial_a - \tfrac{1}{2}x_a I, \quad J_{ab} = x_a\partial_b - x_b\partial_a \ (b < a = 1, \ldots, n), \qquad (2.6)$$

$$D = 2t\partial_t + x_a\partial_a - \tfrac{n}{2}I, \qquad (2.7)$$

$$\Pi = t^2\partial_t + tx_a\partial_a - \left(\tfrac{1}{4}x_a x_a + \tfrac{n}{2}t\right)I, \qquad (2.8)$$

where a summation is assumed from 1 to n over the repeated indices a. In the case $n = 2$, this result was derived almost 100 years ago [124].

The Lie algebra produced by the operators (2.5)–(2.6) is called the Galilei algebra $AG(1.n)$ and its extension by using operator (2.7) is referred to as the extended Galilei algebra $AG_1(1.n)$. Following [108, 109, 114] the algebra generated by the operators (2.5)–(2.8) is called the generalized Galilei algebra $AG_2(1.n)$. On the other hand, following Niederer's paper [188], in which this algebra was rediscovered in the case of the free Schrödinger equation, this algebra is called the Schrödinger algebra in theoretical physics. All the nonzero commutators of the algebra $AG_2(1.n)$ are as follows (see, e.g., [128, 222] and papers cited therein)

$$[\partial_t, G_a] = \partial_a, \quad [\partial_a, G_b] = -\frac{1}{2}\delta_{ab}I, \quad [\partial_a, J_{bc}] = \delta_{ab}\partial_c - \delta_{ac}\partial_b,$$

$$[G_a, J_{bc}] = \delta_{ab}G_c - \delta_{ac}G_b, \quad [J_{ab}, J_{cd}] = \delta_{ac}J_{db} + \delta_{bc}J_{ad} + \delta_{ad}J_{bc} + \delta_{bd}J_{ca},$$

$$[G_a, D] = -G_a, \quad [\partial_a, D] = \partial_a, \quad [\partial_a, \Pi] = G_a,$$

$$[\partial_t, D] = 2\partial_t, \quad [\partial_t, \Pi] = D, \quad [D, \Pi] = 2\Pi,$$

where $\delta_{ab} = \begin{cases} 1, & a = b, \\ 0, & a \neq b. \end{cases}$

Having these commutators, we may note that the algebra $AG_2(1.n)$ contains as subalgebras the classical three-dimensional algebra $sl(2, \mathbb{R})$ with the realization $\langle \partial_t, D, \Pi \rangle$ (see the last line of the commutator relations) and the algebra of rotations $O(n)$ with the standard realization J_{ab}. Notably, all the nonzero commutators for the Galilei algebra $AG(1.n)$ and its extension $AG_1(1.n)$ can be easily extracted from those listed above.

2.2 Symmetry of the nonlinear diffusion equation

The standard generalization of the (1+1)-dimensional linear heat equation reads as

$$u_t = [d(u)u_x]_x \tag{2.9}$$

and is called the nonlinear heat (diffusion) equation. Here $d(u)$ is the diffusion (conductivity) coefficient, which can be an arbitrary nonnegative function. In contrast to the linear equation (2.1), the Lie algebra of invariance of (2.9) is not a fixed algebra but depends on the form of $d(u)$ (hereafter we assume that it is a nonconstant smooth function). It means that the Lie symmetry classification (LSC) problem should be solved, i.e., to describe all possible forms of $d(u)$ corresponding to different algebras of invariance. Ovsiannikov was the first to solve this problem [201], simultaneously he created an approach (hereafter called the Lie–Ovsiannikov algorithm), which helps us to solve such kind of problems [203]. Following Ovsiannikov's works the LSC problems are often called the group classification problems. We think that this terminology is misleading (especially for the reader, who is not specialist in symmetry-based methods) because the algorithm allows us to find Lie algebras for PDEs. Of course, having a complete list of the Lie algebras for a given PDE class, one may easily construct all the corresponding Lie groups.

Here we present the well-known results of the LSC for (2.9) by application of the Lie–Ovsiannikov algorithm. The algorithm is based on the classical Lie method (called also "infinitesimal method") and a set of equivalence transformations (ETs) of the differential equation in question. It is worth noting that there are some recent approaches, which are important for obtaining the so-called canonical list of inequivalent equations admitting nontrivial Lie symmetries and allow us to solve the problem of the LSC in a more efficient way than a formal application of the Lie–Ovsiannikov algorithm. We will discuss this issue in subsequent sections. Here we want to emphasize that nowadays it is widely accepted that the problem of LSC is completely solved for the given PDE class if it has been proved that

i) the Lie symmetry algebras are MAIs of the relevant PDEs from the list obtained;

ii) all PDEs from the list are inequivalent with respect to (w.r.t.) a set of transformations, which are explicitly (or implicitly) presented and, generally speaking, may not form any group;

iii) any other PDE from the class that admits a nontrivial Lie symmetry algebra is reduced by transformations from the set to one of those from the list.

In the case of the Lie–Ovsiannikov algorithm, the second item is specified by using groups of the (continuous) ETs. In other words, the LSC problem should be solved up to the group of the ETs of the PDE class in question according to this algorithm.

Definition 2.1 *[25] A one-parameter group of ETs of the class of m-order PDEs*

$$L(t, x, u, u_1, \ldots, u_m, K_1, \ldots, K_p) = 0 \qquad (2.10)$$

(here $K_1, \ldots, K_p$ are some functions (parameters), which may depend on t, x, u and/or derivatives of u) is a one-parameter Lie group of transformations given by

$$\bar{t} = f(t, x, u, \epsilon), \quad \bar{x}_a = g_a(t, x, u, \epsilon), \quad a = 1, \ldots, n, \quad \bar{u} = h(t, x, u, \epsilon),$$
$$\bar{K}_i = F_i(t, x, u, K_1, \ldots, K_p, \epsilon), \quad i = 1, \ldots, p \qquad (2.11)$$

where $x = (x_1, \ldots, x_n)$ and ϵ is the group parameter, which maps each equation of the form (2.10) into an equation belonging to the same class.

In this definition and in what follows $\underset{1}{u} \equiv (u_t, u_{x_1}, ..., u_{x_n})$, $\underset{2}{u} \equiv (u_{tt}, u_{tx_1}, ..., u_{x_n x_n}), ...$, i.e., it is a totality of derivatives of the corresponding order (the subscripts $t, x_1 ... x_n$, denote differentiation w.r.t. these variables). Obviously, this definition can be easily generalized on multi-parameter groups.

Definition 2.2 *The multi-parameter group $\mathcal{E}$ is called the group of ETs for the PDE class (2.10), if $\mathcal{E}$ contains as subgroups all one-parameter groups of ETs of this class.*

It is worth noting that transformations (2.11) may transform an equation into itself, then such transformations coincide with the relevant one-parameter group of invariance of PDE in question. This observation leads to another important notion.

Definition 2.3 *A Lie group, which is the invariance group for all equations belonging to the PDE class (2.10) and contains as a subgroup any other Lie group that is common for all PDEs from this class, is called the principal Lie group (another terminology is "kernel of main groups"). The corresponding Lie algebra is called the principal algebra of the class in question.*

It can be easily proved using the above definitions that the principal Lie group $\mathcal{E}_0$ must be a subgroup of $\mathcal{E}$ [203].

The technique for constructing the (continuous) group of ETs is well-known (see, e.g., [138, 203]) and is based on a relevant modification of the classical Lie method (called also "infinitesimal method"). It means that the corresponding Lie algebra is calculated instead of $\mathcal{E}$. In the case of the nonlinear heat equation (2.9), this group was established in [203].

Theorem 2.1 *The group of the continuous ETs of the PDE class (2.9) is the 6-parameter Lie group*

$$\bar{t} = e_0 t + t_0, \quad \bar{x} = e_1 x + x_0, \quad \bar{u} = e_2 u + u_0, \quad \bar{d} = \frac{e_1^2}{e_0} d, \qquad (2.12)$$

where $e_0, t_0, e_1, x_0, e_2,$ and u_0 are arbitrary group parameters ($e_i > 0$, $i = 0, 1, 2$).

Obviously, the class of PDEs (2.9) is also invariant with respect to the *discrete* transformations

$$\begin{aligned}(a) \quad & \bar{t} = -t, \quad \bar{d} = -d, \\ (b) \quad & \bar{x} = -x, \\ (c) \quad & \bar{u} = -u,\end{aligned} \qquad (2.13)$$

hence one may assume that $e_i \neq 0$, $i = 0, 1, 2$ without losing a generality. Thus, the group of the ETs of (2.9) is 6-parameter Lie group $\mathcal{E}$. The corresponding

Lie algebra is generated by the three operators of translations ∂_t, ∂_x, ∂_u, the unit operator I and the operators of scaling transformations

$$D_0 = 2t\partial_t + x\partial_x, \quad D_d = x\partial_x + 2d\partial_d. \qquad (2.14)$$

In order to find all possible Lie symmetry operators of the form

$$X = \xi^0(t,x,u)\partial_t + \xi^1(t,x,u)\partial_x + \eta(t,x,u)\partial_u, \qquad (2.15)$$

which an equation of the form (2.9) can admit, one needs to apply the invariance criteria (see Definition 1.2 in Chapter 1)

$$\underset{2}{X}\left(u_t - d(u)u_{xx} - d_u(u)u_x^2\right)\Big|_{\mathcal{M}} = 0, \qquad (2.16)$$

where

$$\mathcal{M} \equiv \{u_t - d(u)u_{xx} - d_u(u)u_x^2 = 0\} \qquad (2.17)$$

(hereafter $d_u(u) \equiv \frac{d}{du}d(u)$) is the manifold in the prolonged space of the variables

$$t, x, u, u_t, u_x, u_{tt}, u_{xx}, u_{xt}$$

(it means that all variables and derivatives are independent in this space). The operator

$$\underset{2}{X} = X + \zeta^0 \partial_{u_t} + \zeta^1 \partial_{u_x} + \sigma^{00}\partial_{u_{tt}} + \sigma^{01}\partial_{u_{tx}} + \sigma^{11}\partial_{u_{xx}} \qquad (2.18)$$

is the second prolongation of the operator X and its coefficients are defined by the formulae (they follow as particular cases from more general formulae (1.21))

$$\begin{aligned}
\zeta^0 &= \eta_t + u_t\eta_u - u_t(\xi^0_t + u_t\xi^0_u) - u_x(\xi^1_t + u_t\xi^1_u),\\
\zeta^1 &= \eta_x + u_x\eta_u - u_t(\xi^0_x + u_x\xi^0_u) - u_x(\xi^1_x + u_x\xi^1_u),\\
\sigma^{00} &= \eta_{tt} + 2u_t\eta_{tu} + u_t^2\eta_{uu} + u_{tt}\eta_u -\\
&\quad - u_t(\xi^0_{tt} + 2u_t\xi^0_{tu} + u_t^2\xi^0_{uu} + u_{tt}\xi^0_u) -\\
&\quad - u_x(\xi^1_{tt} + 2u_t\xi^1_{tu} + u_t^2\xi^1_{uu} + u_{tt}\xi^1_u) -\\
&\quad - 2u_{tt}(\xi^0_t + u_t\xi^0_u) - 2u_{tx}(\xi^1_t + u_t\xi^1_u),\\
\sigma^{01} &= \eta_{tx} + u_x\eta_{tu} + u_t\eta_{ux} + u_t u_x\eta_{uu} + u_{tx}\eta_u -\\
&\quad - u_t(\xi^0_{tx} + u_x\xi^0_{tu} + u_t\xi^0_{ux} + u_t u_x\xi^0_{uu} + u_{tx}\xi^0_u) -\\
&\quad - u_x(\xi^1_{tx} + u_x\xi^1_{tu} + u_t\xi^1_{ux} + u_t u_x\xi^1_{uu} + u_{tx}\xi^1_u) -\\
&\quad - u_{tt}(\xi^0_x + u_x\xi^0_u) - u_{tx}(\xi^0_t + u_t\xi^0_u) -\\
&\quad - u_{tx}(\xi^1_x + u_x\xi^1_u) - u_{xx}(\xi^1_t + u_t\xi^1_u),\\
\sigma^{11} &= \eta_{xx} + 2u_x\eta_{xu} + u_x^2\eta_{uu} + u_{xx}\eta_u -\\
&\quad - u_t(\xi^0_{xx} + 2u_x\xi^0_{xu} + u_x^2\xi^0_{uu} + u_{xx}\xi^0_u) -\\
&\quad - u_x(\xi^1_{xx} + 2u_x\xi^1_{xu} + u_x^2\xi^1_{uu} + u_{xx}\xi^1_u) -\\
&\quad - 2u_{tx}(\xi^0_x + u_x\xi^0_u) - 2u_{xx}(\xi^1_x + u_x\xi^1_u)
\end{aligned} \qquad (2.19)$$

Symmetry of the nonlinear diffusion equation

TABLE 2.1: All possible extensions of the algebra A^{pr} of equations from the PDE class (2.9).

	The form of equation	MAI
1	$u_t = (e^u u_x)_x$	$\langle A^{pr}, \, x\partial_x + 2\partial_u \rangle$
2	$u_t = (u^k u_x)_x, \, k \neq 0, -\frac{4}{3}$	$\langle A^{pr}, \, kx\partial_x + 2u\partial_u \rangle$
3	$u_t = (u^{-\frac{4}{3}} u_x)_x$	$\langle A^{pr}, \, 2x\partial_x - 3u\partial_u, \, x^2\partial_x - 3xu\partial_u \rangle$

Substituting (2.18) with coefficients (2.19) into (2.16), one arrives at a cumbersome expression, which should vanish provided (2.17) takes place. At the next step, the expression obtained can be split into separate parts for the derivatives $u_x, u_x^2, u_x^3, u_{xx}$ and $u_x u_{xx}$ (the time derivative u_t should be expressed via other derivatives using the relation (2.17)). As a result, the following system of determining equations (DEs) is obtained [203]:

$$\xi_x^0 = \xi_u^0 = \xi_u^1 = \eta_{uu} = 0, \qquad (2.20)$$

$$\eta d_u = (2\xi_x^1 - \xi_t^0)d, \qquad (2.21)$$

$$2\eta_x d_u = (\xi_{xx}^1 - 2\eta_{xu})d + \xi_t^1, \qquad (2.22)$$

$$d\eta_{xx} = \eta_t. \qquad (2.23)$$

Obviously, the general solution of this DEs system depends essentially on the form of the diffusivity $d(u)$. All the solutions of the system, which do not depend on $d(u)$, lead to the principal algebra of invariance of the PDE class (2.9). Thus, assuming that $d(u)$ is an arbitrary function, Eqs. (2.21)–(2.23) are reduced to the conditions

$$\eta = \xi_t^1 = \xi_{xx}^1 = 0, \quad 2\xi_x^1 - \xi_t^0 = 0. \qquad (2.24)$$

The general solution of system (2.20) and (2.24) can be easily constructed. As a result, one obtains

$$\xi^0 = 2ct + t_0, \quad \xi^1 = cx + x_0, \quad \eta = 0, \qquad (2.25)$$

where c, t_0 and x_0 are arbitrary parameters. Thus, the three-dimensional principal algebra of the PDE class (2.9) with the basic operators

$$A^{pr} = \langle \partial_t, \, \partial_x, \, D_0 = 2t\partial_t + x\partial_x \rangle \qquad (2.26)$$

is obtained.

A crucial step is to solve the above system of DEs taking into account that some forms of the diffusivity $d(u)$, which should be determined, lead to Lie algebras of invariance of higher dimensionality. In [203], this step is described in detail, hence we present only the result.

Theorem 2.2 *All possible MAIs of four- and of higher dimensionality (up to the group of equivalent transformations $\mathcal{E}$) and the corresponding nonlinear equations from class (2.9) are presented in Table 2.1. Any other equation of the form (2.9) invariant under a nontrivial Lie algebra is reduced to one of those given in Table 2.1 by an ET from $\mathcal{E}$.*

Now we turn to multidimensional nonlinear diffusion equations. As it was demonstrated in Chapter 1, nonlinear diffusion equations of the form

$$u_t = \mathrm{div}\,[d(u)\nabla u] \tag{2.27}$$

(hereafter $u = u(t,x)$ is an unknown function, $x = (x_1,...x_n)$, $\nabla u \equiv (u_{x_1},...,u_{x_n})$) have an enormous number of diverse applications. Accordingly they have been widely studied by the classical Lie method for a long time, in particular, in [95, 186] for $n = 2, 3$. Recently we noted that Lie symmetries of the (1+2)-dimensional nonlinear equation (2.27) have been described for the first time in paper [186], incidentally not cited so often as [95] published 13 years later. Here we summarize the known results in the form of three theorems and present a brief analysis of the algebras of invariance.

Theorem 2.3 *The group $\mathcal{E}$ of continuous ETs of the class of PDEs (2.27) consists of the transformations*

$$\bar{t} = e_0 t + \delta_0, \quad \bar{x}_a = e_1 \gamma_{ab} x_b + \delta_a, \quad \bar{u} = e_2 u + \delta, \quad \bar{d} = \frac{e_1^2}{e_0} d, \tag{2.28}$$

where $e_0 > 0, e_1 > 0, e_2 > 0$, $\delta, \delta_0, \delta_a$ ($a = 1,\ldots,n$), are arbitrary real constants and γ_{ab} ($a, b = 1,\ldots,n$) are real elements of the matrix $(\gamma_{ab}) \in SO(n)$.

Taking into account the similar discrete transformations to those in the (1+1)-dimensional case, one may again assume that e_0, e_1 and e_2 are simply nonzero constants. So, the group of the ETs of (2.27) is the ($\frac{n(n+1)}{2} + 5$)-parameter Lie group $\mathcal{E}$. Similarly to the (1+1)-dimensional case, the corresponding Lie algebra is generated by the operators of translations, the unit operator I and the operators of scaling transformations. However, this algebra also contains the Lie algebra of rotations in $\mathbb{R}^n$ as a subalgebra. In the cases $n = 2$ and $n = 3$, the latter has a clear geometrical interpretation.

In contrast to the linear equation (2.4), Lie symmetry of the multidimensional nonlinear heat equation essentially depends on number n because $n = 2$ is a special case. In fact, $(1+2)$-dimensional diffusion equation with the diffusivity $d = u^{-1}$ is invariant w.r.t. an infinite-dimensional Lie algebra. This exceptional diffusivity has no analogs in the case $n > 2$.

Theorem 2.4 *[95], [186] The principal algebra of invariance of the PDE class*

$$u_t = [d(u)u_{x_1}]_{x_1} + [d(u)u_{x_2}]_{x_2} \tag{2.29}$$

TABLE 2.2: All possible extensions of the algebra A^{pr} of equations from the PDE class (2.29).

	The form of equation	MAI
1	$u_t = (e^u u_{x_1})_{x_1} + (e^u u_{x_2})_{x_2}$	$\langle A^{pr}, x_1\partial_{x_1} + x_2\partial_{x_2} + 2\partial_u \rangle$
2	$u_t = (u^k u_{x_1})_{x_1} + (u^k u_{x_2})_{x_2}, k \neq 0, -1$	$\langle A^{pr}, kx_1\partial_{x_1} + kx_2\partial_{x_2} + 2u\partial_u \rangle$
3	$u_t = (u^{-1} u_{x_1})_{x_1} + (u^{-1} u_{x_2})_{x_2}$	$\langle A^{pr}, A\partial_{x_1} + B\partial_{x_2} - 2A_{x_1}u\partial_u \rangle$

is the five-dimensional Lie algebra A^{pr} with the basic operators

$$\partial_t, \quad \partial_1, \quad \partial_2,$$
$$J_{12} = x_1\partial_2 - x_2\partial_1, \quad D = 2t\partial_t + x_1\partial_1 + x_2\partial_2. \quad (2.30)$$

All possible MAIs of higher dimensionality (up to the group of ETs $\mathcal{E}$ (2.28) with $n = 2$) and the corresponding nonlinear equations from class (2.29) are presented in Table 2.2. Any other equation of the form (2.29) invariant under a six-dimensional (or higher) Lie algebra is reduced to one of those given in Table 2.2 by an ET from $\mathcal{E}$.

In case 3 of Table 2.2, the pair $(A(x_1, x_2), B(x_1, x_2))$ is an arbitrary solution of the Cauchy-Riemann system $A_{x_1} = B_{x_2}$, $A_{x_2} = -B_{x_1}$. Obviously, this system possesses an infinite number of linearly independent solutions and it is a reason why the corresponding operators generate the infinite-dimensional Lie algebra.

Now we present the result of LSC in case $n > 2$.

Theorem 2.5 *The principal algebra of invariance of the class of PDEs (2.27) with $n > 2$ is $(\frac{n(n+1)}{2} + 2)$-dimensional Lie algebra A^{pr} with the basic operators*

$$\partial_t, \quad \partial_a \ (a = 1, \ldots, n),$$
$$J_{ab} = x_a\partial_b - x_b\partial_a \ (b < a = 1, \ldots, n), \quad D = 2t\partial_t + x_a\partial_a. \quad (2.31)$$

All possible MAIs of higher dimensionality (up to the group of equivalent transformations $\mathcal{E}$ (2.28)) and the corresponding nonlinear equations from class (2.27) are presented in Table 2.3. Any other equation of the form (2.27) invariant under a Lie algebra, which is wider than the principal algebra, is reduced to one of those given in Table 2.3 by an ET from $\mathcal{E}$.

It is worth commenting MAI occurring in case 3 of Table 2.3. This Lie algebra is the classical conformal algebra $AC(n)$ (in the space $\mathbb{R}^n$ with the standard metric) extended by the operator of time translation ∂_t and the operator of scale translations D from (2.31). The conformal algebra was studied in an enormous number of papers and books (see, e.g., [127] and the references

TABLE 2.3: All possible extensions of the algebra A^{pr} of equations from the PDE class (2.27).

	The form of equation	MAI
1	$u_t = \text{div}\,(e^u \nabla u)$	$\langle A^{pr},\ x_a \partial_a + 2\partial_u \rangle$
2	$u_t = \text{div}\,(u^k \nabla u),\ k \neq 0, -\frac{4}{n+2}$	$\langle A^{pr},\ kx_a \partial_a + 2u\partial_u \rangle$
3	$u_t = \text{div}\,\left(u^{-\frac{4}{n+2}} \nabla u\right)$	$\langle A^{pr},\ D_1 = x_a \partial_a - \frac{n+2}{2} u \partial_u,$ $K_a = 2x_a D_1 - x_b x_b \partial_a, a = 1,\ldots,n \rangle$

therein) because one appears as the invariance algebra of many equations and models, for example, the classical wave equation and its generalizations [114]. All the nonzero commutators of the $(\frac{n(n+3)}{2} + 1)$-dimensional algebra $AC(n)$ with the basic operators ∂_a, J_{ab}, D_1 and K_a are as follows

$$[\partial_a, J_{bc}] = \delta_{ab}\partial_c - \delta_{ac}\partial_b,\quad [J_{ab}, J_{cd}] = \delta_{ac}J_{db} + \delta_{bc}J_{ad} + \delta_{ad}J_{bc} + \delta_{bd}J_{ca},$$

$$[\partial_a, D_1] = \partial_a,\quad [\partial_a, K_b] = 2(\delta_{ab}D_1 - J_{ab}),$$

$$[J_{ab}, K_c] = \delta_{bc}K_a - \delta_{ac}K_b,\quad [D_1, K_a] = K_a.$$

Two additional operators occurring in case 3 of Table 2.3 produce the following nonzero commutators:

$$[\partial_t, D] = 2\partial_t,\quad [\partial_a, D] = \partial_a,\quad [D, K_a] = K_a.$$

The exponent $-\frac{4}{n+2}$ arising in the diffusion coefficient is often called conformal. Interestingly, this conformal exponent is still the same if one considers some generalizations of (2.27) like systems of multidimensional reaction-diffusion (RD) equations (see for detail [67]).

Two cases, $n = 1$ and $n = 2$, are particularly noteworthy. In fact, the first one corresponds to conformal exponent $-\frac{4}{3}$ and the conformal algebra $AC(1)$ with the basic operators (see Table 2.1)

$$\partial_x,\quad x\partial_x - \frac{3}{2}u\partial_u,\quad x^2\partial_x - 3xu\partial_u. \tag{2.32}$$

It is a special case because the three above operators are also a realization of the algebra $sl(2, \mathbb{R})$.

The second case corresponds to conformal exponent -1 and we already know that the corresponding algebra of invariance is infinite-dimensional (see Table 2.2). However, the conformal algebra $AC(2)$ can be extracted as a subalgebra using appropriate solutions of the Cauchy-Riemann system. In fact, the particular solution (x_1, x_2) of this system leads to the operator of scale transformations

$$D_1 = x_1 \partial_1 + x_2 \partial_2 - 2u\partial_u \tag{2.33}$$

while the solutions $(x_1^2-x_2^2, 2x_1x_2)$ and $(2x_1x_2, x_2^2-x_1^2)$ generate the operators of conformal transformations

$$K_a = 2x_a D_1 - x_b x_b \partial_a, \quad a = 1, 2. \tag{2.34}$$

Now one may check via calculating commutator relations that the Lie algebra with the basic operators $\partial_1, \partial_2, J_{12}$ from (2.30) and operators (2.33)–(2.34) is nothing else but the conformal algebra $AC(2)$.

2.3 Equivalence transformations and form-preserving transformations

Now we turn to the class of reaction-diffusion-convection (RDC) equations

$$u_t = [A(u)u_x]_x + B(u)u_x + C(u), \tag{2.35}$$

where A, B and C are arbitrary smooth functions. Equations from this class arise in a wide range of real world applications and several examples were presented in Chapter 1.

The class of RDC equations (2.35) contains three arbitrary functions and the Lie symmetry of its different representatives depends essentially on the form of the triplet (A, B, C). Thus, the LSC problem arises. In Section 2.2, it was shown that the Lie–Ovsiannikov algorithm, which is based on the classical Lie scheme and a set of equivalence transformations (ETs) of the PDE class in question, can be applied for solving such kind of problems. However this algorithm may lead to a very long list of equations with nontrivial Lie symmetry provided the given class of PDEs (systems of PDEs) contains several arbitrary functions. In particular, it will be shown in subsequent sections that class (2.35) contains 30 equations, which admit nontrivial Lie algebras of invariance and are inequivalent w.r.t. the group of ETs of (2.35).

During the last two decades new approaches for solving the LSC problems were developed, which are important for obtaining the so-called canonical list of inequivalent equations admitting nontrivial Lie symmetry algebras and allow us to solve this problem in a more efficient way than a formal application of the Lie–Ovsiannikov algorithm. Here we use the algorithm based on so-called form-preserving transformations (FPTs) [152, 153], which were used initially for finding locally-equivalent PDEs, especially those nonlinear PDEs that are linearizable by a point transformation (i.e., a nongenerating transformation, which does not involve derivatives of unknown function(s)). It is worth noting that these transformations are called also "admissible transformations" following the 1992 paper [122], in which they were used to classify the Lie symmetries of a class of variable coefficient Korteweg–de Vries equations. Interestingly such transformations were implicitly used much earlier in the 1978 paper [189] in order to find all possible heat equations with nonlinear sources that admit the Lie symmetry either of the linear heat equation or the Burgers equation.

It was shown later that the FPTs allow us an essential reduction of the number of cases obtained via the Lie–Ovsiannikov algorithm (see, e.g., extensive discussions on this matter in [65, 66, 82]). For example, it was proved using a set of FPTs that the canonical list of inequivalent two-component systems of RD equations (with a nonconstant diffusivity) consists of 10 systems only [66] (not about 30 systems derived by the Lie–Ovsiannikov algorithm in [155]). It turns out that the canonical list of inequivalent RDC equations consists of 16 equations, while 30 equations are obtained by the direct applications of the Lie–Ovsiannikov algorithm. In order to prove this we apply both algorithms mentioned above to the class of RDC equations (2.35).

In subsequent sections, both algorithms are united in a single one consisting of the following separate steps.

1. Construction of the group of ETs $\mathcal{E}$ for the class of RDC equations (2.35).

2. Construction of FPTs for this class.

3. Applying the classical Lie method for deriving the system of DEs.

4. Finding the principal algebra A^{pr} by solving the system of DEs in a special case.

5. Deriving necessary conditions for possible extensions of A^{pr} (i.e., existence of nontrivial Lie symmetry).

6. Finding sufficient conditions for extensions of A^{pr} and deriving LSC for class (2.35) using the group of ETs $\mathcal{E}$.

7. Deriving LSC for class (2.35) using FPTs.

2.3.1 The group of equivalence transformations

In order to find the group of ETs $\mathcal{E}$ for class (2.35), we apply the standard technique, which was used by Ovsiannikov for the first time and later was formalized in [3, 138] (see also [163] for detail).

Theorem 2.6 *The group of the continuous ETs of the class of RDC equations (2.35) is the 7-parameter Lie group*

$$\bar{t} = e_0 t + t_0, \quad \bar{x} = e_1 x + gt + x_0, \quad \bar{u} = e_2 u + u_0,$$
$$\bar{A} = \tfrac{e_1^2}{e_0} A, \quad \bar{B} = \tfrac{e_1}{e_0} B, \quad \bar{C} = \tfrac{1}{e_0} C, \tag{2.36}$$

where $e_0, t_0, e_1, x_0, g, e_2,$ and u_0 are arbitrary group parameters ($e_i > 0, i = 0, 1, 2$).

Equivalence transformations and form-preserving transformations 31

Proof As it was pointed out above, the technique for constructing the group of ETs (continuous) is based on a modification of the classical Lie method. In the case of class (2.35), we should start from the infinitesimal operator

$$E = \xi^0(t,x,u)\partial_t + \xi^1(t,x,u)\partial_x + \eta(t,x,u)\partial_u + \\ +\zeta^1(t,x,u,A,B,C)\partial_A + \zeta^2(t,x,u,A,B,C)\partial_B + \zeta^3(t,x,u,A,B,C)\partial_C \quad (2.37)$$

being ξ^0, ξ^1, η, ζ^1, ζ^2 and ζ^3 to-be-determined functions. The coefficients ξ^0, ξ^1 and η do not depend on the function triplet (A,B,C) because any ETs cannot involve these functions for transformations of the variables t, x and u.

In order to find the operator E, we should apply Lie's invariance criteria to the system of equations consisting of (2.35) and a set of differential consequences of the triplet (A,B,C) w.r.t. the variables t, x and the first-order derivatives u_t and u_x. Of course, each consequence is equal to zero. As a result, we obtain a multicomponent system of PDEs, however one mostly consists of primitive equations excepting Eq. (2.35) (the algorithm for search ETs implies that it is no longer the class of PDEs but a single equation involving three additional variables). Thus, the invariance criteria takes the form to the operator E

$$\underset{2}{E}\bigl(u_t - (Au_x)_x + Bu_x + C\bigr)\Big|_{\mathcal{M}} = 0, \quad \underset{2}{E}S^i\Big|_{\mathcal{M}} = 0, \; i = 1,\dots,12, \quad (2.38)$$

where $\underset{2}{E}$ is the second-order prolongation of operator (2.37) and $\mathcal{M} = \{u_t = (Au_x)_x + Bu_x + C$, $S^1 \equiv A_t = 0$, $S^2 \equiv A_x = 0$, $S^3 \equiv B_t = 0$, $S^4 \equiv B_x = 0$, $S^5 \equiv C_t = 0$, $S^6 \equiv C_x = 0$, $S^7 \equiv A_{u_t} = 0$, $S^8 \equiv A_{u_x} = 0$, $S^9 \equiv B_{u_t} = 0$, $S^{10} \equiv B_{u_x} = 0$, $S^{11} \equiv C_{u_t} = 0$, $S^{12} \equiv C_{u_x} = 0\}$.

After the straightforward calculations using formulae (2.38), which are omitted here in order to avoid cumbersome expressions, one arrives at the following system of PDEs for the functions ξ^0, ξ^1, η, ζ^a, $a = 1,2,3$:

$$\xi^0_x = \xi^0_u = \xi^1_u = \eta_t = \eta_x = \eta_{uu} = \zeta^a_t = \zeta^a_x = 0, \quad (2.39)$$

$$\zeta^1 = (2\xi^1_x - \xi^0_t)A, \quad (2.40)$$

$$\zeta^2 = (\xi^1_x - \xi^0_t)B + \xi^1_{xx}A - \xi^1_t, \quad (2.41)$$

$$\zeta^3 = (\eta_u - \xi^0_t)C. \quad (2.42)$$

Differentiating Eqs. (2.40)–(2.42) w.r.t. the variable u and taking into account (2.39), one obtains

$$\zeta^a_u = 0, \; a = 1,2,3 \quad (2.43)$$

Thus, the functions ζ^a do not depend on t, x and u.

Similarly differentiating Eqs. (2.40)–(2.42) w.r.t. x and t, and using (2.39), one obtains the linear second-order PDEs

$$\xi^0_{tt} = \xi^1_{tt} = \xi^1_{xx} = \xi^1_{tx} = 0. \quad (2.44)$$

Integrating (2.44) and the first three equations from (2.39), one easily finds the functions ξ^0, ξ^1 and η. Having the latter, the functions ζ^a can be immediately calculated using Eqs. (2.40)–(2.42). As a result, the following coefficients of the infinitesimal operator (2.37) are derived

$$\xi^0 = k_0 t + d_0, \quad \xi^1 = k_1 x + gt + d_1, \quad \eta = k_2 u + d_2,$$
$$\zeta^1 = (2k_1 - k_0)A, \quad \zeta^2 = (k_1 - k_0)B - g, \quad \zeta^3 = (k_2 - k_0)C, \quad (2.45)$$

where k_μ, d_μ, $\mu = 0, 1, 2$ and g are arbitrary parameters. Thus, operator (2.37) with the coefficients (2.45) generate 7-dimensional Lie algebra. At the final step, having the above algebra, one easily finds the corresponding Lie group, which is nothing else but the group of ETs $\mathcal{E}$ given by formulae (2.36).

The proof is now completed. □

Remark 2.1 *Each equation belonging to the RDC equation class (2.35) with $B(u) = \lambda$ can be reduced to an equation belonging to the RD equation class (i.e., (2.35) with $B(u) = 0$) (2.51) by the ET $x^* = x + \lambda t$.*

2.3.2 Form-preserving transformations

Now we turn to FPTs. Roughly speaking, a given FPT is a local substitution, which reduces some PDE from the given class to another PDE belonging to the same class. The rigorous definition can be as follows.

Definition 2.4 *A point transformation given by*

$$\bar{t} = f(t, x, u), \quad \bar{x}_a = g_a(t, x, u), \quad \bar{u} = h(t, x, u), \quad (a = 1, \ldots, n), \quad (2.46)$$

which maps at least one equation of the form (2.10) into an equation belonging to the same class, is called the FPTs for the PDE class (2.10).

Comparing this definition with Definition 2.2, one immediately notes that each ET from $\mathcal{E}$ is automatically a FPT but not vice versa. In contrast to the ETs, a set of all possible FPTs for the given class of PDEs usually do not form a Lie group. However, a subset of FPTs may generate a group of ETs on a subclass of the given class [66]. This is a reason why FPTs are also called additional ETs.

Now we present an example, which highlights very nicely the difference between ET and FPT.

Example 2.1 *Consider the RDC equation*

$$u_t = u_{xx} + \lambda_1 u u_x + \lambda_0, \quad (2.47)$$

(here λ_0 and λ_1 are arbitrary constants), which coincides with the well-known Burgers equation in the case $\lambda_0 = 0$.

Equivalence transformations and form-preserving transformations 33

One may easily check that the transformation

$$\bar{t} = t, \quad \bar{x} = x + \epsilon t, \quad \bar{u} = u - \epsilon \tag{2.48}$$

(here ϵ is an arbitrary parameter) maps (2.47) into itself, hence it is an example of FPT. On the other hand, transformation (2.48) belongs to the group of ETs of the class of RDC equations (2.35) (see Theorem 2.6). Thus, it is not a pure FPT.

It was found in [80] that the transformation

$$\bar{t} = t, \quad \bar{x} = x - \frac{1}{2}\lambda_0\lambda_1 t^2, \quad \bar{u} = u - \lambda_0 t \tag{2.49}$$

maps (2.47) into the Burgers equation for $\bar{u}(\bar{t}, \bar{x})$. Since both equations belong to the class of RDC equations (2.35), one concludes that (2.49) is FPT. However, transformation (2.49) does not belong to the group $\mathcal{E}$ for this class (see Theorem 2.6). Thus, it is a pure FPT.

Remark 2.2 *Transformation (2.49) was missed in [189] where all semi-linear parabolic equations have been constructed that are reducible to the Burgers equation by local substitutions.*

In what follows we will use the notion of the pure FPT in order to underline that it is not ET for the PDE class in question. Now we present the main result of this section, that is the theorem describing a general form of FPTs for the class of RDC equations (2.35).

Theorem 2.7 *[83] An arbitrary RDC equation of the form (2.35) can be reduced to another equation of the same form*

$$w_\tau = [F(w)w_y]_y + G(w)w_y + H(w) \tag{2.50}$$

by the local transformation

$$\tau = a(t, x, u), \quad y = b(t, x, u), \quad w = c(t, x, u) \tag{2.51}$$

with the correctly-specified smooth functions a, b and c if and only if these functions are of the form

$$a = a(t), \quad b = b(t, x), \quad c = \alpha(t, x)u + \beta(t, x), \tag{2.52}$$

and the following equalities take place

$$b_x^2 A(u) = a_t F(\alpha u + \beta), \tag{2.53}$$

$$-2\frac{b_x \alpha_x}{\alpha}\frac{d}{du}[uA(u)] - 2\frac{b_x \beta_x}{\alpha}A_u(u) + b_{xx}A(u) + b_x B(u) = a_t G(\alpha u + \beta) + b_t, \tag{2.54}$$

$$\frac{\alpha_x^2}{\alpha}\frac{d}{du}[u^2 A(u)] + 2\frac{\alpha_x \beta_x}{\alpha}\frac{d}{du}[uA(u)] + \frac{\beta_x^2}{\alpha}A_u(u) - (\alpha_{xx}u + \beta_{xx})A(u) - \\ -(\alpha_x u + \beta_x)B(u) + \alpha C(u) + \alpha_t u + \beta_t = a_t H(\alpha u + \beta) \tag{2.55}$$

provided

$$a_t b_x \alpha \neq 0. \tag{2.56}$$

Proof Firstly we note that any FPT (2.51) must be nondegenerate, i.e., its Jacobian is nonvanish:

$$J(t, x, u) = \begin{vmatrix} a_t & a_x & a_u \\ b_t & b_x & b_u \\ c_t & c_x & c_u \end{vmatrix} \neq 0. \qquad (2.57)$$

Having transformation (2.51) one can express the derivatives of the function u by the well-known formulas (see, e.g., [157])

$$u_t = -\frac{w_\tau a_t + w_y b_t - c_t}{w_\tau a_u + w_y b_u - c_u}, \quad u_x = -\frac{w_\tau a_x + w_y b_x - c_x}{w_\tau a_u + w_y b_u - c_u},$$

$$u_{xx} = -\frac{w_{\tau\tau}(a_x+a_u u_x)^2 + 2w_{\tau y}(a_x+a_u u_x)(b_x+b_u u_x) + w_{yy}(b_x+b_u u_x)^2}{w_\tau a_u + w_y b_u - c_u} - \qquad (2.58)$$

$$-\frac{w_\tau(a_{xx}+2a_{xu}u_x+a_{uu}u_x^2) + w_y(b_{xx}+2b_{xu}u_x+b_{uu}u_x^2) - (c_{xx}+2c_{xu}u_x+c_{uu}u_x^2)}{w_\tau a_u + w_y b_u - c_u}.$$

Substituting (2.58) into (2.35) one arrives at a rather cumbersome expression. Let us assume that (2.51) is a FPT. So, the expression obtained must be reduced to an equation of the form (2.50). In the particular case, the coefficient next to the second-order derivative $w_{\tau\tau}$ should vanish and the coefficients next to the derivatives u_{xx} and w_{yy} must be equal, hence one obtains the system

$$\begin{aligned} a_x + a_u u_x &= 0, \\ (b_x + b_u u_x)^2 A(u) &= a_t F(w). \end{aligned} \qquad (2.59)$$

Since the functions $A(u)$ and $F(w)$ do not depend on any derivatives, system (2.59) immediately leads to the conditions

$$a_x = a_u = b_u = 0, \qquad (2.60)$$

$$b_x^2 A(u) = a_t F(w). \qquad (2.61)$$

Obviously, (2.61) is exactly Eq. (2.53) arising in the theorem.

Taking into account (2.60), we can now simplify the condition (2.57) to the form (2.56).

Substituting (2.60)–(2.61) into formulas (2.58), we establish that those can be essentially simplified:

$$u_t = \frac{w_\tau a_t + w_y b_t - c_t}{c_u}, \quad u_x = \frac{w_y b_x - c_x}{c_u},$$

$$u_{xx} = \frac{w_{yy} b_x^2 + w_y b_{xx} - c_{xx}}{c_u} - 2\frac{c_{xu}}{c_u^2}(w_y b_x - c_x) - \frac{c_{uu}}{c_u^3}(w_y b_x - c_x)^2. \qquad (2.62)$$

Now we substitute (2.62) into (2.35) and obtain the equation

$$w_\tau a_t + w_y b_t - c_t = A_u(u)\frac{1}{c_u}(w_y b_x - c_x)^2 + B(u)(w_y b_x - c_x) + C(u)c_u +$$
$$+ A(u)[w_{yy} b_x^2 + w_y b_{xx} - c_{xx} - 2\frac{c_{xu}}{c_u^2}(w_y b_x - c_x) - \frac{c_{uu}}{c_u^3}(w_y b_x - c_x)^2], \qquad (2.63)$$

which must coincide with (2.50). Taking into account (2.61), one can easily

… check that (2.63) is reduced to (2.50) if and only if

$$-A(u)\frac{c_{uu}}{c_u^3}b_x^2 + A_u(u)\frac{b_x^2}{c_u} = a_t\frac{dF(w)}{dw}, \qquad (2.64)$$

$$-b_t + A(u)(b_{xx} - 2\frac{c_{xu}}{c_u}b_x) - 2A_u(u)\frac{b_xc_x}{c_u} + B(u)b_x = a_tG(w), \qquad (2.65)$$

$$c_t - A(u)(c_{xx} - 2\frac{c_{xu}}{c_u}c_x) + A_u(u)\frac{c_x^2}{c_u} - B(u)c_x + C(u)c_u = a_tH(w). \qquad (2.66)$$

The last step is to define the function $c(t, x, u)$. Differentiating (2.61) with respect to (w.r.t.) u and substituting the expression obtained into (2.64), one arrives at $b_x c_{uu} = 0$. It gives

$$c_{uu} = 0 \qquad (2.67)$$

because of condition (2.56).

Thus, integrating (2.60) and (2.67), we obtain the FPTs (2.52). Finally substituting $w = \alpha(t, x)u + \beta(t, x)$ into (2.64)–(2.66) one sees that (2.64) is nothing else but a differential consequence of (2.61) while (2.65) and (2.66) are equivalent to (2.54) and (2.55), respectively.

The proof is now completed. □

Remark 2.3 *Nevertheless the PDE class (2.35) is a particular case of one arising in Theorems 4.2a and 4.2b [153], our theorem cannot be obtained formally as a simple corollary of those theorems. Of course, one may substitute the right-hand sides of (2.35) and (2.50) into condition (4.3)[153] and try to derive conditions (2.53)–(2.55), however this is rather a long way.*

Having this theorem, one may derive the group of ETs of (2.35) using formulae (2.52)–(2.55). Because this group was already derived using the standard technique, we present here how the group $\mathcal{E}$ of the nonlinear heat equation (2.9) can be easily obtained. In fact, assuming that transformation (2.51) with (2.52) belongs to $\mathcal{E}$, one concludes that equality (2.53) takes place for an *arbitrary* smooth function F. Since the function A is related with F via (2.53) and both functions do not depend on independent variables t and x, we immediately arrive at the conditions

$$a_t(t) = eb_x^2(t, x), \quad \alpha(t, x) = e_2, \quad \beta(t, x) = u_0 \qquad (2.68)$$

where e, e_2 and u_0 are arbitrary constants. Taking into account that the function $a(t)$ does not depend on x, we immediately observe that $b(t, x)$ must be linear w.r.t. x. Moreover, substituting conditions (2.68) into (2.54) with $B = G = 0$ (i.e., no convective terms in the nonlinear heat equation), one obtains

$$b_t(t, x) = 0. \qquad (2.69)$$

As a result, we arrive at the function

$$b(t, x) = e_1 x + x_0, \qquad (2.70)$$

where e_1 and x_0 are arbitrary constants.

Now one realizes that the last relation (2.55) with $B = C = H = 0$ (i.e., no convective and reactive terms) is automatically satisfied provided conditions (2.68) take place.

Finally, combining formulae (2.68)–(2.70), using (2.52) and substituting the derived functions a, b and c into (2.51), one obtains

$$\tau = e_0 t + t_0, \quad y = e_1 x + x_0, \quad w = e_2 u + u_0, \quad F = \frac{e_1^2}{e_0} A, \quad e_0 = e e_1^2 \quad (2.71)$$

with the restriction $e e_1 e_2 \neq 0$ (because of restriction (2.56)). On the other hand, formulae (2.71) coincide with those (2.12) up to notations. Thus, we proved Theorem 2.1 using FPTs in a much simpler way than it was done via the standard technique [203].

2.4 Determining equations for reaction-diffusion-convection equations

Here we construct a system of DEs for the class of reaction-diffusion-convection (RDC) equations (2.35). As it was shown in Chapter 1, a wide range of processes in physics, biology, chemistry etc. can be described by equations of the form (2.35). The most common equations belonging to this class are heat (diffusion) equations and their Lie symmetries were already discussed above.

In the case $B = 0$, we obtain the class of heat equations with sources

$$u_t = [A(u)u_x]_x + C(u), \quad (2.72)$$

which are known also as RD equations. Lie symmetries of the class of equations (2.72) were completely described in [94].

In the case $C = 0$, the class of heat equations with convective terms (another terminology is "diffusion-convection equations")

$$u_t = [A(u)u_x]_x + B(u)u_x, \quad (2.73)$$

is obtained from (2.50). The first attempt to describe completely Lie symmetries of this class was done in [200]. However, the results were incomplete, while the complete classification was derived much later in [215]. Notably, the Lie symmetry of the most representative equation from (2.73), the Burgers equation

$$u_t = u_{xx} + \lambda u u_x,$$

was found for the first time in [145].

The most general case of (2.35) with $B \neq 0$ was successfully studied in papers [80, 82] with aim to derive a complete LSC. It should be noted that the first attempt was done independently in [228] and [18], however the results obtained therein were incomplete.

Finally, all the known results about Lie symmetries of RDC equations were combined and systematized in [83]. Here and in the subsequent sections we present in detail how the problem of LSC is solved using the Lie–Ovsiannikov algorithm and a relatively new approach based on FPTs.

From the very beginning, we construct a system of DEs according to the classical Lie scheme. So, the class of equations (2.35) is considered as the manifold

$$\mathcal{M} \equiv \{u_t - Au_{xx} - A_u u_x^2 - Bu_x - C = 0\} \tag{2.74}$$

in the prolonged space of the variables

$$t, x, u, u_t, u_x, u_{tt}, u_{tx}, u_{xx}.$$

An equation of the form (2.35) is invariant under the transformations generated by the infinitesimal operator (2.15) when the following invariant criteria is satisfied:

$$\underset{2}{X} \left(u_t - Au_{xx} - A_u u_x^2 - Bu_x - C \right) \Big|_{\mathcal{M}} = 0. \tag{2.75}$$

Applying the second prolongation (2.18) of X to the equation, we obtain

$$\underset{2}{X} \left(u_t - Au_{xx} - A_u u_x^2 - Bu_x - C \right) = \\ = A\sigma_{11} + (2A_u u_x + B)\zeta_1 - \zeta_0 + \eta(A_u u_{xx} + A_{uu} u_x^2 + B_u u_x + C_u). \tag{2.76}$$

Taking into account (2.76) and substituting the coefficients from (2.19) into (2.75), one obtains a cumbersome expression. Because the expression should vanish on the manifold $\mathcal{M}$, the time derivative u_t can be excluded using (2.74). As a result, the following expression is obtained:

$$A\Big[\eta_{xx} + 2u_x \eta_{xu} + u_x^2 \eta_{uu} + u_{xx}\eta_u - 2u_{tx}(\xi_x^0 + u_x \xi_u^0) - \\ - u_x(\xi_{xx}^1 + 2u_x \xi_{xu}^1 + u_x^2 \xi_{uu}^1 + u_{xx}\xi_u^1) - 2u_{xx}(\xi_x^1 + u_x \xi_u^1)\Big] + \\ + \eta(A_u u_{xx} + A_{uu} u_x^2 + B_u u_x + C_u) + \\ + (2A_u u_x + B)\Big[\eta_x + u_x \eta_u - u_x(\xi_x^1 + u_x \xi_u^1)\Big] - \eta_t + u_x \xi_t^1 + \\ + \xi_u^0(Au_{xx} + A_u u_x^2 + Bu_x + C)^2 - \\ - (Au_{xx} + A_u u_x^2 + Bu_x + C)\Big[A(\xi_{xx}^0 + 2u_x \xi_{xu}^0 + u_x^2 \xi_{uu}^0 + u_{xx}\xi_u^0) + \\ + (2A_u u_x + B)(\xi_x^0 + u_x \xi_u^0) + \eta_u - \xi_t^0 - u_x \xi_u^1\Big] = 0 \tag{2.77}$$

Since the functions ξ^0, ξ^1 and η do not depend on derivatives of u, expression (2.77) can be split into separate parts, which must independently vanish.

In particular, the coefficients in front of u_{tx}, $u_x u_{tx}$, and $u_x u_{xx}$ should be zero, hence

$$\xi_x^0 = \xi_u^0 = \xi_u^1 = 0. \tag{2.78}$$

Substituting (2.78) into (2.77), we obtain

$$\left[\eta A_u + (\xi_t^0 - 2\xi_x^1)A\right]u_{xx} + \left[\eta_{uu}A + \eta A_{uu} + (\eta_u + \xi_t^0 - 2\xi_x^1)A_u\right]u_x^2 +$$
$$+ \left[\eta B_u + (\xi_t^0 - \xi_x^1)B + 2\eta_x A_u + (2\eta_{xu} - \xi_{xx}^1)A + \xi_t^1\right]u_x +$$
$$+ \eta C_u + (\xi_t^0 - \eta_u)C + \eta_{xx}A + \eta_x B - \eta_t = 0$$

Now one can again split the above expression w.r.t. the derivatives u_x^2, u_{xx}, and u_x, so that the equations

$$\eta_{uu}A + \eta A_{uu} + (\eta_u + \xi_t^0 - 2\xi_x^1)A_u = 0 \tag{2.79}$$

$$\eta A_u = (2\xi_x^1 - \xi_t^0)A; \tag{2.80}$$

$$\eta B_u = (\xi_x^1 - \xi_t^0)B - 2\eta_x A_u + (\xi_{xx}^1 - 2\eta_{xu})A - \xi_t^1; \tag{2.81}$$

$$\eta C_u = (\eta_u - \xi_t^0)C - \eta_{xx}A - \eta_x B + \eta_t \tag{2.82}$$

are obtained. It can be noted that Eq. (2.79) is simplified to the form

$$\eta_{uu} = 0 \tag{2.83}$$

because of (2.80). Thus, the system of DEs for the class of RDC equations (2.35) consists of Eqs. (2.78), (2.80)–(2.83). Obviously, Eqs. (2.78) and (2.83) can be easily integrated, hence

$$\xi^0 = \xi^0(t), \quad \xi^1 = \xi^1(t, x), \quad \eta = \alpha(t, x)u + \beta(t, x), \tag{2.84}$$

where the functions arising in right-hand sides are arbitrary smooth functions at the moment.

Finally, substituting (2.84) into (2.80)–(2.82), one arrives at a three-component system of PDEs, which are called classification equations. The general solution of the system obtained essentially depends on the form of the triplet (A, B, C). It turns out that constructing all possible triplets leading to different solutions of the classification equations is a highly nontrivial problem. It is worth noting that the relevant system of DEs in the case of the nonlinear heat equation (2.9) is much simpler, so that its solving is straightforward. In order to complete the algorithm of LSC presented in Section 2.3 for the class of RDC equations (2.35), we need to implement steps 4–7. Section 2.5 is devoted to this task.

2.5 Complete description of Lie symmetries of reaction-diffusion-convection equations

2.5.1 Principal algebra of invariance

In order to obtain LSC, one needs to construct the principal Lie algebra of the class in question. The algebra A^{pr} of the class of RDC equations (2.35) can be derived from the system of DEs obtained above under the natural assumption that A, B and C are arbitrary smooth functions, i.e., it is maximal algebra of invariance (MAI) of Eq. (2.35) with arbitrary given coefficients.

Theorem 2.8 *The principal Lie algebra of the class of RDC equations (2.35) is the two-dimensional Abelian algebra*

$$A^{pr} = <\partial_t, \partial_x>. \tag{2.85}$$

Proof To prove this statement we solve the system of DEs (2.80)–(2.82) assuming that A, B and C are arbitrary smooth functions. Substituting (2.84) into (2.80)–(2.82) we obtain the system

$$(\alpha u + \beta) A_u = (2\xi_x^1 - \xi_t^0) A, \tag{2.86}$$

$$(\alpha u + \beta) B_u = (\xi_x^1 - \xi_t^0) B - 2(\alpha_x u + \beta_x) A_u + (\xi_{xx}^1 - 2\alpha_x) A - \xi_t^1, \tag{2.87}$$

$$(\alpha u + \beta) C_u = (a - \xi_t^0) C - (\alpha_{xx} u + \beta_{xx}) A - (\alpha_x u + \beta_x) B + \alpha_t u + \beta_t. \tag{2.88}$$

Because the arbitrary given functions A, B and C depend on u while unknown functions do not depend on u, we may split Eqs. (2.86)–(2.88) w.r.t. A, B, C, A_u, B_u and C_u. As a result, the following system of conditions and linear equations

$$\alpha = \beta = 0, \tag{2.89}$$

$$\xi_t^0 = \xi_x^1 = \xi_t^1 = 0 \tag{2.90}$$

is obtained. Integrating the above equations, we conclude that

$$\xi^0 = d_0, \quad \xi^1 = d_1, \quad \eta = 0, \tag{2.91}$$

where d_0, d_1 are arbitrary parameters. Operator (2.15) with coefficients (2.91) generates the Abelian algebra (2.85).

The proof is now completed. □

2.5.2 Necessary conditions for nontrivial Lie symmetry

Here we are searching for necessary conditions, which are needed for extension of the principal Lie algebra (2.85). In other words, we need to establish all possible triplets (A, B, C) leading to extension of Lie symmetry of the relevant RDC equations of the form (2.35). The main result of this section can be formulated as follows.

Theorem 2.9 *If an arbitrary equation belonging to the class of RDC equations (2.35) admits MAI of a higher dimensionality than algebra (2.85), then the triplets (A, B, C) and the corresponding equations have one of the following nine forms:*

1. $(A, B, C) = (A(u), \lambda_1, 0)$,

$$u_t = [A(u)u_x]_x + \lambda_1 u_x;$$

2. $(A, B, C) = (e^{ku}, \lambda_1 + \lambda_2 e^{mu}, \lambda_3 e^{(2m-k)u})$,

$$u_t = \lambda_0(e^{ku}u_x)_x + (\lambda_1 + \lambda_2 e^{mu})u_x + \lambda_3 e^{(2m-k)u}, \ m \neq 0, \frac{k}{2}, k;$$

3. $(A, B, C) = (e^{ku}, \lambda_1 + \lambda_2 u, \lambda_3 e^{-ku})$,

$$u_t = \lambda_0(e^{ku}u_x)_x + (\lambda_1 + \lambda_2 u)u_x + \lambda_3 e^{-ku}, \ k \neq 0;$$

4. $(A, B, C) = (e^{ku}, \lambda_1 + \lambda_2 e^{\frac{k}{2}u} + \lambda_3 e^{ku}, \lambda_4 + \lambda_5 e^{\frac{k}{2}u} + \lambda_6 e^{ku})$,

$$u_t = \lambda_0(e^{ku}u_x)_x + (\lambda_1 + \lambda_2 e^{\frac{k}{2}u} + \lambda_3 e^{ku})u_x + \lambda_4 + \lambda_5 e^{\frac{k}{2}u} + \lambda_6 e^{ku}, k \neq 0;$$

5. $(A, B, C) = ((u+\theta)^k, \lambda_1 + \lambda_2(u+\theta)^m, \lambda_3(u+\theta)^{2m-k+1})$,

$$u_0 = \lambda_0[(u+\theta)^k u_x]_x + [\lambda_1 + \lambda_2(u+\theta)^m]u_x + \lambda_3(u+\theta)^{2m-k+1}, \ m \neq 0, \frac{k}{2}, k;$$

6. $(A, B, C) = ((u+\theta)^k, \lambda_1 + \lambda_2 \ln(u+\theta), \lambda_3(u+\theta)^{-k+1})$,

$$u_0 = \lambda_0[(u+\theta)^k u_x]_x + [\lambda_1 + \lambda_2 \ln(u+\theta)]u_x + \lambda_3(u+\theta)^{-k+1}, \ k \neq 0;$$

7. $(A, B, C) = ((u+\theta)^k, \lambda_1 + \lambda_2(u+\theta)^{\frac{k}{2}} + \lambda_3(u+\theta)^k,$
   $[\lambda_4 + \lambda_5(u+\theta)^{\frac{k}{2}} + \lambda_6(u+\theta)^k](u+\theta))$,

$$u_0 = \lambda_0[(u+\theta)^k u_x]_x + [\lambda_1 + \lambda_2(u+\theta)^{\frac{k}{2}} + \lambda_3(u+\theta)^k]u_x +$$
$$+[\lambda_4 + \lambda_5(u+\theta)^{\frac{k}{2}} + \lambda_6(u+\theta)^k](u+\theta), \ k \neq 0;$$

8. $(A, B, C) = (\lambda_0, \lambda_1 + \lambda_2 u, \lambda_3 + \lambda_4 u)$,

$$u_0 = \lambda_0 u_{xx} + (\lambda_1 + \lambda_2 u)u_x + \lambda_3 + \lambda_4 u;$$

9. $(A, B, C) = (\lambda_0, \lambda_1 + \lambda_2 \ln(u+\theta), [\lambda_3 + \lambda_4 \ln(u+\theta) + \lambda_5 \ln^2(u+\theta)](u+\theta))$,

$$u_0 = \lambda_0 u_{xx} + [\lambda_1 + \lambda_2 \ln(u+\theta)]u_x + [\lambda_3 + \lambda_4 \ln(u+\theta) + \lambda_5 \ln^2(u+\theta)](u+\theta).$$

Hereafter $m, k, \theta, \lambda_0 \neq 0, \lambda_i (i = 1, \ldots, 6)$ *are arbitrary constants.*

Proof The system of DEs for the class of RDC equations (2.35) consists of Eqs.(2.80)–(2.83), which can be simplified to the form (2.86)–(2.88). System (2.86)–(2.88) consists of three nonlinear PDEs w.r.t. to the functions ξ^0, ξ^1, α, β, A, B, C. From the formal point of view it is more complicated object than (2.35). However, unknown functions in (2.86)–(2.88) depend on different variables (e.g., ξ^0 depends on t while A depends on u only) and it is a common peculiarity of such type systems, which allows us to work out an algorithm for their solving. Unfortunately, this algorithm usually is quite cumbersome and consists of examination of several inequivalent cases. In the case of system (2.86)–(2.88), these cases arise naturally from the first equation. In fact, depending on zero and/or nonzero values of α, β and $2\xi_x^1 - \xi_t^0$, one may distinguish 5 inequivalent cases as follows.

1) $\alpha = 0$, $\beta = 0$, $2\xi_x^1 - \xi_t^0 = 0$;
2) $\alpha = 0$, $\beta \neq 0$, $2\xi_x^1 - \xi_t^0 \neq 0$;
3) $\alpha \neq 0$, $2\xi_x^1 - \xi_t^0 \neq 0$;
4) $\alpha = 0$, $\beta \neq 0$, $2\xi_x^1 - \xi_t^0 = 0$;
5) $\alpha \neq 0$, $2\xi_x^1 - \xi_t^0 = 0$.

Generally speaking, one obtains 2^3 cases, however the sixth and seventh cases were included in **3)** and **5)**, while the last case $\alpha = 0$, $\beta = 0$, $2\xi_x^1 - \xi_t^0 \neq 0$ leads immediately to the contradiction $A = 0$.

Now we examine each case in order to construct the most general form of the triplet (A, B, C), which solves system (2.86)–(2.88) and leads to an extension of the principal Lie algebra A^{pr}.

Case 1). In this case the first equation vanishes (i.e., A is an arbitrary function) and system (2.86)–(2.88) reduces to the form

$$\xi_x^1 B + \xi_t^1 = 0, \quad \xi_t^0 C = 0.$$

The second equation immediately gives $C = 0$ (otherwise $\xi_t^0 = 0$ and the principal Lie algebra A^{pr} can be obtained only). On the other hand, the first equation in the above system can be satisfied by $B = \lambda_1$ only (otherwise $\xi_x^1 = \xi_t^1 = 0$ leading to A^{pr}). As a result, one readily obtains the general solution $(A, B, C) = (A(u), \lambda_1, 0)$ arising above in the theorem, hence the corresponding RDC equation has the form

$$u_t = [A(u)u_x]_x + \lambda_1 u_x.$$

Case 2). In this case, system (2.86)–(2.88) takes the form

$$A_u = \frac{2\xi_x^1 - \xi_t^0}{\beta} A,$$
$$B_u = \frac{\xi_x^1 - \xi_t^0}{\beta} B - 2\frac{\beta_x}{\beta} A_u + \frac{\xi_{xx}^1}{\beta} A - \frac{\xi_t^1}{\beta}, \qquad (2.92)$$
$$C_u = -\frac{\xi_t^0}{\beta} C - \frac{\beta_{xx}}{\beta} A - \frac{\beta_x}{\beta} B + \frac{\beta_t}{\beta}.$$

In contrast to the functions α, β, ξ^0 and ξ^1, the function triplet (A, B, C) does depend on the variable x and it allows us to introduce so-called structural constants of the form

$$\frac{2\xi_x^1 - \xi_t^0}{\beta} = k, \qquad (2.93)$$

$$\frac{\xi_x^1 - \xi_t^0}{\beta} = m, \qquad (2.94)$$

$$-\frac{\xi_t^1}{\beta} = m_2, \qquad (2.95)$$

$$\frac{\beta_t}{\beta} = m_4, \quad -\frac{\beta_x}{\beta} = m_1, \quad -\frac{\beta_{xx}}{\beta} = m_3. \qquad (2.96)$$

Substituting structural constants into (2.92), one obtains

$$\begin{aligned} A_u &= kA, \\ B_u &= mB + m_1(2A_u - \tfrac{k}{2}A) + m_2, \\ C_u &= (2m-k)C + m_3 A + m_1 B + m_4. \end{aligned} \qquad (2.97)$$

Obviously the form of solutions of system (2.97) essentially depends on the coefficients k, m, m_1, m_2, m_3 and m_4. In fact, Eqs. (2.93) and (2.94) lead to the equation

$$(2m-k)\xi_x^1 = (m-k)\xi_t^0.$$

Differentiating this equation w.r.t. x, one obtains

$$(2m-k)\xi_{xx}^1 = 0. \qquad (2.98)$$

On the other hand, differentiation of (2.94) w.r.t. x gives $\xi_{xx}^1 = m\beta_x$, hence (2.98) is equivalent to

$$(2m-k)m\beta_x = 0. \qquad (2.99)$$

Differentiating (2.93) and (2.94) with respect to (w.r.t.) t and (2.95) w.r.t. x and using (2.99), we find the compatibility constraint

$$m(2m-k)(m-k)\beta_t = 0. \qquad (2.100)$$

The compatibility constraint (2.100) leads to four different subcases that can be considered step by step. Namely, the following different subcases should be examined: (2i) $m \neq 0, \tfrac{k}{2}, k$, (2ii) $m = 0$, (2iii) $k = m \neq 0$ and (2iv) $k = 2m \neq 0$.

We start from the most general one (2i). In this case, making rather simple calculations using Eqs. (2.96) and (2.99)–(2.100), one arrives at the restrictions $m_1 = m_3 = m_4 = 0$. Because three coefficients vanish in (2.97), its general solution can be immediately constructed. As a result, one obtains

$$A = \lambda_0 e^{ku}, \quad B = \lambda_1 + \lambda_2 e^{mu}, \quad C = \lambda_3 e^{(2m-k)u}.$$

Thus, the triplet (A, B, C), which is presented in the second item of the theorem, is derived. The corresponding RDC equation has the form

$$u_t = \lambda_0(e^{ku}u_x)_x + (\lambda_1 + \lambda_2 e^{mu})u_x + \lambda_3 e^{(2m-k)u}, \quad m \neq 0, \frac{k}{2}, k; k \neq 0.$$

(It will be shown below that the restriction $k \neq 0$ can be omitted.)

Now we turn to the subcase (2ii) $m = 0$. In quite similar way as it was done in case (2i), one may extract the restrictions $m = m_1 = m_3 = m_4 = 0$ from Eqs. (2.93)–(2.99). Now the general solution of system (2.97) can be easily constructed and we obtain

$$A = \lambda_0 e^{ku}, \quad B = \lambda_1 + \lambda_2 u, \quad C = \lambda_3 e^{-ku}.$$

Thus, the triplet (A, B, C) from the third item of theorem is identified. The corresponding RDC equation has the form

$$u_t = \lambda_0(e^{ku}u_x)_x + (\lambda_1 + \lambda_2 u)u_x + \lambda_3 e^{-ku}.$$

In subcase (2iii), constraint (2.99) leads to the restrictions $m_1 = m_3 = 0$. However, the general solution (2.97) with these restrictions is a particular case of (2iv) because one is obtainable by setting $\lambda_2 = \lambda_5 = 0$ in (2.101).

Completing examination of case **2)**, we construct the general solution of (2.97) in subcase (2iv) $m = \frac{k}{2} \neq 0$. As a result, we obtain

$$A = \lambda_0 e^{ku}, \quad B = \lambda_1 + \lambda_2 e^{\frac{k}{2}u} + \lambda_3 e^{ku}, \quad C = \lambda_4 + \lambda_5 e^{\frac{k}{2}u} + \lambda_6 e^{ku}, \quad (2.101)$$

what is nothing else but the triplet (A, B, C) from the fourth item of theorem. Simultaneously Eq. (2.35) takes the form

$$u_t = \lambda_0(e^{ku}u_x)_x + (\lambda_1 + \lambda_2 e^{\frac{k}{2}u} + \lambda_3 e^{ku})u_x + \lambda_4 + \lambda_5 e^{\frac{k}{2}u} + \lambda_6 e^{ku}.$$

Thus, analysis of case **2)** is now completed.

Now we turn to case **3)** $\alpha \neq 0$, $2\xi_x^1 - \xi_t^0 \neq 0$, which is the most complicated. It is convenient to rewrite Eqs.(2.86)–(2.88) in the form

$$(u + \tfrac{\beta}{\alpha})A_u = \tfrac{2\xi_x^1 - \xi_t^0}{\alpha} A,$$

$$(u + \tfrac{\beta}{\alpha})B_u = \tfrac{\xi_x^1 - \xi_t^0}{\alpha} B - 2(\tfrac{\alpha_x}{\alpha}u + \tfrac{\beta_x}{\alpha})A_u + \tfrac{\xi_{xx}^1 - 2\alpha_x}{\alpha}A - \tfrac{\xi_t^1}{\alpha}, \quad (2.102)$$

$$(u + \tfrac{\beta}{\alpha})C_u = (1 - \tfrac{\xi_t^0}{\alpha})C - (\tfrac{\alpha_{xx}}{\alpha}u + \tfrac{\beta_{xx}}{\alpha})A - (\tfrac{\alpha_x}{\alpha}u + \tfrac{\beta_x}{\alpha})B + \tfrac{\alpha_t}{\alpha}u + \tfrac{\beta_t}{\alpha},$$

and to introduce the structural constants

$$\frac{2\xi_x^1 - \xi_t^0}{\alpha} = k, \quad (2.103)$$

$$\frac{\xi_x^1 - \xi_t^0}{\alpha} = m, \quad (2.104)$$

$$-\frac{\xi_t^1}{\alpha} = m_2, \quad (2.105)$$

$$\frac{\beta}{\alpha} = \theta, \quad \frac{\alpha_t}{\alpha} = m_4, \quad -\frac{\alpha_x}{\alpha} = m_1, \quad -\frac{\alpha_{xx}}{\alpha} = m_3. \quad (2.106)$$

Substituting the structural constants (2.103)–(2.106) into system (2.102), we arrive at the system

$$(u+\theta)A_u = kA,$$
$$(u+\theta)B_u = mB + m_1[2(u+\theta)A_u + (2-\tfrac{k}{2})A] + m_2, \qquad (2.107)$$
$$(u+\theta)C_u = (2m-k+1)C + m_3(u+\theta)A + m_1(u+\theta)B + m_4(u+\theta).$$

Eqs. (2.103) and (2.104) produce the condition

$$(2m-k)\xi_x^1 = (m-k)\xi_t^0.$$

Its differential consequence w.r.t. the variable x gives

$$(2m-k)\xi_{xx}^1 = 0. \qquad (2.108)$$

The latter together with the differential consequence (w.r.t. x) of (2.104) allows us to obtain the equation

$$(2m-k)m\alpha_x = 0. \qquad (2.109)$$

On the other hand, differentiating Eqs. (2.103) and (2.104) w.r.t. t and Eq. (2.105) w.r.t. x, we arrive at the equation

$$(2m-k)(k-m)m\alpha_t = 0. \qquad (2.110)$$

Now one may say that (2.109) and (2.110) are compatibility conditions for system (2.103)–(2.106).

Having conditions (2.109) and (2.110), one realizes that the same subcases should be analyzed in order to solve system (2.107).

We again start from the most general one (3i): $m \neq 0, \tfrac{k}{2}, k$. Having Eqs. (2.106), (2.108)–(2.110), one arrives at the restrictions $m_1 = m_3 = m_4 = 0$. Because three coefficients vanish in (2.107), its general solution can be immediately constructed. As a result, one obtains

$$A = \lambda_0(u+\theta)^k, \quad B = \lambda_1 + \lambda_2(u+\theta)^m, \quad C = \lambda_3(u+\theta)^{2m-k+1}.$$

Thus, the triplet (A, B, C), which is presented in the fifth item of the theorem, is derived. The corresponding RDC equation has the form

$$u_t = \lambda_0[(u+\theta)^k u_x]_x + [\lambda_1 + \lambda_2(u+\theta)^m]u_x + \lambda_3(u+\theta)^{2m-k+1}, \quad m \neq 0, \frac{k}{2}, k.$$

In the next subcase (3ii) $m = 0$, the restrictions $m_1 = m_3 = m_4 = 0$ again are derived. Thus, the general solution of (2.107) can be immediately constructed, however, its form is different (because $m = 0$), namely

$$A = \lambda_0(u+\theta)^k, \quad B = \lambda_1 + \lambda_2\ln(u+\theta), \quad C = \lambda_3(u+\theta)^{-k+1}.$$

The corresponding RDC equation has the form
$$u_t = \lambda_0[(u+\theta)^k u_x]_x + [\lambda_1 + \lambda_2 \ln(u+\theta)]u_x + \lambda_3(u+\theta)^{-k+1}, \quad k \neq 0.$$

Thus, the sixth item of the theorem is identified.

Now we turn to subcase (3iii) $m = \frac{k}{2}$. Making straightforward calculations, the general solution of system (2.107) can be again constructed, hence the triplet (A, B, C) and the corresponding RDC equation have the forms

$$A = \lambda_0(u+\theta)^k, B = \lambda_1 + \lambda_2(u+\theta)^{\frac{k}{2}} + \lambda_3(u+\theta)^k,$$
$$C = [\lambda_4 + \lambda_5(u+\theta)^{\frac{k}{2}} + \lambda_6(u+\theta)^k](u+\theta),$$

and

$$u_t = \lambda_0[(u+\theta)^k u_x]_x + [\lambda_1 + \lambda_2(u+\theta)^{\frac{k}{2}} + \lambda_3(u+\theta)^k]u_x +$$
$$+ [\lambda_4 + \lambda_5(u+\theta)^{\frac{k}{2}} + \lambda_6(u+\theta)^k](u+\theta), \quad k \neq 0,$$

which arise in item 7 of the theorem are obtained.

Completing examination of the third case, we consider subcase (3iv) $m = k$. However, the analogous calculations show that the general solution of system (2.107) is

$$A = \lambda_0(u+\theta)^k, \quad B = \lambda_1 + \lambda_2(u+\theta)^k, \quad C = [\lambda_3 + \lambda_4(u+\theta)^k](u+\theta)$$

and it means that a particular case of subcase (3iii) is derived.

Let us consider case **4)**. In this case

$$\alpha = 0, \quad 2\xi_x^1 - \xi_t^0 = 0. \tag{2.111}$$

Introducing again the structural constants, system (2.86)–(2.88) takes the form
$$A_u = 0,$$
$$B_u = mB + m_1, \tag{2.112}$$
$$C_u = 2mC + m_4 A + m_2 B + m_3,$$

where

$$-\xi_t^0 = 2m\beta, \quad -\xi_t^1 = m_1\beta, \tag{2.113}$$
$$\beta_t = m_3\beta, \quad -\beta_x = m_2\beta, \quad -\beta_{xx} = m_4\beta. \tag{2.114}$$

Obviously, we should consider separately subcases $m \neq 0$ and $m = 0$.

Assuming $m \neq 0$, the differential consequences of Eqs. (2.113) lead to the conditions
$$\beta_x = \beta_t = 0. \tag{2.115}$$

Thus, we obtain $m_2 = m_3 = m_4 = 0$ from (2.114), otherwise $\beta = 0$ and the trivial Lie algebra is obtained. The general solution of (2.112) with $m_2 = m_3 = m_4 = 0$ has the form

$$A = \lambda_0, \quad B = \lambda_1 + \lambda_2 e^{mu}, \quad C = \lambda_3 e^{2mu},$$

where $\lambda_1 = -\frac{m_1}{m}$. The relevant RDC equation

$$u_t = \lambda_0 u_{xx} + (\lambda_1 + \lambda_2 e^{mu})u_x + \lambda_3 e^{2mu}$$

can be united with subcase (2i), i.e., the latter is valid also for $k = 0$, hence one obtains the second item of the theorem.

Assuming $m = 0$, one easily notes that $m_1 \beta_x = 0$ (see Eqs. (2.113)). If $\beta_x = 0$ then $m_2 = m_4 = 0$ (see (2.114)) and the general solution of (2.112) has the form

$$A = \lambda_0, \quad B = \lambda_1 + \lambda_2 u, \quad C = \lambda_3 + \lambda_4 u,$$

where $\lambda_2 = m_1$ and $\lambda_4 = m_3$. The relevant RDC equation is listed in item 8 of the theorem. If $m_1 = 0$ then $(A, B, C) = (\lambda_0, \lambda_1, \lambda_3 + \lambda_4 u)$ is obtained, leading to a particular case of item 8.

Thus, analysis of case **4)** is now completed.

Finally, we examine case **5)**. Since $\alpha \neq 0$ and $2\xi_x^1 - \xi_t^0 = 0$ system (2.86)–(2.88) can be rewritten as

$$A_u = 0,$$
$$(u + \tfrac{\beta}{\alpha})B_u = -\tfrac{\xi_t^0}{2\alpha}B - 2\tfrac{\alpha_x}{\alpha}A - \tfrac{\xi_t^1}{\alpha}, \qquad (2.116)$$
$$(u+\tfrac{\beta}{\alpha})C_u = (1-\tfrac{\xi_t^0}{\alpha})C - (\tfrac{\alpha_{xx}}{\alpha}u+\tfrac{\beta_{xx}}{\alpha})A - (\tfrac{\alpha_x}{\alpha}u+\tfrac{\beta_x}{\alpha})B + \tfrac{\alpha_t}{\alpha}u + \tfrac{\beta_t}{\alpha},$$

which is equivalent to the system

$$A_u = 0,$$
$$(u+\theta)B_u = mB + 2m_1 A + m_2, \qquad (2.117)$$
$$(u+\theta)C_u = (1+2m)C + m_3(u+\theta)A + m_1(u+\theta)B + m_4(u+\theta),$$

where

$$-\tfrac{\xi_t^0}{2\alpha} = m, \quad \tfrac{\beta}{\alpha} = \theta, \quad -\tfrac{\alpha_x}{\alpha} = m_1,$$
$$-\tfrac{\xi_t^1}{\alpha} = m_2, \quad -\tfrac{\alpha_{xx}}{\alpha} = m_3, \quad \tfrac{\alpha_t}{\alpha} = m_4 \qquad (2.118)$$

are the structural constants. In a quite similar way to case **4)** the restrictions

$$m\alpha_x = m\alpha_t = 0 \qquad (2.119)$$

can be derived from (2.118). Taking into account restrictions (2.119), one needs to analyze two possibilities, $m \neq 0$ and $m = 0$ in order to solve (2.117).

Having $m \neq 0$, one easily obtains from (2.118) the further restrictions $m_1 = m_3 = m_4 = 0$. The above restrictions allow us to integrate system (2.117). As a result, the following triplet (A, B, C) is obtained

$$A = \lambda_0, \quad B = \lambda_1 + \lambda_2(u+\theta)^m, \quad C = \lambda_3(u+\theta)^{2m+1}.$$

Thus, the corresponding RDC equation takes the form

$$u_t = \lambda_0 u_{xx} + [\lambda_1 + \lambda_2(u+\theta)^m]u_x + \lambda_3(u+\theta)^{2m+1}.$$

Now we can unite the above equation with that derived in subcase (3i), hence the equation

$$u_t = \lambda_0[(u+\theta)^k u_x]_x + [\lambda_1 + \lambda_2(u+\theta)^m]u_x + \lambda_3(u+\theta)^{2m-k+1}$$

(here k is an arbitrary parameter) listed in item 5 of the theorem is derived.

Having $m = 0$, we note that restrictions (2.119) are automatically fulfilled. Straightforward calculations show that the general solution of (2.117) with $m = 0$ has the form

$$A = \lambda_0, \quad B = \lambda_1 + \lambda_2 \ln(u+\theta),$$
$$C = [\lambda_3 + \lambda_4 \ln(u+\theta) + \lambda_5 \ln^2(u+\theta)](u+\theta).$$

Thus, the corresponding RDC equation is

$$u_t = \lambda_0 u_{xx} + [\lambda_1 + \lambda_2 \ln(u+\theta)]u_x + [\lambda_3 + \lambda_4 \ln(u+\theta) + \lambda_5 \ln^2(u+\theta)](u+\theta),$$

which coincides with that listed in item 9 of the theorem.

The proof is now completed. □

Now we can apply the set of equivalence transformations (ETs) (2.36) in order to simplify the equations listed in Theorem 2.9. The relevant transformations that remove the parameters λ_0, λ_1, θ and k from these equations can be easily determined and are presented in the second column of Table 2.4, while the equations obtained are shown in the last column. Now we note that each RDC equation listed in the last column of Table 2.4 contains at least one parameter less than one listed in Theorem 2.9.

Remark 2.4 *The coefficients $\tilde{\lambda}_i$ arising in the last column of Table 2.4 are arbitrary constants (because they are expressed via arbitrary constants arising in the equations numbered in the first column), while A is an arbitrary smooth function.*

Remark 2.5 *Table 2.4 is important for those researchers who are dealing with a RDC equation with the specified nonlinearities and aiming to construct particular solutions using the Lie machinery. The first step is to find the equation under question among those listed in Table 2.4. The Lie method can be successfully applied only for such equations.*

Thus, necessary conditions, which are needed for extension of the principal Lie algebra (2.85) of the RDC equations from class (2.35) are derived, i.e., the fifth step of the algorithm (see Section 2.3) is completed. Hereinafter we deal with equations arising in the last column of Table 2.4 only and use the old variables t, x and u for simplicity.

TABLE 2.4: Simplification of the RDC equations from Theorem 2.9 using ETs (2.36)

Equations from Theorem 2.9	Equivalence transformations	Simplified equations
Eq. 1	$\tau = t,$ $y = x + \lambda_1 t,$ $w = u$	$w_\tau = [A(w)w_y]_y$
Eq. 2	$\tau = \lambda_0 t,$ $y = x + \lambda_1 t,$ $w = u$	$w_\tau = (e^{kw}w_y)_y + \tilde{\lambda}_2 e^{mw} w_y +$ $+ \tilde{\lambda}_3 e^{(2m-k)w}$
Eq. 3	$\tau = \lambda_0 t,$ $y = x + \lambda_1 t,$ $w = ku$	$w_\tau = (e^w w_y)_y +$ $+ \tilde{\lambda}_2 w w_y + \tilde{\lambda}_3 e^{-w}$
Eq. 4	$\tau = \lambda_0 t,$ $y = x + \lambda_1 t,$ $w = ku$	$w_\tau = (e^w w_y)_y + (\tilde{\lambda}_2 e^{\frac{1}{2}w} +$ $+ \tilde{\lambda}_3 e^w) w_y + \tilde{\lambda}_4 + \tilde{\lambda}_5 e^{\frac{1}{2}w} + \tilde{\lambda}_6 e^w$
Eq. 5	$\tau = \lambda_0 t,$ $y = x + \lambda_1 t,$ $w = u + \theta$	$w_\tau = (w^k w_y)_y +$ $+ \tilde{\lambda}_2 w^m w_y + \tilde{\lambda}_3 w^{2m-k+1}$
Eq. 6	$\tau = \lambda_0 t,$ $y = x + \lambda_1 t,$ $w = u + \theta$	$w_\tau = (w^k w_y)_y +$ $+ \tilde{\lambda}_2 \ln w \, w_y + \tilde{\lambda}_3 w^{-k+1}$
Eq. 7	$\tau = \lambda_0 t,$ $y = x + \lambda_1 t,$ $w = u + \theta$	$w_\tau = (w^k w_y)_y +$ $+ (\tilde{\lambda}_2 w^{\frac{k}{2}} + \tilde{\lambda}_3 w^k) w_y +$ $+ (\tilde{\lambda}_4 + \tilde{\lambda}_5 w^{\frac{k}{2}} + \tilde{\lambda}_6 w^k) w$
Eq. 8	$\tau = \lambda_0 t,$ $y = x + \lambda_1 t,$ $w = u$	$w_\tau = w_{yy} + \tilde{\lambda}_2 w w_y + \tilde{\lambda}_3 + \tilde{\lambda}_4 w$
Eq. 9	$\tau = \lambda_0 t,$ $y = x + \lambda_1 t,$ $w = u + \theta$	$w_\tau = w_{yy} + \tilde{\lambda}_2 \ln w \, w_y +$ $+ (\tilde{\lambda}_3 + \tilde{\lambda}_4 \ln w + \tilde{\lambda}_5 \ln^2 w) w$

2.5.3 Lie symmetry classification via the Lie–Ovsiannikov algorithm

Looking on the algorithm presented in the beginning of Section 2.3, one notes that the first five steps are already realized. Finding sufficient conditions needed for extension of the principal Lie algebra (2.85) is based on examination of nine equations from the last column of Table 2.4. Having this done and taking into account the group of ETs $\mathcal{E}$, we can complete LSC of the class of the RDC equations (2.35). As a result, the sixth step of our plan will be realized.

To obtain LSC of class (2.35) via the Lie–Ovsiannikov algorithm, we need to find all possible Lie symmetries of each of those nine equations. It should be noted that only the first equation among those contains an arbitrary function A. However, it is the well-known nonlinear heat equation and its classification has been done in the seminal work [201]. The other eight equations contain no arbitrary functions hence their maximal algebras of invariance (MAIs) can be found by straightforward calculations. At the present time it is a rather simple task, hence we present below the main theorem and a sketch of its proof.

Theorem 2.10 *[83] All possible MAI of RDC equations of the form (2.35) depending on the function triplet (A, B, C) are presented in Table 2.5. Any other equation of the form (2.35) with nontrivial Lie symmetry (i.e., its MAI is of dimensionality three and higher) is reduced by an ET from $\mathcal{E}$ (2.36) to one of 30 equations listed in Table 2.5.*

TABLE 2.5: The complete LSC of equations of the form (2.35) using the group of ETs $\mathcal{E}$

	RDC equations	MAI	Constraints
1	$u_t = u_{xx}$	$<\partial_t, \partial_x, G, I, D, \Pi, Q_1^\infty>$	
2	$u_t = (u^{-\frac{4}{3}} u_x)_x$	$<\partial_t, \partial_x, D, D_1, \mathcal{K}>$	$k = -\frac{4}{3}$
3	$u_t = (u^k u_x)_x$	$<\partial_t, \partial_x, D, D_1>$	$k \neq 0, -\frac{4}{3}$
4	$u_t = (e^u u_x)_x$	$<\partial_t, \partial_x, D, D_2>$	
5	$u_t = [A(u) u_x]_x$	$<\partial_t, \partial_x, D>$	
6	$u_t = u_{xx} + p$	$<\partial_t, \partial_x, G + \frac{p}{2} tx \partial_u,$ $I - pt \partial_u, D + 2pt \partial_u,$ $\Pi + \frac{p}{4} t(x^2 + 6t) \partial_u, Q_1^\infty>$	
7	$u_t = u_{xx} + pu$	$<\partial_t, \partial_x, G, I, D + 2ptI$ $\Pi + pt^2 I, Q_2^\infty>$	
8	$u_t = (u^{-\frac{4}{3}} u_x)_x + pu^{-\frac{1}{3}}$	$<\partial_t, \partial_x, D_3, \mathcal{K}_1, \mathcal{K}_2>$	$k=-\frac{4}{3}$,
9	$u_t = (u^{-\frac{4}{3}} u_x)_x +$ $+(pu^{-\frac{4}{3}} + \lambda_6)u$	$<\partial_t, \partial_x, T_1, \mathcal{K}_1, \mathcal{K}_2>$	$k=-\frac{4}{3}$
10	$u_t = u_{xx} + pu \ln u$	$<\partial_t, \partial_x, \mathcal{G}, e^{pt} I>$	
11	$u_t = (u^k u_x)_x + pu$	$<\partial_t, \partial_x, T_1, D_1>$	$k \neq 0$
12	$u_t = (e^u u_x)_x + p$	$<\partial_t, \partial_x, T_2, D_2>$	
13	$u_t = u_{xx} + uu_x$	$<\partial_t, \partial_x, G_0, D-I, \Pi_0>$	
14	$u_t = u_{xx} + uu_x + p$	$<\partial_t, \partial_x, G_0, D-I-$ $-\frac{3p}{2} t(t \partial_x - 2 \partial_u),$ $\Pi_0 - \frac{p}{2} t^2 (t \partial_x - 3 \partial_u)>$	
15	$u_t = (u^k u_x)_x + u^k u_x +$ $+\frac{2(k+2)}{(3k+4)^2} u^{k+1}$	$<\partial_t, \partial_x, D_3, X_1>$	$k \neq 0, -\frac{4}{3}$
16	$u_t = (u^k u_x)_x + u^k u_x +$ $+[p + \frac{2(k+2)}{(3k+4)^2} u^k] u$	$<\partial_t, \partial_x, T_1, X_1>$	$k \neq 0, -\frac{4}{3}$

Continuation of Table 2.5

	RDC equations	MAI	Constraints
17	$u_t = (e^u u_x)_x + e^u u_x + \frac{2}{9}e^u$	$<\partial_t, \partial_x, D_4, X_2>$	
18	$u_t = (e^u u_x)_x + e^u u_x + p + \frac{2}{9}e^u$	$<\partial_t, \partial_x, T_2, X_2>$	
19	$u_t = u_{xx} + u u_x + pu$	$<\partial_t, \partial_x, \mathcal{G}_1>$	
20	$u_t = u_{xx} + \ln u\, u_x +$ $+\lambda_4 u \ln u$	$<\partial_t, \partial_x, \mathcal{G}_2>$	$\lambda_4 \neq 0$
21	$u_t = u_{xx} + \ln u\, u_x + \lambda_3 u$	$<\partial_t, \partial_x, G_1>$	
22	$u_t = u_{xx} + \ln u\, u_x +$ $+(\lambda_3 + \frac{1}{4}\ln^2 u)u$	$<\partial_t, \partial_x, Y>$	
23	$u_t = (u^k u_x)_x + \lambda_2 u^m u_x +$ $+\lambda_3 u^{2m-k+1}$	$<\partial_t, \partial_x, (k-2m)D+D_1>$	$|\lambda_2|+|\lambda_3| \neq 0$, $|k|+|m| \neq 0$
24	$u_t = (u^k u_x)_x + \lambda_3 u^k u_x +$ $+(\lambda_6 u^k + p)u$	$<\partial_t, \partial_x, T_1>$	$k \neq 0$, $|\lambda_3|+|\lambda_6| \neq 0$
25	$u_t = (u^k u_x)_x + (\lambda_2 u^{\frac{k}{2}} + u^k)u_x +$ $+[\lambda_4 + \frac{2\lambda_2}{3k+4}u^{\frac{k}{2}} + \frac{2(k+2)}{(3k+4)^2}u^k]u$	$<\partial_t, \partial_x, X_1>$	$k \neq 0, -\frac{4}{3}$, $\lambda_2 \neq 0$
26	$u_t = (u^k u_x)_x + \ln u\, u_x +$ $+\lambda_3 u^{-k+1}$	$<\partial_t, \partial_x, Z_1>$	$k \neq 0$,
27	$u_t = (e^{ku} u_x)_x + \lambda_2 e^{mu} u_x +$ $+\lambda_3 e^{(2m-k)u}$	$<\partial_t, \partial_x, (k-2m)D+D_2>$	$|\lambda_2|+|\lambda_3| \neq 0$, $|k|+|m| \neq 0$
28	$u_t = (e^u u_x)_x + \lambda_3 e^u u_x +$ $+\lambda_6 e^u + p$	$<\partial_t, \partial_x, T_2>$	$|\lambda_3|+|\lambda_6| \neq 0$, $\lambda_6 \neq \frac{2}{9}\lambda_3^2$
29	$u_t = (e^u u_x)_x + (\lambda_2 e^{\frac{1}{2}u} + e^u)u_x +$ $+\lambda_4 + \frac{2\lambda_2}{3}e^{\frac{1}{2}u} + \frac{2}{9}e^u$	$<\partial_t, \partial_x, X_2>$	$\lambda_2 \neq 0$
30	$u_t = (e^u u_x)_x + u u_x + \lambda_3 e^{-u}$	$<\partial_t, \partial_x, Z_2>$	

Remark 2.6 In Table 2.5, the following designations for Lie symmetry operators are introduced:
$Z_1 = k(t\partial_t + x\partial_x) - t\partial_x + u\partial_u$, $Z_2 = t\partial_t + x\partial_x - t\partial_x + \partial_u$,
$X_1 = e^{-\frac{k}{3k+4}x}(\partial_x - \frac{2}{3k+4}u\partial_u)$, $X_2 = e^{-\frac{1}{3}x}(\partial_x - \frac{2}{3}\partial_u)$,
$T_1 = e^{-pkt}(\partial_t + pu\partial_u)$, $T_2 = e^{-pt}(\partial_t + p\partial_u)$,
$\mathcal{G}_1 = e^{pt}(\partial_x - p\partial_u)$, $\mathcal{G}_2 = e^{\lambda_4 t}(\partial_x - \lambda_4 u\partial_u)$,
$\mathcal{G} = e^{pt}(\partial_x - \frac{p}{2}xu\partial_u)$,
$G_0 = t\partial_x - \partial_u$, $G_1 = t\partial_x - u\partial_u$,
$G = t\partial_x - \frac{1}{2}xu\partial_u$, $I = u\partial_u$,
$D = 2t\partial_t + x\partial_x$,
$D_4 = t\partial_t - \partial_u$, $D_3 = kt\partial_t - u\partial_u$,
$D_2 = x\partial_x + 2\partial_u$, $D_1 = kx\partial_x + 2u\partial_u$,
$\Pi_0 = t^2\partial_t + tx\partial_x - (x + tu)\partial_u$,
$\Pi = t^2\partial_t + tx\partial_x - \frac{1}{2}(\frac{x^2}{2} + t)u\partial_u$,
$K = x^2\partial_x - 3xu\partial_u$,
$Y = e^{\frac{1}{4}t - \frac{1}{2}x}u\partial_u$,
$Q_a^\infty = \beta^a(t,x)\partial_u$, $a = 1, 2$, where $\beta^1(t,x)$ and $\beta^2(t,x)$ are arbitrary solutions of the linear heat equations $\beta_t^1 = \beta_{xx}^1$ and $\beta_t^2 = \beta_{xx}^2 + p\beta^2$, respectively; $K_a = \varphi^a(x)\partial_x - \dot\varphi^a u\partial_u$, where $\varphi^1(x)$ and $\varphi^2(x)$ form a fundamental system of solutions of the linear ODE $\ddot\varphi = p\varphi$, $a = 1, 2$.

Remark 2.7 All the parameters arising in Table 2.5 are arbitrary constants provided they satisfy the restrictions from the last column. However, the parameter p is nonzero and is reducible to ± 1 by an appropriate ET. An arbitrary parameter $k \neq 0$ in case 27 is reducible to $k = 1$, while $k = 0$ allows us to make $m = 1$ in this case.

Proof is based on examination of nine equations from the last column of Table 2.4. It means that all possible Lie symmetries of each of those nine equations should be found.

The first equation containing an arbitrary function A is nothing else but the well-known nonlinear heat equation and its classification has been done in [201] (see also Sections 2.1 and 2.2 above). Thus, the examination of the first equation immediately leads to the equations and MAIs listed in cases 1–5 of Table 2.5.

MAIs of the other eight equations can be found by straightforward calculations using the classification equations (2.86)–(2.88). The scheme of examination for each of those equations is the same, so that we present the relevant details only for two equations.

Consider Equation 5 listed in the last column of Table 2.4 (we remind the reader that the "old" notations t, x and u are used hereafter)

$$u_t = (u^k u_x)_x + \lambda_2 u^m u_x + \lambda_3 u^{2m-k+1}, \tag{2.120}$$

where $|\lambda_2| + |\lambda_3| \neq 0$. Here we assume additionally that $m \neq 0, \frac{k}{2}, k$ because Eq. (2.120) with $m = 0$ is a particular case of Equation 6 (see the last column of Table 2.4), while that with $m = \frac{k}{2}$ and $m = k$ leads to a particular case of Equation 7, which will be examined after Equation 5.

Substituting the triplet (A, B, C) from (2.120) into Eqs. (2.86)–(2.88), we arrive at the system of PDEs

$$k(\alpha u + \beta) u^{k-1} = (2\xi_x^1 - \xi_t^0) u^k, \tag{2.121}$$

$$\lambda_2 m(\alpha u + \beta) u^{m-1} = \lambda_2(\xi_x^1 - \xi_t^0) u^m - 2k(\alpha_x u + \beta_x) u^{k-1} + \\ + (\xi_{xx}^1 - 2\alpha_x) u^k - \xi_t^1, \tag{2.122}$$

$$(2m - k + 1)\lambda_3(\alpha u + \beta) u^{2m-k} = \lambda_3(\alpha - \xi_t^0) u^{2m-k+1} - \\ - (\alpha_{xx} u + \beta_{xx}) u^k - \lambda_2(\alpha_x u + \beta_x) u^m + \alpha_t u + \beta_t. \tag{2.123}$$

Now we use the very common trick for solving such kind of PDE systems. Because all unknown functions in Eqs.(2.121)–(2.123) do not depend on u, one can split these equations w.r.t. different exponents of the function u. As a result, the system

$$\xi_t^1 = 0, \quad \alpha_t = \alpha_x = \beta = 0,$$
$$k\alpha = 2\xi_x^1 - \xi_t^0, \quad \lambda_2 m\alpha = \lambda_2(\xi_x^1 - \xi_t^0), \tag{2.124}$$
$$\lambda_3[\xi_t^0 + (2m - k)\alpha] = 0$$

is obtained. Because (2.124) is the overdetermined system consisting of 6 (or 5 under the restriction $\lambda_2 \lambda_3 = 0$) linear equations for the functions α, ξ^0 and ξ^1, its solving is rather trivial. As a result, we arrive at the general solution

$$\xi^0 = (2m - k)C_1 t + C_2, \quad \xi^1 = (m - k)C_1 x + C_3, \quad \alpha = -C_1, \quad \beta = 0$$

(hereafter C with the lower indices are arbitrary constants). The infinitesimal operator X (see (2.15)) with the above coordinates (taking into account also (2.84)) generates the three-dimensional MAI of Eq. (2.120):

$$< \partial_t, \ \partial_x, \ (k - 2m)D + D_1 >,$$

where $(k - 2m)D + D_1 = (2m - k)t\partial_t + (m - k)x\partial_x - u\partial_u$. Thus, case 23 (under the restriction $m \neq 0, \frac{k}{2}, k$) of Table 2.5 is derived.

Now we turn to Equation 7 listed in the last column of Table 2.4, namely

$$u_t = (u^k u_x)_x + (\lambda_2 u^{\frac{k}{2}} + \lambda_3 u^k)u_x + (\lambda_4 + \lambda_5 u^{\frac{k}{2}} + \lambda_6 u^k)u, \quad k \neq 0. \quad (2.125)$$

Examination of this equation is the most complicated among others.

First of all this equation was investigated under the restrictions $\lambda_2 = \lambda_3 = 0$, i.e., the above equation does not involve a convection term. Simultaneously we should assume that $|\lambda_4| + |\lambda_5| + |\lambda_6| \neq 0$ (otherwise a subcase of the first equation from Table 2.4 is obtained). Under two above restrictions, analysis of the classification equations (2.86)–(2.88) with the triplet (A, B, C) from (2.125) is rather simple. As a result, five inequivalent subcases were derived leading to the corresponding cases(or subcases) in Table 2.5. They are listed below.

1) If $\lambda_5 = \lambda_6 = 0$ then Eq. (2.125) admits the four-dimensional MAI

$$< \partial_t, \ \partial_x, \ T_1, \ D_1 >,$$

i.e., case 11 of Table 2.5 is derived.

2) If $k = -\frac{4}{3}$, $\lambda_4 = \lambda_5 = 0$ then Eq. (2.125) admits the five-dimensional MAI

$$< \partial_t, \ \partial_x, \ D_3, \ \mathcal{K}_1, \ \mathcal{K}_2 >.$$

Thus, case 8 of Table 2.5 is obtained.

3) If $k = -\frac{4}{3}$, $\lambda_5 = 0$ and $\lambda_4 \lambda_6 \neq 0$ then Eq. (2.125) admits the five-dimensional MAI with another representation, namely

$$< \partial_t, \ \partial_x, \ T_1, \ \mathcal{K}_1, \ \mathcal{K}_2 >.$$

This equation and MAI arise in case 9 of Table 2.5.

4) If $k \neq -\frac{4}{3}$, $\lambda_4 = \lambda_5 = 0$ then Eq. (2.125) admits the three-dimensional MAI

$$< \partial_t, \ \partial_x, \ kD - D_1 >,$$

It can be noted that a subcase ($m = k$) of case 23 of Table 2.5 is obtained.

5) If $\lambda_5 = 0$ and $\lambda_4 \lambda_6 \neq 0$ then Eq. (2.125) admits the three-dimensional MAI

$$< \partial_t, \ \partial_x, \ T_1 >.$$

It means that a particular case ($\lambda_3 = 0$) of case 24 from Table 2.5 is obtained.

Now we examine Eq. (2.125) under the restriction $|\lambda_2| + |\lambda_3| \neq 0$, i.e., the convective term arises in the equation.

Substituting the triplet (A, B, C) from (2.125) into Eqs. (2.86)–(2.88), we arrive at the system of PDEs

$$k(\alpha + \frac{\beta}{u}) = 2\xi_x^1 - \xi_t^0, \quad (2.126)$$

$$k\alpha(\tfrac{\lambda_2}{2} u^{\frac{k}{2}} + \lambda_3 u^k) = (\xi_x^1 - \xi_t^0)(\lambda_2 u^{\frac{k}{2}} + \lambda_3 u^k) - \\ -2\alpha_x k u^k + (\xi_{xx}^1 - 2\alpha_x)u^k - \xi_t^1, \quad (2.127)$$

$$\alpha[\lambda_5(\tfrac{k}{2}+1)u^{\frac{k}{2}} + \lambda_6(k+1) + \lambda_4] = (\alpha - \xi_t^0)(\lambda_4 + \lambda_5 u^{\frac{k}{2}} + \lambda_6 u^k) - \\ -\alpha_{xx} u^k - \alpha_x(\lambda_2 u^{\frac{k}{2}} + \lambda_3 u^k) + \alpha_t. \tag{2.128}$$

Splitting Eqs. (2.126)-(2.128) w.r.t. different exponents of the function u, we derive the linear system of PDEs

$$\xi_t^1 = \beta = 0, \tag{2.129}$$

$$\lambda_2 \xi_t^0 = 0, \tag{2.130}$$

$$k\alpha = 2\xi_x^1 - \xi_t^0, \tag{2.131}$$

$$\xi_{tt}^0 + k\lambda_4 \xi_t^0 = 0, \tag{2.132}$$

$$(3k+4)\xi_{xx}^1 + k\lambda_3 \xi_x^1 = 0, \tag{2.133}$$

$$\xi_{xxx}^1 + \lambda_3 \xi_{xx}^1 + k\lambda_6 \xi_x^1 = 0, \tag{2.134}$$

$$4\lambda_2 \xi_{xx}^1 + k\lambda_5(\xi_t^0 + 2\xi_x^1) = 0. \tag{2.135}$$

Taking into account the Eq. (2.135) structure, it is convenient to consider separately four cases depending on zero and nonzero values of the parameters λ_2 and λ_5.

Case 1) $\lambda_2 \neq 0$ and $\lambda_5 \neq 0$. It is the most general case. Under above restrictions, Eqs.(2.129), (2.130) and (2.135) immediately give

$$\xi^0 = C_1, \quad \xi^1 = C_2 e^{-\frac{k}{2}\frac{\lambda_5}{\lambda_2} x} + C_3. \tag{2.136}$$

Substituting (2.136) into Eqs. (2.133)–(2.134), we derive that only the restrictions

$$\lambda_5 = \frac{2\lambda_2 \lambda_3}{3k+4}, \quad k \neq -\frac{4}{3}, \quad \lambda_3 \neq 0, \quad \lambda_6 = \frac{2(k+2)}{(3k+4)^2}\lambda_3^2,$$

can lead to an extension of the principal algebra A^{pr}. Having these restrictions, we immediately obtain

$$\alpha = -C_2 \frac{2\lambda_3}{3k+4} e^{-\frac{\lambda_3 k}{3k+4} x} \tag{2.137}$$

from Eqs. (2.131) and (2.136). As a result, the infinitesimal operator X with coordinates (2.129), (2.136) and (2.137) produces the three-dimensional MAI

$$< \partial_t, \partial_x, X_1 = e^{-\frac{\lambda_3 k}{3k+4} x} \left(\partial_x - \frac{2\lambda_3}{3k+4} u \partial_u \right) > .$$

The corresponding RDC equation takes the form

$$u_t = (u^k u_x)_x + (\lambda_2 u^{\frac{k}{2}} + \lambda_3 u^k) u_x + \left[\lambda_4 + \frac{2\lambda_2 \lambda_3}{3k+4} u^{\frac{k}{2}} + \frac{2(k+2)}{(3k+4)^2} \lambda_3^2 u^k \right] u,$$

where $k \neq 0, -\frac{4}{3}, \lambda_3 \neq 0$. Using ET

$$\lambda_3^2 t \to t, \quad \lambda_3 x \to x, \quad u \to u \tag{2.138}$$

we obtain case 25 of Table 2.5.

Case 2) $\lambda_2 \neq 0, \lambda_5 = 0$. Eqs. (2.130), (2.131) and (2.135) can be easily integrated, so that

$$\xi^0 = C_1, \quad \xi^1 = C_2 x + C_3, \quad \alpha = \frac{2}{k} C_2. \tag{2.139}$$

Substituting the functions ξ^0, ξ^1 and α from (2.139) into (2.133)–(2.134), one notes that the principal algebra A^{pr} admits an extension only under the additional restrictions

$$\lambda_3 = \lambda_6 = 0.$$

Thus, we arrive at the equation

$$u_t = (u^k u_x)_x + \lambda_2 u^{\frac{k}{2}} u_x + \lambda_4 u,$$

which is invariant w.r.t. the three-dimensional algebra

$$< \partial_t, \ \partial_x, \ D_1 = kx\partial_x + 2I >.$$

It means that case 23 from Table 2.5 is obtained under the restriction $m = \frac{k}{2}$ (we remind the reader that the most general case, $m \neq \frac{k}{2}$, k was obtained by examination of Equation 5 from the last column of Table 2.4).

Case 3) $\lambda_2 = 0, \lambda_5 \neq 0$ does not lead to any new extension of the principal algebra. In fact, Eq. (2.135) takes the form

$$\xi^0_t = -2\xi^1_x \tag{2.140}$$

and taking into account (2.84), we obtain from (2.140)

$$2\xi^1_{xx} = 0. \tag{2.141}$$

Substituting (2.141) into Eq. (2.133) we conclude that Eq. (2.125) with $\lambda_3 \neq 0$ does not allow any extension of A^{pr}.

In case 4) $\lambda_2 = \lambda_5 = 0$, Eqs. (2.130) and (2.135) vanish, while Eqs. (2.133) and (2.134) produce the compatibility condition

$$[(3k+4)^2 \lambda_6 - 2(k+2)\lambda_3^2]\xi^1_x = 0.$$

So, taking into account Eq. (2.132), we can distinguish four different cases in order to solve the system (2.131)–(2.134).

Subcase 4a) $\lambda_4 \neq 0, (3k+4)^2 \lambda_6 - 2(k+2)\lambda_3^2 \neq 0$. The general solution of Eqs.(2.129), (2.131)–(2.133) can be readily constructed by straightforward calculations:

$$\xi^0 = C_1 e^{-k\lambda_4 t} + C_2, \quad \xi^1 = C_3 \tag{2.142}$$

and

$$\alpha = C_1 \lambda_4 e^{-k\lambda_4 t}, \quad \beta = 0. \tag{2.143}$$

The infinitesimal operator X with coordinates (2.142) and (2.143) produces the three-dimensional MAI

$$< \partial_t, \partial_x, T_1 = e^{-k\lambda_4 t}(\partial_t + \lambda_4 I) > .$$

Notably, the restriction $(3k+4)^2\lambda_6 - 2(k+2)\lambda_3^2 \neq 0$ can be skipped because $\xi_x^1 = 0$ in the above Lie algebra.

The corresponding RDC equation is

$$u_t = (u^k u_x)_x + \lambda_3 u^k u_x + (\lambda_4 + \lambda_6 u^k)u.$$

Taking into account ET (2.138) with $\lambda_3 \to \lambda_4$, one realizes that the above Lie algebra and equation coincide with those listed in case 24.

Subcase 4b) $\lambda_4 \neq 0, (3k+4)^2\lambda_6 - 2(k+2)\lambda_3^2 = 0$. The second restriction implies $k \neq -\frac{4}{3}$ (otherwise the convection term vanishes). The general solution of Eqs.(2.129), (2.131)–(2.133) was found in the same way as in the above subcase:

$$\xi^0 = C_1 e^{-k\lambda_4 t} + C_2, \quad \xi^1 = C_3 e^{-\frac{k\lambda_3}{3k+4}x} + C_4 \qquad (2.144)$$

and

$$\alpha = -2C_3 \frac{\lambda_3}{3k+4} e^{-\frac{k\lambda_3}{3k+4}x} + C_1 \lambda_4 e^{-k\lambda_4 t}, \quad \beta = 0, \qquad (2.145)$$

where $\lambda_6 = \frac{2(k+2)}{(3k+4)^2}\lambda_3^2$. The corresponding Lie algebra is the four-dimensional MAI

$$< \partial_t, \partial_x, T_1 = e^{-k\lambda_4 t}(\partial_t + \lambda_4 I), X_1 = e^{-\frac{\lambda_3 k}{3k+4}x}\left(\partial_x - \frac{2\lambda_3}{3k+4}I\right) >$$

of the RDC equation

$$u_t = (u^k u_x)_x + \lambda_3 u^k u_x + \left[\lambda_4 + \frac{2(k+2)}{(3k+4)^2}\lambda_3^2 u^k\right]u,$$

where $\lambda_4 \neq 0, k \neq -\frac{4}{3}$. Using ET (2.138) with $\lambda_3 \to \lambda_4$ we obtain case 16 from Table 2.5.

Subcase 4c) $\lambda_4 = 0, (3k+4)^2\lambda_6 - 2(k+2)\lambda_3^2 = 0$. The second restriction again implies $k \neq -\frac{4}{3}$, while the first that leads to the linear function ξ^0 (see Eq. (2.131)):

$$\xi^0 = C_1 t + C_2 \qquad (2.146)$$

in (2.144), while ξ^1 is still the same. So, Eq. (2.131) with (2.146) leads to

$$\alpha = -\frac{2\lambda_3}{3k+4} C_3 e^{-\frac{k\lambda_3}{3k+4}x} - \frac{C_1}{k}. \qquad (2.147)$$

As a result, the four-dimensional Lie algebra is obtained:

$$< \partial_t, \partial_x, D_3 = kt\partial_t - I, X_1 = e^{-\frac{\lambda_3 k}{3k+4}x}\left(\partial_x - \frac{2\lambda_3}{3k+4}I\right) >,$$

which is MAI of the RDC equation

$$u_t = (u^k u_x)_x + \lambda_3 u^k u_x + \frac{2(k+2)}{(3k+4)^2}\lambda_3^2 u^{k+1},$$

where $k \neq 0; -\frac{4}{3}$. Using ET (2.138) we obtain case 15 from Table 2.5.

The last possible subcase, $\lambda_4 = 0$, $(3k+4)^2 \lambda_6 - 2(k+2)\lambda_3^2 \neq 0$, immediately produces

$$\xi^0 = c_1 t + c_2, \quad \xi^1 = c_3, \quad \alpha = -\frac{c_1}{k}. \tag{2.148}$$

Formulae (2.148) lead to the Lie algebra

$$< \partial_t, \ \partial_x, \ D_3 = kt\partial_t - I >$$

and the corresponding RDC equation

$$u_t = (u^k u_x)_x + \lambda_3 u^k u_x + \lambda_6 u^{k+1},$$

which are nothing else but the particular case, $m = k$, of case 23 from Table 2.5.

Examination of other equations from Table 2.4 is simpler than that of Equation 7, so that we present briefly only the sketch of how other cases from Table 2.5 have been derived.

Equation 6

$$u_t = (u^k u_x)_x + \lambda_2 \ln u u_x + \lambda_3 u^{-k+1}, \quad k \neq 0 \tag{2.149}$$

admits the three-dimensional MAI

$$< \partial_t, \ \partial_x, \ Z_1 = k(t\partial_t + x\partial_x) - \lambda_2 t \partial_x + I >,$$

where $\lambda_2 \neq 0$. Using ET (2.138) with $\lambda_3 \to \lambda_2$, we obtain case 26 from Table 2.5.

If $\lambda_2 = 0$ then a particular case of the 23rd case of Table 2.5 is obtained.

Equation 8

$$u_t = u_{11} + \lambda_2 u u_x + \lambda_3 + \lambda_4 u, \quad |\lambda_2| + |\lambda_3| + |\lambda_4| \neq 0 \tag{2.150}$$

depending on values of λ_2, λ_3 and λ_4, admits four different Lie algebras.

If $\lambda_2 = \lambda_4 = 0$ then the classical heat equation with a constant source (sink) is obtained, which admits infinite-dimensional MAI

$$< \partial_t, \ \partial_x, \ G + \frac{\lambda_3}{2}tx\partial_u, \ I - \lambda_3 t \partial_u, \ D + 2\lambda_3 t \partial_u, \ \Pi + \frac{\lambda_3}{4}t(x^2 + 6t)\partial_u, Q_1^\infty >.$$

One concludes that this equation and the Lie algebra are listed in case 6 of Table 2.5.

If $\lambda_2 = 0$ and $\lambda_4 \neq 0$ then the term λ_3 can be removed by the known ET

$u + \frac{\lambda_3}{\lambda_4} \to u$, which was not used in case 19 of Table 2.5. Having this done, the classical heat equation with linear source (sink)

$$u_t = u_{xx} + \lambda_4 u$$

is obtained, which, of course, admits infinite-dimensional MAI

$$< \partial_t, \ \partial_x, \ G, \ I, \ D + 2\lambda_4 tI, \ \Pi + \lambda_4 t^2 I, \ Q_2^\infty > .$$

Thus, case 7 of Table 2.5 is identified (λ_4 can be reduced to $p = \pm 1$ by an appropriate ET).

If $\lambda_2 \neq 0$ and $\lambda_3 = \lambda_4 = 0$ then the well-known Burgers equation is obtained. The Burgers equation and its MAI is listed in case 13 of Table 2.5.

If $\lambda_2 \lambda_3 \neq 0$ and $\lambda_4 = 0$ then the Burgers equation with a constant source (sink) is obtained. After simple calculations the five-dimensional MAI of this equation can be easily found. As a result, case 14 of the table was identified.

Finally, if $\lambda_2 \lambda_3 \lambda_4 \neq 0$ then Eq. (2.150) can be simplified by applying an appropriate ET to that with $\lambda_2 = 1$, $\lambda_3 = 0$ and $\lambda_4 = \pm 1$. The equation obtained admits the three-dimensional MAI

$$< \partial_t, \ \partial_x, \ \mathcal{G}_1 > .$$

Thus, case 19 of Table 2.5 is identified.

The last equation (see the third column of Table 2.4) is

$$u_t = u_{xx} + \lambda_2 \ln u u_x + (\lambda_3 + \lambda_4 \ln u + \lambda_5 \ln^2 u)u. \tag{2.151}$$

Depending on values of four coefficients (at least one of them must be nonzero), six subcases were examined.

If $\lambda_2 = \lambda_4 = \lambda_5 = 0$ then the classical heat equation with a constant source (sink) is obtained, which was already examined above.

If $\lambda_2 = \lambda_5 = 0$ and $\lambda_4 \neq 0$ then Eq. (2.151) admits the four-dimensional MAI. On the other hand, the equation can be simplified by the scale ET

$$|\lambda_4| t \to t, \quad \sqrt{|\lambda_4|} x \to x, \quad e^{\frac{\lambda_3}{\lambda_4}} u \to u,$$

hence, $\lambda_4 = \pm 1$. As a result, case 10 from Table 2.5 was identified.

If $\lambda_2 = \lambda_4 = 0$ and $\lambda_5 \neq 0$ then Eq. (2.151) admits only the algebra A^{pr}.

If $\lambda_2 \neq 0$ then Eq. (2.151) contains a convective term and its symmetry depends on the coefficient λ_5. Assuming $\lambda_5 = 0$ and $\lambda_4 \neq 0$, one arrives at the equation

$$u_t = u_{xx} + \lambda_2 \ln u u_x + (\lambda_3 + \lambda_4 \ln u)u, \tag{2.152}$$

which is invariant w.r.t. the three-dimensional MAI

$$< \partial_t, \ \partial_x, \ e^{\lambda_4 t}(\partial_x - \lambda_4 I) > . \tag{2.153}$$

Eq. (2.152) and Lie algebra (2.153) are listed in case 20 of Table 2.5. It should be noted that Eq. (2.151) with $\lambda_2 \neq 0$ can be simplified by setting $\lambda_2 = 1$. The appropriate ET is

$$\lambda_2^2 t \to t, \quad \lambda_2 x \to x + \frac{\lambda_3}{\lambda_4} t, \quad e^{\frac{\lambda_3}{\lambda_4}} u \to u.$$

Thus, the coefficient λ_2 is skipped in case 20 of Table 2.5 (as well as in cases 21 and 22).

Assuming $\lambda_5 = \lambda_4 = 0$, one obtains the Galilei-invariant equation

$$u_t = u_{xx} + \lambda_2 \ln u u_x + \lambda_3 u.$$

Using ET (2.138) with $\lambda_3 \to \lambda_2$ we obtain case 21 from Table 2.5.

The last possibility is to assume $\lambda_5 \neq 0$. Using the ET

$$\lambda_2^2 t \to t, \quad \lambda_2 x \to x + \frac{\lambda_2^2 \lambda_4}{2\lambda_5} t, \quad e^{\frac{\lambda_4}{2\lambda_5}} u \to u$$

Eq. (2.151) reduces to the form

$$u_t = u_{xx} + \ln u u_x + (\lambda_3 + \lambda_5 \ln^2 u)u. \tag{2.154}$$

Eq. (2.154) possesses a nontrivial Lie symmetry only provided $\lambda_5 = \frac{1}{4}$. As a result, case 22 of Table 2.5 has been derived.

Finally, three equations involving exponential functions, Equations 2–4 from Table 2.4, were examined. It was proved that Equation 4 leads only to case 29 of Table 2.5, while the other two equations produce cases 12, 17, 18, 27, 28 and 30 of Table 2.5.

The sketch of proof is now completed. □

2.5.4 Application of form-preserving transformation

In this subsection, we present the Lie symmetry classification (LSC) for class (2.35) based on form-preserving transformations (FPTs). In order to derive such LSC, we need to use the results of the Lie–Ovsiannikov classification presented in Table 2.5 and Theorem 2.7 describing FPTs. The main result can be formulated as follows.

Theorem 2.11 *1. There are exactly 15 equations in Table 2.5, which are reducible to other equations from the same table by an appropriate FPT of the form (2.51)–(2.52). All the equations and the corresponding transformations are presented in cases 1–15 of Table 2.6. Case 16 of Table 2.6 shows that the equation arising in case 21 of Table 2.5 can be simplified by the highly nontrivial FPT presented in the third column of Table 2.6.*
2. All possible RDC equations of the form (2.35) admitting nontrivial Lie symmetries are reduced to one of the 15 canonical equations listed in the second column of Table 2.7 by the relevant FPTs presented in Theorem 2.7. The relevant maximal algebras of invariance (MAIs) of the canonical RDC equations are listed in the third column of Table 2.7.

TABLE 2.6: Simplification of the RDC equations from Table 2.5 by means of FPTs

	RDC equation	FPT	Canonical form of RDC equation
1	$u_t = u_{xx} + p$	$\tau = t,$ $y = x,$ $w = u - pt$	$w_\tau = w_{yy}$
2	$u_t = u_{xx} + pu$	$\tau = t,$ $y = x,$ $w = ue^{pt}$	$w_\tau = w_{yy}$
3	$u_t = (u^{-\frac{4}{3}} u_x)_x + pu^{-\frac{1}{3}}$	$\tau = t,$ $y = \varphi(x),$ $w = \psi^3(x)u,$ $\ddot{\psi} = \frac{p}{3}\psi, \dot{\varphi} = \psi^{-2}$	$w_\tau = (w^{-\frac{4}{3}} w_y)_y$
4	$u_t = (u^{-\frac{4}{3}} u_x)_x + pu + \lambda_6 u^{-\frac{1}{3}}$	$\tau = \frac{-3}{4p} e^{-\frac{4p}{3}t},$ $y = \varphi(x),$ $w = ue^{-pt}\psi^3(x),$ $\ddot{\psi} = \frac{\lambda_6}{3}\psi, \dot{\varphi} = \psi^{-2}$	$w_\tau = (w^{-\frac{4}{3}} w_y)_y$
5	$u_t = (u^k u_x)_x + pu,$ $k \neq 0$	$\tau = \frac{e^{kpt}}{kp},$ $y = x,$ $w = ue^{-pt}$	$w_\tau = (w^k w_y)_y$
6	$u_t = (e^u u_x)_x + p$	$\tau = \frac{e^{pt}}{p},$ $y = x,$ $w = u - pt$	$w_\tau = (e^w w_y)_y$
7	$u_t = u_{xx} + uu_x + p$	$\tau = t,$ $y = x + \frac{p}{2}t^2,$ $w = u - pt$	$w_\tau = w_{yy} + ww_y$
8	$u_t = (u^k u_x)_x + u^k u_x + \frac{2(k+2)}{(3k+4)^2} u^{k+1},$ $k \neq 0, -\frac{4}{3}$	$\tau = t, y = \frac{e^{k_0 x}}{k_0},$ $w = ue^{\frac{2k_0}{k}x},$ $k_0 = \frac{k}{3k+4}$	$w_\tau = (w^k w_y)_y$

Continuation of Table 2.6

	RDC equation	FPT	Canonical form of RDC equation
9	$u_t = (u^k u_x)_x + u^k u_x +$ $+ \left[\lambda_4 + \frac{2(k+2)}{(3k+4)^2} u^k\right] u,$ $k \neq 0, -\frac{4}{3},\ \lambda_4 \neq 0$	$\tau = \frac{e^{k\lambda_4 t}}{k\lambda_4},$ $y = \frac{e^{k_0 x}}{k_0},$ $w = u e^{-\lambda_4 t + \frac{2k_0}{k} x},$ $k_0 = \frac{k}{3k+4}$	$w_\tau = (w^k w_y)_y$
10	$u_t = (u^k u_x)_x + \lambda_3 u^k u_x +$ $+ (\lambda_6 u^k + p) u,$ $k \neq 0,\ |\lambda_3| + |\lambda_6| \neq 0$	$\tau = \frac{e^{kpt}}{kp},$ $y = x,$ $w = u e^{-pt}$	$w_\tau = (w^k w_y)_y +$ $+ \lambda_3 w^k w_y + \lambda_6 w^{k+1},$
11	$u_t = (u^k u_x)_x + (\lambda_2 u^{\frac{k}{2}} + u^k) u_x +$ $+ \left[\lambda_4 + \frac{2\lambda_2}{3k+4} u^{\frac{k}{2}} + \frac{2(k+2)}{(3k+4)^2} u^k\right] u,$ $k \neq 0, -\frac{4}{3},\ |\lambda_2| + |\lambda_4| \neq 0$	$\tau = t,\ y = \frac{e^{k_0 x}}{k_0},$ $w = u e^{\frac{2k_0}{k} x},$ $k_0 = \frac{k}{3k+4}$	$w_\tau = (w^k w_y)_y +$ $+ \lambda_2 w^{\frac{k}{2}} w_y + \lambda_4 w,$
12	$u_t = (e^u u_x)_x + e^u u_x + \frac{2}{9} e^u$	$\tau = t,$ $y = 3 e^{\frac{1}{3} x},$ $w = u + \frac{2}{3} x$	$w_\tau = (e^w w_y)_y$
13	$u_t = (e^u u_x)_x + e^u u_x + \lambda_4 + \frac{2}{9} e^u,$ $\lambda_4 \neq 0$	$\tau = \frac{e^{\lambda_4 t}}{\lambda_4},$ $y = 3 e^{\frac{1}{3} x},$ $w = u - \lambda_4 t + \frac{2}{3} x$	$w_\tau = (e^w w_y)_y$
14	$u_t = (e^u u_x)_x + \lambda_3 e^u u_x +$ $+ \lambda_6 e^u + p,\ \lambda_6 \neq \frac{2}{9} \lambda_3^2,$ $|\lambda_3| + |\lambda_6| \neq 0$	$\tau = \frac{e^{pt}}{p},$ $y = x,$ $w = u - pt$	$w_\tau = (e^w w_y)_y +$ $+ \lambda_3 e^w w_y + \lambda_6 e^w,$
15	$u_t = (e^u u_x)_x + (\lambda_2 e^{\frac{1}{2} u} + e^u) u_x +$ $+ \lambda_4 + \frac{2\lambda_2}{3} e^{\frac{1}{2} u} + \frac{2}{9} e^u,$ $|\lambda_2| + |\lambda_4| \neq 0$	$\tau = t,$ $x = 3 e^{\frac{1}{3} x},$ $w = u + \frac{2}{3} x$	$w_\tau = (e^w w_y)_y +$ $+ \lambda_2 e^{\frac{1}{2} w} w_y + \lambda_4,$
16	$u_t = u_{xx} + \ln u\, u_x + \lambda_3 u$	$\tau = t,$ $y = x + \frac{\lambda_3}{2} t^2,$ $w = e^{-\lambda_3 t} u$	$w_\tau = w_{yy} + \ln w\, w_y$

TABLE 2.7: The complete LSC of equations of the form (2.35) using FPTs

	the RDC equation	MAI
1	$u_t = u_{xx}$	$\langle \partial_t, \partial_x, G = t\partial_x - \frac{1}{2}xI,$ $I = u\partial_u,\ D = 2t\partial_t + x\partial_x,$ $\Pi = t^2\partial_t + tx\partial_x - \frac{1}{2}(\frac{x^2}{2} + t)I,$ $Q_1^\infty = \beta^1(t,x)\partial_u\rangle,\ \beta_t^1 = \beta_{xx}^1$
2	$u_t = (u^{-\frac{4}{3}}u_x)_x$	$\langle \partial_t, \partial_x, D, D_1 = kx\partial_x + 2I,$ $K = x^2\partial_x - 3xI\rangle,\ k = -\frac{4}{3}$
3	$u_t = (u^k u_x)_x,\ k \neq 0, -\frac{4}{3}$	$\langle \partial_t, \partial_x, D, D_1\rangle$
4	$u_t = (e^u u_x)_x$	$\langle \partial_t, \partial_x, D, D_2 = kx\partial_x + 2\partial_u\rangle,$ $k=1$
5	$u_t = [A(u)u_x]_x,\ A(u) \neq u^k, e^u$	$\langle \partial_t, \partial_x, D\rangle$
6	$u_t = u_{xx} + pu\ln u$	$\langle \partial_t, \partial_x, \mathcal{G} = e^{pt}(\partial_x - \frac{p}{2}xI), e^{pt}I\rangle$
7	$u_t = u_{xx} + uu_x$	$\langle \partial_t, \partial_x,\ G_0 = t\partial_x - \partial_u, D - I,$ $\Pi_0 = t^2\partial_t + tx\partial_x - (x + tu)\partial_u\rangle$
8	$u_t = u_{xx} + uu_x + pu$	$\langle \partial_t, \partial_x, \mathcal{G}_1 = e^{pt}(\partial_x - p\partial_u)\rangle$
9	$u_t = u_{xx} + \ln uu_x + r_1 u\ln u,\ r_1 \neq 0$	$\langle \partial_t, \partial_x, \mathcal{G}_2 = e^{r_1 t}(\partial_x - r_1 I)\rangle$
10	$u_t = u_{xx} + \ln uu_x$	$\langle \partial_t,\ \partial_x, G_1 = t\partial_x - I\rangle$
11	$u_t = u_{xx} + \ln uu_x + u(\frac{1}{4}\ln^2 u + r_1),$	$\langle \partial_t, \partial_x, Y = e^{\frac{1}{4}t - \frac{1}{2}x}I\rangle$
12	$u_t = (u^k u_x)_x + qu^m u_x + ru^{2m-k+1},$ $\|m\| + \|k\| \neq 0,\ \|q\| + \|2m - k\| \neq 0$	$\langle \partial_t, \partial_x, (k - 2m)D + D_1\rangle$
13	$u_t = (u^k u_x)_x + \ln uu_x + r_1 u^{1-k},$ $k \neq 0$	$\langle \partial_t,\ \partial_x,$ $Z_1 = k(t\partial_t + x\partial_x) - t\partial_x + I\rangle$
14	$u_t = (e^{ku}u_x)_x + qe^{mu}u_x + re^{(2m-k)u},$ $\|m\| + \|k\| \neq 0,\ \|q\| + \|2m - k\| \neq 0$	$\langle \partial_t, \partial_x, (k - 2m)D + D_2\rangle$
15	$u_t = (e^u u_x)_x + uu_x + r_1 e^{-u}$	$\langle \partial_t, \partial_x,$ $Z_2 = t\partial_t + x\partial_x - t\partial_x + \partial_u\rangle$

Remark 2.8 *All the parameters arising in Table 2.7 are arbitrary constants provided they satisfy the restrictions given therein. However, the parameter $p = \pm 1$, while the pair $(q, r) = (0, \pm 1)$ (no convection term) otherwise $q = 1$ and r is an arbitrary constant. An arbitrary parameter $k \neq 0$ in case 14 is reducible to $k = 1$, while $k = 0$ allows us to set $m = 1$ in the case 14.*

Proof First of all, we stress that two equations can be locally equivalent only under condition that their MAI have the same dimensionality or are infinity-dimensional. Obviously the linear heat equations listed in cases 6 and 7 of Table 2.5 can be mapped to the first one. The relevant FPTs are presented in Table 2.6 (cases 1 and 2).

Consider five equations arising in cases 2, 8, 9, 13 and 14 of Table 2.5, admitting five-dimensional MAI, i.e., the Lie algebras of the highest finite dimensionality. The equation listed in case 14 is mapped to the Burgers equation by the highly nontrivial transformation (see case 7 of Table 2.6). This transformation is belonging to the set of FPTs and was found for the first time in [80] (see also Example 2.1 above).

The equations listed in the eighth and ninth cases of Table 2.5 are mapped to that from the second case by the the corresponding FPTs (see cases 3 and 4 of Table 2.6). These transformations were earlier found in [94].

In order to complete examination of RDC equations with five-dimensional MAI, we note that the nonlinear heat equation $u_t = (u^{-\frac{4}{3}} u_x)_x$ (see case 2 of Table 2.5) cannot be reduced to the Burgers equation because condition (2.53) from Theorem 2.7 is not valid. In fact, this condition says that the diffusivities of two locally equivalent RDC equations must be the same functions (up to constant coefficients). Thus, there are only two RDC equations of the form (2.35), which are invariant under five-dimensional MAI.

Now we examine 9 equations admitting four-dimensional MAI (cases 3–4, 10–12, 15–18 of Table 2.5). Let us show that only three among them are locally inequivalent, while 6 other equations are reducible to those three equations by FPTs. Obviously, the equations listed in cases 3, 4 and 10 cannot be mapped one to another because condition (2.53) is not valid. However other equations are reduced to those listed in cases 3 and 4.

Indeed, equations 11, 15 and 16 from Table 2.5 can be united as follows

$$u_t = (u^k u_x)_x + \lambda_3 u^k u_x + \left[\lambda_4 + \frac{2(k+2)}{(3k+4)^2}\lambda_3^2 u^k\right] u, \quad k \neq 0, -\frac{4}{3}. \quad (2.155)$$

Let us construct a FPT reducing (2.155) to the equation

$$w_\tau = (w^k w_y)_y, \quad k \neq 0, -\frac{3}{4}. \quad (2.156)$$

Condition (2.53) takes the form

$$b_x^2 u^k = a_t (\alpha u + \beta)^k$$

Complete description of Lie symmetries of RDC equations

so that
$$\beta = 0, \quad b_x^2 = a_t \alpha^k. \tag{2.157}$$

Conditions (2.54) and (2.55) take the forms
$$b_t = 0, \quad b_{xx} + \left[\lambda_3 - 2(k+1)\frac{\alpha_x}{\alpha}\right] b_x = 0, \tag{2.158}$$

and
$$\alpha_t = -\lambda_4 \alpha, \quad \alpha \frac{\alpha_{xx} + \lambda_3 \alpha_x}{k+2} = \alpha_x^2 + \frac{2\lambda_3^2}{(3k+4)^2}\alpha^2. \tag{2.159}$$

Eqs. (2.159) are compatible and have the particular solution (hereafter we are not looking for general solutions in order to find an appropriate FPT)
$$\alpha = e^{-\lambda_4 t + \frac{2\lambda_3}{3k+4}x}. \tag{2.160}$$

Substituting (2.160) into (2.157) and (2.158) one obtains
$$b_x^2 = a_t e^{-k\lambda_4 t + \frac{2k\lambda_3}{3k+4}x}, \quad b_{xx} - \frac{k\lambda_3}{3k+4}b_x = 0. \tag{2.161}$$

Solutions of system (2.161) depend on the values λ_3 and λ_4. There are three different cases, which correspond to the equations listed in cases 11, 15 and 16 of Table 2.5 and those lead to the relevant FPTs

$$\tau = \frac{e^{\lambda_4 k t}}{\lambda_4 k}, \quad y = \frac{e^{k_0 x}}{k_0}, \quad w = u e^{-\lambda_4 t + \frac{2k_0}{k}x}, \quad k_0 = \frac{k}{3k+4} \tag{2.162}$$

if $\lambda_3 = 1$ and $\lambda_4 \neq 0$;

$$\tau = t, \quad y = \frac{e^{k_0 x}}{k_0}, \quad w = u e^{\frac{2k_0}{k}x}, \quad k_0 = \frac{k}{3k+4} \tag{2.163}$$

if $\lambda_3 = 1$ and $\lambda_4 = 0$;

$$\tau = \frac{e^{\pm k t}}{\pm k}, \quad y = x, \quad w = u e^{\mp t} \tag{2.164}$$

if $\lambda_3 = 0$ and $\lambda_4 = \pm 1$. The above transformations and the corresponding equations are listed in cases 5, 8 and 9 of Table 2.6.

Similarly, equations 12, 17 and 18 from Table 2.5 can be united as
$$u_t = (e^u u_x)_x + \lambda_3 e^u u_x + \lambda_4 + \frac{2}{9}\lambda_3^2 e^u, \tag{2.165}$$

which is reduced to the equation
$$w_\tau = (e^w w_y)_y. \tag{2.166}$$

In this case, condition (2.53) takes the form
$$b_x^2 e^u = a_t e^{\alpha u + \beta}. \tag{2.167}$$

Since the functions a, b, α and β do not depend on the variable u and $a_t b_x \neq 0$ (see condition (2.56)), we immediately arrive at

$$\alpha = 1, \quad b_x^2 = a_t e^{\beta} \tag{2.168}$$

So, conditions (2.54) and (2.55) are reduced to the forms

$$(-2b_x \beta_x + b_{xx} + \lambda_3 b_x)e^u = b_t, \tag{2.169}$$

and

$$(-\beta_{xx} + \beta_x^2 - \lambda_3 \beta_x + \frac{2}{9}\lambda_3^2)e^u + \lambda_4 + \beta_t = 0, \tag{2.170}$$

respectively. Splitting Eqs. (2.169) and (2.170) with respect to e^u, we arrive at the overdetermined system

$$b_t = 0, \quad b_{xx} + (\lambda_3 - 2\beta_x)b_x = 0, \tag{2.171}$$

$$\beta_t = -\lambda_4, \quad \beta_{xx} - \beta_x^2 + \lambda_3 \beta_x - \frac{2}{9}\lambda_3^2 = 0. \tag{2.172}$$

System (2.172) has, for example, the solution

$$\beta = -\lambda_4 t + \frac{2}{3}\lambda_3 x. \tag{2.173}$$

Substituting (2.173) into (2.168) and (2.171) one obtains the system

$$b_x^2 = a_t e^{-\lambda_4 t + \frac{2}{3}\lambda_3 x}, \tag{2.174}$$

$$b_t = 0, \quad b_{xx} - \frac{\lambda_3}{3}b_x = 0 \tag{2.175}$$

to find the functions a and b. The above system can be easily integrated because Eqs. (2.175) are linear. As a result, solutions of (2.174)–(2.175) essentially depend on the values λ_3 and λ_4. There are three different cases, which correspond to equations 12, 17 and 18 from Table 2.5 and those lead to the relevant FPTs. They are listed in cases 6, 12 and 13 of Table 2.6. Now examination of equations admitting four-dimensional MAI is complete.

Finally, we should analyze 13 equations admitting three-dimensional MAI listed in cases 5 and 19–30 of Table 2.5. Of course, none of the equations arising in cases 19–30 of Table 2.5 is reducible to that from case 5. However, using Theorem 2.7 in the quite similar way as it was done above, we have found that four of them are reducible to other by the relevant FPTs listed in Table 2.6 (see cases 10–11 and 14–15). Moreover, the equation listed in case 21 of Table 2.5 can be simplified by the nontrivial transformation presented in case 16 of Table 2.6. Thus, 9 equations only admitting three-dimensional MAI have been obtained.

Here we present some details for the most complicated equation arising in case 25 of Table 2.5, namely

$$u_t = (u^k u_x)_x + (\lambda_2 u^{\frac{k}{2}} + u^k)u_x + \left[\lambda_4 + \frac{2\lambda_2}{3k+4}u^{\frac{k}{2}} + \frac{2(k+2)}{(3k+4)^2}u^k\right]u, \tag{2.176}$$

where $\lambda_2 k(3k+4) \neq 0$. Let us show that Eq. (2.176) is reducible to the equation
$$w_\tau = (w^k w_x)_x + \lambda_2 w^{\frac{k}{2}} w_x + \lambda_4 w, \qquad (2.177)$$
listed in case 11 of Table 2.6. Substituting the known triplets (A, B, C) and (F, G, H) from Eqs. (2.176)–(2.177) into system (2.53)–(2.55) and taking into account (2.52), one obtains a very cumbersome expression. However, one can be split w.r.t. different exponents of the variable u. As a result, one immediately finds the function $\beta = 0$ and derives the system

$$\begin{aligned} & b_t = 0, \quad a_t + \lambda_4(1 - a_t)\alpha = 0, \\ & \lambda_2(\alpha_x - \tfrac{2}{3k+4}\alpha) = 0, \quad b_x^2 = a_t \alpha^k, \\ & b_{xx} + [1 - 2(k+1)\tfrac{\alpha_x}{\alpha}]b_x = 0, \\ & \alpha_{xx} + [1 - (k+2)\tfrac{\alpha_x}{\alpha}]\alpha_x - 2\tfrac{k+2}{(3k+4)^2}\alpha = 0 \end{aligned} \qquad (2.178)$$

for finding the functions a, b and α. Because finding the general solution of system (2.178) is a nontrivial task, we restrict ourselves by the search for a particular solution. Assuming that α does not depend on the time t, the second equation with $\lambda_4 \neq 0$ (the subcase $\lambda_4 = 0$ can be treated in the same way) immediately gives $a = t + t_0$ (the parameter t_0 can be skipped) while the third equation produces $\alpha = \alpha_0 e^{\frac{k}{3k+4}x}$ (α_0 is an arbitrary constant at the moment). Having this, other equations of system (2.178) can be easily solved, hence the particular solution

$$a = t, \quad b = \frac{3k+4}{k} e^{\frac{k}{3k+4}x}, \quad \alpha = e^{\frac{2}{3k+4}x}, \quad \beta = 0$$

was found. Thus, Eq. (2.176) is reducible to Eq. (2.177) via the following FPT

$$\tau = t, \quad y = \frac{3k+4}{k} e^{\frac{k}{3k+4}x}, \quad w = u e^{\frac{2}{3k+4}x}.$$

So, case 11 of Table 2.5 is derived.

The last step is to show that those nine equations admitting three-dimensional MAI are not reducible one to another.

Consider the four equations listed in cases 19–22 of Table 2.5, which have the same diffusivity $A(u) = 1$. Taking any two equations from this set, we see that condition (2.53) can be satisfied if $b_x^2 = a_t$. Let us show that condition (2.54) of Theorem 2.7 cannot be satisfied if one wants to reduce the equation listed in case 19 to any other equation from this set of equations. Taking into account the structure of the equations listed in cases 20–22, it is easily seen that condition (2.54) takes the same form for all of them:

$$-2\frac{b_x \alpha_x}{\alpha} + b_{xx} + b_x u = a_t \ln(\alpha u + \beta) + b_t. \qquad (2.179)$$

In expression (2.179), two terms only depend on u which leads to the constraints $b_x = a_t = 0$ and this contradicts condition (2.56).

Consider the equation listed in case 21 and assume that there is a FPT, which reduces this equation to the equation

$$w_\tau = w_{yy} + \ln w w_y + \lambda_5 w \ln w, \quad \lambda_5 \neq 0 \qquad (2.180)$$

from case 20. In this case, conditions (2.53) and (2.54) take the form

$$b_x^2 = a_t, \qquad (2.181)$$

$$-2\frac{b_x \alpha_x}{\alpha} + b_{xx} + b_x \ln u = a_t \ln(\alpha u + \beta) + b_t. \qquad (2.182)$$

Eq. (2.182) immediately leads to the constraint $b_x = a_t$. Thus, Eq. (2.181) takes the form

$$a_t = b_x = 1 \qquad (2.183)$$

and then condition (2.55) from Theorem 2.7 can be easily reduced to the overdetermined system

$$2\frac{\alpha_x^2}{\alpha} - \alpha_{xx} + \lambda_7 \alpha + \alpha_t = \lambda_5 \alpha \ln \alpha, \quad \alpha_x = -\lambda_5 \alpha \qquad (2.184)$$

to find the function α. System (2.184) is compatible only under the condition $\lambda_5 = 0$ which contradicts the constraint in (2.180). Hence the equation listed in case 21 of the Table 2.5 is locally inequivalent to one from case 20. Simultaneously, one sees that this equation can be reduced to (2.180) with $\lambda_5 = 0$ and it is exactly case 16 from Table 2.6.

In quite similar way one proves that the equation listed in case 22 is not reducible either to equation from case 20 or one from case 21 of Table 2.5.

We also omit the similar analysis of equations with power and exponential diffusivities equations (cases 23–24 and 26–30 in Table 2.5), which leads to the FPTs and the relevant equivalent equations listed in cases 10 and 14–15 of Table 2.6.

At the final step, we exclude 15 locally equivalent equations from Table 2.5 using Table 2.6 and take into account case 16 from Table 2.6.

As a result, we obtain Table 2.7 containing 15 canonical equations and their MAIs (constant lambda-s with indices were renamed).

The proof is now completed. □

Thus, we have fully implemented the algorithm formulated in Section 2.3 for the general class of RDC equations (2.35). The main results are summarized in Tables 2.5–2.7. It is worth commenting on the results obtained. The Lie–Ovsiannikov classification method applied to RDC equations of the form (2.35) leads to 30 equations (listed in Table 2.5) admitting nontrivial Lie symmetry (i.e., MAI is of dimensionality three and higher), while the LSC method based on FPTs leads to the 15 equations only (they are listed in Table 2.7). These 15 equations are inequivalent w.r.t. to any point transformations.

2.6 Nonlinear equations arising in applications and their Lie symmetry

In Section 2.5, all possible RDC equations admitting a nontrivial Lie symmetry were described and all possible point (local) transformations mapping a given equation into another RDC equation with the same symmetry were identified. Here we consider some specified equations belonging to the class of RDC equations (2.35), which are used in real world applications, in order to establish their Lie symmetry. We restrict ourselves on some well-known examples only and the reader may examine many other nonlinear RDC arising in applications whether they admit a nontrivial Lie symmetry using Tables 2.4–2.7.

2.6.1 Heat (diffusion) equations with power-law nonlinearity

It is well-known that a typical derivation (in one-dimensional space approximation) of the classical heat (diffusion) equation leads to the equation

$$\rho c T_t = \Lambda T_{xx}, \tag{2.185}$$

where ρ, c, and Λ are density, heat capacity, and heat conductivity, respectively, and $T(t,x)$ is unknown temperature. Obviously, Eq. (2.185) is reducible to Eq. (2.1) by a scaling transformation (actually, one is ET belonging to (2.36)). As one may note, the linear heat equation (2.1) is the simplest RDC equation from class (2.35) and possesses the widest Lie symmetry (see case 1 of Table 2.7). The relevant Lie algebra contains as a subalgebra the generalized Galilei algebra, which was discussed in Section 2.1.

However, if the temperature range is sufficiently wide then one should assume that the above heat coefficients are nonnegative functions depending on T. As a result, one obtains the nonlinear equation

$$\rho(T)c(T)T_t = [\Lambda(T)T_x]_x \tag{2.186}$$

instead of Eq. (2.185). This nonlinear equation can be essentially simplified using the Goodman substitution [91, 125, 158]

$$u = \phi(T) \equiv \int_0^T \rho(\zeta)c(\zeta)\,\mathrm{d}\zeta. \tag{2.187}$$

Thus, the standard nonlinear heat equation (2.9) is obtained with the diffusivity

$$d(u) = \frac{\Lambda[\phi^{-1}(u)]}{\rho[\phi^{-1}(u)]c[\phi^{-1}(u)]}, \tag{2.188}$$

where ϕ^{-1} is the inverse function to ϕ. An enormous number of papers and books (see, e.g., the above cited books and references therein) is devoted to equations of the form (2.186) arising in real world applications. Here we mention only some of them, which arise in a wide range of applications and possess a nontrivial Lie symmetry.

The most typical form of the coefficient $d(u)$ arising in applications is the power-law function u^k, i.e., the equation

$$u_t = \left(u^k u_x\right)_x \tag{2.189}$$

is obtained.

The classical example when such power-law function occurs is the heat transfer within Storm's metals. A material is called Storm's metal if its heat coefficients satisfy the relations [236](see also [158])

$$\Lambda(T)\rho(T)c(T) = \kappa_0, \quad \frac{d}{dT}\log\frac{\rho(T)c(T)}{\Lambda(T)} = \kappa_1, \tag{2.190}$$

where κ_0 and κ_1 are correctly-specified parameters. It can be shown that the above relation simplifies Eq. (2.186) to the form

$$u_t = \left(u^{-2} u_x\right)_x. \tag{2.191}$$

Notably, Eq. (2.191) is linearizable by an integral substitution, which was firstly found in [236] and later rediscovered in [28, 223]. The substitution can be rewritten in the following nonlocal form

$$dy = u\,dx + u^{-2}\frac{\partial u}{\partial x}dt, \quad w = u^{-1} \tag{2.192}$$

and one reduces Eq. (2.191) to the classical heat equation $w_t = w_{yy}$.

Substitution (2.192) can be presented also as the superposition of the transformations

$$t = t, \ x = x, \ u = v_x,$$

and

$$t = t, \ x = w, \ v = y.$$

Although relation (2.190) is valid for a wide range of metals, some widely used in engineering metals do not possess this property. For example, $\Lambda(T)$ is a constant and $\rho(T)c(T)$ is a linear function for aluminium [74]. However, the resulting equation is again Eq. (2.189) with a negative exponent k.

In the case of the power-law diffusivity with a positive exponent k, Eq. (2.9) also takes the form (2.189), which is known in literature as the equation with slow diffusion. In particular, Eq. (2.189) with $k = 1$ is often called the porous media equation or the Boussinesq equation because J. Boussinesq was the first to introduce this equation as a model for water filtration in soil. The paper [19] was pioneering that, in which Eq. (2.189) with the diffusivity u^k, $k \geq 1$ was studied in order to model unsteady motion of gas through porous media.

Because of the importance of Eq. (2.189) for real world application, its exact solutions were studied extensively since the 1950s [19, 20, 33, 104, 105, 207, 233, 250]. Moreover, some new solutions were derived in the 1970s–1980s and almost all exact solutions of Eq. (2.189) were summarized in Hill's excellent work [131]. We note that plane wave solutions (especially traveling fronts playing an important role in real world models), which form a simplest subclass of invariant solutions, can be found for arbitrary diffusivity (at least in implicit form)[55] and several examples of such solutions are presented in [55, 73, 123]. On the other hand, Eq. (2.189) admits *a nontrivial Lie symmetry* (see cases 3 and 4 of Table 2.7). In the case $k \neq 0, -\frac{4}{3}$, the equation admits the four-dimensional MAI generated by two operators of time and space translations and two operators of scale transformations D_0 and D_1. Eq. (2.189) with $k = -\frac{4}{3}$ admits an extension of the above algebra via the operator of the conformal transformations K and the corresponding Lie algebra was discussed in Section 2.2. It was pointed out in [131] that *all the known solutions (at that time) can be derived using the above Lie symmetry operators of Eq. (2.189)*. This fact underlines importance and powerfulness of the Lie method for searching exact solutions of nonlinear PDEs.

In conclusion of this subsection, we note that some new exact solutions of Eq. (2.189) with correctly-specified exponents $k = -\frac{4}{3}; -\frac{3}{2}; -\frac{1}{2}$ and $k = -1$ were found in the 1990s [16, 117, 120, 148, 149] (these special cases are discussed in Chapters 4 and 5), which are not obtainable via Lie symmetries. Notably authors of some recently published papers present exact solutions of Eq. (2.189) and they are refereed as new, although these solutions were found much earlier in the above cited papers.

2.6.2 Diffusion equations with a convective term

The equation listed in case 2 of Table 2.7 is nothing else but the famous Burgers equation

$$u_t = u_{xx} + uu_x, \qquad (2.193)$$

which occurs in mathematical description of a wide range of processes. The equation was proposed by Burgers for modeling the fluid flow with turbulence effects in 1948 and the author summarized his results about Eq. (2.193) in monograph [39]. It is widely accepted that the Burgers equation is a one-dimensional approximation of the Navier–Stokes equations. Nowadays the Burgers equation occurs in modeling processes, which are quite far from fluid dynamics, in particular the transport traffic (see e.g., [184]).

It is well-known that the Burgers equation is linearizable. The corresponding substitution is nonlocal and has the form

$$u = 2\frac{v_x}{v}. \qquad (2.194)$$

The resulting equation is the linear diffusion equation $v_t = v_{xx}$. This substitution is called the Cole–Hopf transformation because they derived (2.194) independently each from other [89, 134].

As it follows from Table 2.7, MAI of the Burgers equation is the five-dimensional Lie algebra with the basic operators

$$\partial_t,\ \partial_x,\ G_0 = t\partial_x - \partial_u,$$
$$D = 2t\partial_t + x\partial_x - u\partial_u, \qquad (2.195)$$
$$\Pi_0 = t^2\partial_t + tx\partial_x - (x+tu)\partial_u,$$

which were found for the first time in [145]. One may note that this algebra has a similar structure to that of the linear heat equations (we do not take into account the infinite-dimensional part of MAI of the latter) and the operators ∂_t, ∂_x, G_0, D and Π_0 generate transformations with the same geometrical interpretation as those from (2.2). The only difference is absence of the unit operator I (see (2.2)). As a result, the commutator $[\partial_x, G_0] = 0$ while $[\partial_x, G] = -I$ in the case of the generalized Galilei algebra (2.2). All other Lie brackets for operators from (2.195) are the same as for those from (2.2). Thus, the algebra with the basic operators (2.195) can be called the generalized Galilei algebra without the unit operator.

Of course, all the operators from (2.195) can be used for construction of exact solutions of the Burgers equation. However, the solutions obtained are useful in exceptional cases only because the Cole–Hopf transformation (2.194) allows us to transform each solution of the linear heat equation into a solution of Eq. (2.193), so that it is a nontrivial problem how to construct invariant solutions, which are not obtainable by the Cole–Hopf transformation.

A natural generalization of the Burgers equation reads as

$$u_t = \left(u^k u_x\right)_x + u^m u_x. \qquad (2.196)$$

This equation takes into account the most typical form of nonconstant diffusivity and assumes that the term describing the convective transport can be more complicated. For example, Eq. (2.196) with $k \geq 0$, $m \geq 1$ arises in the theory of infiltration of water under gravity through porous medium [21, 209](see also [37, 242] for more references). Eq. (2.196) with $k = m = -2$ occurs also for describing the liquid transport in porous media and one is linearizable [36, 102].

Eq. (2.196) is a particular case of the RDC equation arising in case 8 of Table 2.7, hence one is invariant w.r.t. three-dimensional MAI generated by the operators

$$\partial_t,\ \partial_x,\ D = (k-2m)t\partial_t + (k-m)x\partial_x + u\partial_u. \qquad (2.197)$$

The structure of this algebra essentially depends on the exponents k and m and three different cases occur, namely $k = 2m$, $k = m$, and $k \neq m, 2m$. It means that different types of exact solutions for Eq. (2.196) can be constructed using the Lie algebra (2.197).

Now we present the following observation. It is widely accepted that fundamental equations in physics should be invariant w.r.t. Galilei transformations or Lorentz transformations (in addition to the space translations and rotations). In particular, it should be Galilean invariance in the case of classical (Newtonian) mechanics, fluid dynamics and classical thermodynamics. As it was mentioned in Section 2.1, the classical Newton equation admits the Galilei operator $t\partial_x$, which generate the invariance transformations

$$x \to x + vt. \tag{2.198}$$

Formula (2.198) is a mathematical reflection of the Galilean relativistic principle, being v the inertial reference system velocity (obviously it is a vector if one considers the relevant process in n-dimensional space). It can be easily calculated that the linear heat equation (2.1) and the Burgers equation (2.193) are compatible with the Galilean relativistic principle because the operators G and G_0 generate invariance transformations (2.198) (simultaneously the function u should be transformed accordingly). Thus, both equations can be considered as generic models in heat transfer theory and fluid dynamics theory. It is worth noting that there is the third equation among RDC equations of the form (2.35), which admits the Galilean transformations (2.198) and is inequivalent to (2.1) and (2.193). In fact, the equation listed in case 10 of Table 2.7, which can be rewritten in the equivalent form (in order to avoid a singularity for $u = 0$)

$$u_t = u_{xx} + \ln(1+u)u_x, \tag{2.199}$$

admits the Lie symmetry operator $t\partial_x - (1+u)\partial_u$. This operator generates transformations (2.198) and $u \to e^{-v}(u+1) - 1$. Surprisingly, in contrast to the linear heat equation and the Burgers equation, Eq. (2.199) is not known in applications. However, we assume that one should consider this equation as a mathematical expression of some physical (chemical, biological) laws.

Finally, it is worth noting that the problem of construction of Galilei-invariant equations was an important task for a long time since the remarkable Levi-Leblond work was published [165]. In particular, all possible scalar second-order PDEs, which are invariant under Galilei algebra and its typical extensions, were described in several authors' papers. Nonlinear PDEs admitting Galilei algebra with unit operator (nonzero "mass") were constructed in [114] (see Chapter 3), while those with Galilean symmetry with zero "mass" were described in [52, 107].

2.6.3 Nonlinear equations describing three types of transport mechanisms

RDC equations of the form (2.35) with $ABC \neq 0$ occur in mathematical modeling many processes in physics (biology, ecology, chemistry, etc.), which involve three different types of energy transport. A vast literature is devoted to such equations, especially papers and books devoted to models arising in biomedical applications (see, e.g., books [160, 180, 181, 245]).

The simplest equation of such type is

$$u_t = u_{xx} + \lambda u_x + \lambda_1 u + \lambda_0, \qquad (2.200)$$

where λ, λ_1 and λ_0 are arbitrary constants. Although the linear equation (2.200) seems to be an important example among RDC equations (see an example of its application in [245]), one does not possess a special status from the symmetry point of view. In fact, the equation is reducible to the classical heat equation (2.1) by a combination of equivalence transformations (ETs) from the set (2.36) and FPTs listed either in case 1 ($\lambda_1 = 0$) or case 2 ($\lambda_1 \neq 0$) of Table 2.6.

The simplest RDC equation, which involves three transport mechanisms and is not reducible to a RD equation reads as

$$u_t = u_{xx} + \lambda u u_x + \lambda_1 u, \quad \lambda\lambda_1 \neq 0. \qquad (2.201)$$

Eq. (2.201) is the Burgers equation with the linear term $\lambda_1 u$, which can be thought as a distributed source (sink) of energy. This equation is equivalent to one listed in case 8 of Table 2.7. Thus, Eq. (2.201) admits three-dimensional Lie algebra. Another example of a generalization of the Burgers equation possessing a nontrivial Lie symmetry is

$$u_t = u_{xx} + \lambda u u_x + \lambda_1 u^3, \quad \lambda\lambda_1 \neq 0. \qquad (2.202)$$

Eq. (2.202) is a special case ($k = 0$) of the RDC equation listed in case 12 of Table 2.7, hence one admits the three-dimensional MAI generated by

$$\partial_t, \ \partial_x, \ D = 2t\partial_t + x\partial_x - u\partial_u. \qquad (2.203)$$

The well-known generalization of the Burgers equation

$$u_t = u_{xx} + \lambda u u_x + \lambda_1 u - \lambda_2 u^2, \quad \lambda_1 > 0, \lambda_2 > 0, \qquad (2.204)$$

occurring in mathematical biology was extensively studied by Murray [179, 180, 181] and can be refereed as the Murray equation [80]. As it follows from the results of Section 2.5 (see Tables 2.5 and 2.7), Eq. (2.204) admits the trivial Lie symmetry (the operators space and time translations) only, however, it will be shown in Chapter 2 that the Murray equation admits a conditional symmetry. Interestingly, a simple generalization of Eq. (2.204) via introducing the porous diffusivity, namely

$$u_t = (uu_x)_x + \lambda u u_x + \lambda_1 u - \lambda_2 u^2 \qquad (2.205)$$

possesses a nontrivial Lie symmetry (see case 24 of Table 2.5 with $k = 1$). Notably Eq. (2.205) is reduced to that (with $k = m = 1$) in case 12 of Table 2.7 by FPT from case 10 of Table 2.6. Moreover, the further generalization of the Murray equation

$$u_t = (u^k u_x)_x + \lambda u^k u_x + \lambda_1 u - \lambda_2 u^{k+1} \qquad (2.206)$$

still admits the operator $e^{-\lambda_1 kt}(\partial_t + \lambda_1 u \partial_u)$ (see case 24 of Table 2.5).

It is worth pointing out that density-dependent RDC equations play an essential role in mathematical modeling tumor growth processes. In particular, the recently proposed model for the tumor cells density has the form [160, 235]

$$u_t = [d(u)u_x]_x - \nu u_x + \lambda_1 u - \lambda_2 u^2. \qquad (2.207)$$

Here $u(t,x)$ is the density of tumor cells, $d(u)$ is the tumor cell diffusivity, which may essentially depend on u (see, e.g., P.90 in [160]), ν is the degree at which cells migrate away from the tumor core and the last two terms express the logistic law of growth of tumor cells. If the coefficient ν is a constant then Eq. (2.207) is reducible to the relevant RD equation (as it was shown above). If one naturally assumes that ν depends on u then Eq. (2.207) with the simplest function $\nu = -\lambda u$ and the porous diffusivity $d(u) = u$ is nothing else but Eq. (2.205). Thus, we have shown that the generic equation for modeling tumor growth admits three-dimensional Lie algebra provided two coefficients are correctly-specified.

Chapter 3

Conditional symmetries of reaction-diffusion-convection equations

3.1	Conditional symmetry of differential equations: historical review, definitions and properties	77
3.2	Q–conditional symmetry of the nonlinear heat equation	83
3.3	Determining equations for finding Q-conditional symmetry of reaction-diffusion-convection equations	87
3.4	Q–conditional symmetry of reaction-diffusion-convection equations with constant diffusivity	92
3.5	Q–conditional symmetry of reaction-diffusion-convection equations with power-law diffusivity	100
	3.5.1 The case of proportional diffusion and convection coefficients	101
	3.5.2 The case of different diffusion and convection coefficients	107
3.6	Q–conditional symmetry of reaction-diffusion-convection equations with exponential diffusivity	111
	3.6.1 Solving the nonlinear system (3.166)	118
	3.6.2 Solving the nonlinear system (3.169)	125
3.7	Nonlinear equations arising in applications and their conditional symmetry	129

3.1 Conditional symmetry of differential equations: historical review, definitions and properties

In 1969, Bluman and Cole [26] introduced an essential generalization of the Lie symmetry notion using the (1+1)-dimensional linear heat equation

$$u_t = u_{xx}$$

(we remind the reader that the subscripts t and x denote differentiation with respect to (w.r.t.) these variables).

The crucial idea used for introducing the notion of new type of symmetries, called later nonclassical symmetry, is to change the classical criteria of Lie symmetry by inserting into the criteria an additional equation, which is produced by the symmetry in question. As a result, the system of determining equations (DEs) for finding the nonclassical symmetry becomes nonlinear (in contrast to that for finding Lie symmetry). Integration of the system of DEs is the most difficult task for finding nonclassical symmetry of PDE in question. It was the main reason why a new type of non-Lie symmetry introduced in [26] was forgotten for almost 20 years. To the best of our knowledge, nonclassical symmetry was briefly described and applied to the Burgers equation only in Ames's monograph [6] until the 1980s when Olver and Rosenau [198] and independently Fushchych and Tsyfra [116] brought attention of researchers to nonclassical symmetry. As a result, in the end of the 1980s and in the 1990s, the theory of nonclassical symmetry was essentially developed and applied to a wide range of PDEs (especially, evolution equations) [13, 16, 17, 49, 79, 80, 113, 115, 116, 121, 164, 191, 192, 193, 198, 197, 199, 217, 218, 219, 229, 252, 253] (see also an extensive overview in [224]).

A new generalization of Lie symmetry, conditional symmetry, was suggested by Fushchych and his collaborators [112], [114, Section 5.7], [227]. Note that the notion of nonclassical symmetry can be derived as a particular case from conditional symmetry but not vice versa (see highly nontrivial examples in the end of the section and in [63, 64]). In the 1980s–1990s, the notions of weak symmetry [198, 199, 218], potential symmetry [29, 30, 159], generalized conditional symmetry [101, 173, 220, 251] were introduced and each of those symmetries is not the classical Lie symmetry. Thus, there are several types of non-Lie symmetries at the present time and each of them can be called a nonclassical symmetry. Taking this into account, to avoid any misunderstanding we continuously use the terminology Q-conditional symmetry introduced instead of nonclassical symmetry in [112, 114].

In the 2000s–2010s, a further development of the Q-conditional symmetry theory has been done and the theory was applied for search symmetries for new classes of PDEs such as nonlinear RDC equations with variable (in space) coefficients [34, 211, 234, 243] and systems of RDC equations [14, 54, 56, 57, 58, 59, 81, 126, 178, 210] (see also the relevant references in [25]). Notably, it was shown in [54] that the Q-conditional symmetry definition can be modified in such a way that a given system of PDEs can admit a hierarchy of conditional symmetry operators.

Now we present a definition of Q-conditional symmetry for an arbitrary PDE with smooth coefficients. Let us consider the k-order PDE

$$L\left(t, x, u, \underset{1}{u}, \ldots, \underset{k}{u}\right) = 0, \quad k \geq 1, \tag{3.1}$$

where $u = u(t,x)$ is an unknown function, $\underset{s}{u}$ means a totality of s-order derivatives of $u(t,x)$ ($s = 1, 2, \ldots, k$) and L is a given smooth function.

Similarly to Lie symmetries, Q-conditional symmetries are constructed in the form of the first-order differential operators

$$Q = \xi^0(t,x,u)\partial_t + \xi^1(t,x,u)\partial_x + \eta(t,x,u)\partial_u, \quad (\xi^0)^2 + (\xi^1)^2 \neq 0, \quad (3.2)$$

where the operator coefficients $\xi^0(t,x,u), \xi^1(t,x,u)$ and $\eta(t,x,u)$ should be found using the corresponding criteria. Throughout the book we use the notation $Q(u) = 0$ for the first-order PDE

$$\xi^0 u_t + \xi^1 u_x - \eta = 0,$$

which is generated by the Q-conditional symmetry (3.2). Rigorously speaking, one should write the invariance surface condition (see, e.g., [24] for details)

$$Q[u - u_0(t,x)]\Big|_{u=u_0(t,x)} = 0,$$

which leads exactly to the above first-order PDE provided an arbitrary solution $u = u_0(t,x)$ is given.

Definition 3.1 *Operator (3.2) is called Q-conditional (nonclassical) symmetry of PDE (3.1) if the following invariance criteria is satisfied:*

$$\underset{k}{Q}(L)\Big|_{\mathcal{M}} = 0, \qquad (3.3)$$

where the differential operator $\underset{k}{Q}$ is the k-order prolongation of operator (3.2) and the manifold $\mathcal{M}$ is defined by the system of equations

$$L = 0, \quad Q(u) = 0, \quad \frac{\partial^{p+q} Q(u)}{\partial t^p x^q} = 0, \quad 1 \leq p + q \leq k - 1$$

in the prolonged space of the variables

$$t, \ x, \ u, \ \underset{1}{u}, \ \ldots, \ \underset{k}{u}.$$

Remark 3.1 *The operator $\underset{k}{Q}$ is the kth-order prolongation of the operator Q. Its coefficients are expressed via the functions ξ^0, ξ^1 and η by the well-known formulae, which are presented in Section 1.3.*

The definition generalizes the classical definition of Lie symmetry (see Definition 1.2 in Chapter 1). We remind the reader that the latter is defined by the same criteria, however the simpler manifold $\mathcal{M}_L = \{L = 0\}$ is used instead of $\mathcal{M}$. In particular, each Lie symmetry is automatically a Q-conditional symmetry.

The above definition allows us to formulate the main properties of Q-conditional symmetries.

Property 1 *If operator (3.2) is Q-conditional symmetry of PDE (3.1) then this symmetry can be multiplied by an arbitrary smooth function $M(t,x,u)$ and the operator obtained is again Q-conditional symmetry of the same PDE.*

Proof To avoid cumbersome calculations, we prove the property in the case of an arbitrary second-order PDE, i.e., Eq. (3.1) with $k = 2$:

$$L(t, x, u, \underset{1}{u}, \underset{2}{u}) = 0. \qquad (3.4)$$

The proof for the arbitrary order k can be realized in the same way.

Let us assume that operator (3.2) is a Q-conditional symmetry of Eq. (3.4). Thus, according to Definition 3.1, the condition

$$Q\underset{2}{L}\Big|_{\mathcal{M}} = 0, \quad \mathcal{M} = \{L = 0, \ Qu = 0, \ D_t(Qu) = 0, \ D_x(Qu) = 0\} \qquad (3.5)$$

is fulfilled. Hereafter we use the notations $Qu = \xi^0 u_t + \xi^1 u_x - \eta$, $D_t = \partial_t + u_t \partial_u + u_{tt}\partial_{u_t} + u_{tx}\partial_{u_x}$, $D_x = \partial_x + u_x \partial_u + u_{tx}\partial_{u_t} + u_{xx}\partial_{u_x}$ and $D = (D_t, D_x)$. Let us show, that the operator

$$Y = M(t, x, u)Q, \qquad (3.6)$$

being M an arbitrary smooth function, is also a Q-conditional symmetry of Eq. (3.4), i.e., one satisfies the criteria

$$Y\underset{2}{L}\Big|_{\mathcal{M}^*} = 0, \quad \mathcal{M}^* = \{L = 0, \ Yu = 0, \ D_t(Yu) = 0, \ D_x(Yu) = 0\}. \qquad (3.7)$$

Obviously the manifold $\mathcal{M}^*$ coincides with the manifold $\mathcal{M}$ in the prolonged space of variables, because

$$D_t(Yu) = MD_t(Qu) + QuD_tM, \quad D_x(Yu) = MD_x(Qu) + QuD_xM.$$

According to formulae (1.21), the second prolongations of the operators Q and Y have the forms

$$\underset{2}{Q} = Q + \underset{1}{\eta} \cdot \partial_{\underset{1}{u}} + \underset{2}{\eta} \cdot \partial_{\underset{2}{u}}$$

and

$$\underset{2}{Y} = Y + (M\underset{1}{\eta}) \cdot \partial_{\underset{1}{u}} + (M\underset{2}{\eta}) \cdot \partial_{\underset{2}{u}},$$

where $\partial_{\underset{1}{u}} = (\partial_{u_t}, \partial_{u_x})$ and $\partial_{\underset{2}{u}} = (\partial_{u_{tt}}, \partial_{u_{tx}}, \partial_{u_{xx}})$.

The coefficient of the operators Q can be written in a shortened way as follows [203]

$$\underset{1}{\eta} = (\zeta^0, \zeta^1) = D\eta - uD\xi, \quad \xi = (\xi^0, \xi^1),$$
$$\underset{2}{\eta} = (\sigma^{00}, \sigma^{01}, \sigma^{11}) = D\underset{1}{\eta} - \underset{1}{u}D\xi \qquad (3.8)$$

(the explicit expressions see in (1.23)).

Now we calculate those of Y:

$$(M\underset{1}{\eta}) = D(M\eta) - \underset{1}{u}D(M\xi) = M(D\eta - \underset{1}{u}D\xi) + DM(\eta - \underset{1}{u}\xi) = M\underset{1}{\eta} - DMQu,$$

$$(M\underset{2}{\eta}) = D(M\underset{1}{\eta}) - \underset{2}{u}D(M\xi) = D(M\underset{1}{\eta} - DMQu) - \underset{2}{u}D(M\xi) = M(D\underset{1}{\eta} - \underset{1}{u}D\xi) +$$
$$+ DM(\underset{1}{\eta} - \underset{2}{u}\xi) - D^2MQu - DMD(Qu) = M\underset{2}{\eta} - 2DMD(Qu) - D^2MQu.$$

In the above formulae we used the equalities

$$Qu = -(\eta - u\underset{1}{\xi}), \quad D(Qu) = -(\eta - u\underset{2}{\xi}),$$

which follows from notations (3.8).

Now one notes that

$$\underset{2}{Y} = Y + (M\eta - DMQu)\partial_{\underset{1}{u}} + (M\underset{2}{\eta} - 2DMD(Qu) - D^2 MQu)\partial_{\underset{2}{u}}$$

$$= MQ + M\eta\partial_{\underset{1}{u}} + M\underset{2}{\eta}\partial_{\underset{2}{u}} - DMQu\partial_{\underset{1}{u}} - (2DMD(Qu) + D^2 MQu)\partial_{\underset{2}{u}}.$$

Thus, we arrive at

$$\underset{2}{Y} = MQ - DMQu\partial_{\underset{2}{u}} - (2DMD(Qu) + D^2 MQu)\partial_{\underset{2}{u}}. \tag{3.9}$$

Substituting operator (3.9) into condition (3.7), we obtain

$$MQL\Big|_{\underset{2}{\mathcal{M}}} - (DML_u + D^2 ML_{\underset{2}{u}})Qu\Big|_{\mathcal{M}} - 2DML_{\underset{2}{u}}D(Qu)\Big|_{\mathcal{M}} \equiv 0,$$

because of (3.5).

The proof is now completed. □

Property 2 *In contrast to Lie symmetries, all possible Q-conditional symmetries of PDE (3.1) form a set, which is not a Lie algebra in the general case.*

Property 3 *Similarly to Lie symmetries, each Q-conditional symmetry of PDE (3.1) guarantees reduction of this equation to ODE.*

Let us consider a class of the k-order evolution equations

$$u_t = F\left(t, x, u, u_x, \ldots, u_x^{(k)}\right), \quad k \geq 1, \tag{3.10}$$

where F is an arbitrary smooth function and $u = u(t,x)$, $u_x^{(s)} = \frac{\partial^s u}{\partial x^s}$, $s = 1, 2, \ldots, k$. In order to find Q-conditional symmetry (3.2) of this equation, one needs to examine two essentially different cases:

(a) $\xi^0 \neq 0$,   (b) $\xi^0 = 0$, $\xi^1 \neq 0$.

In fact, taking into account Property 1, the symmetry in question has essentially different form in these cases, namely:

$$Q = \partial_t + \xi^1(t,x,u)\partial_x + \eta(t,x,u)\partial_u \tag{3.11}$$

in case (a) and

$$Q = \partial_x + \eta(t,x,u)\partial_u \tag{3.12}$$

in case *(b)*.

It turns out that systems of DEs for finding Q-conditional symmetries (3.11) and (3.12) are essentially different. Hereafter we concentrate ourselves mostly on case *(a)*. The natural reason to avoid examination of case *(b)* (so-called no-go case) follows from the well-known fact (firstly proved in [252]) that a complete description of Q-conditional symmetries with $\xi^0 = 0$ for scalar evolution equations is equivalent to solving the equation in question. In case *(a)*, Definition 3.1 can be simplified (in contrast to case *(b)* !). In fact, one notes that differential consequences of equation $Q(u) = 0$ w.r.t. the variables t and x lead to the second- and higher-order PDEs involving the time derivatives u_{tt}, u_{ttt}, ... and the mixed derivatives u_{tx}, u_{txx}, ..., which do not occur in any evolution equation. It means that the PDEs obtained do not play any role in criteria (3.3), hence the definition can be reformulated as follows.

Definition 3.2 *Operator (3.11) is called Q-conditional symmetry for an evolution equation of the form (3.10) if the following invariance criteria is satisfied:*

$$\underset{k}{Q}(u_t - F)\Big|_{\mathcal{M}} = 0,$$

where the manifold $\mathcal{M}$ is formed by two equations $\{u_t = F,\ Q(u) = 0\}$.

It should be noted that differential consequences of the equation $Q(u) = 0$ w.r.t. the variable x do not involve the time and mixed derivatives in the case *(b)*. Thus, Definition 3.1 should be used provided $\xi^0 = 0$.

Finally, a generalization of Q-conditional symmetry, the notion of conditional symmetry, introduced by Fushchych and his collaborators [112], [114, Section 5.7] should be mentioned.

Definition 3.3 *[114, Section 5.7] The k-order PDE (3.1) is said to be conditionally invariant under operator (3.2) if it is invariant in Lie's sense under this operator only together with the additional condition in the form of a m-order PDE*

$$L_Q\left(t, x, u, \underset{1}{u}, \ldots, \underset{m}{u}\right) = 0. \tag{3.13}$$

It means that the overdetermined system (3.1) and (3.13) admits (3.2) as a Lie symmetry.

Obviously, each Q-conditional (nonclassical) symmetry can be derived as a particular case from conditional symmetry by setting $L_Q = Q(u)$, however there are many nontrivial examples when the conditional symmetries derived for some PDEs cannot be found via the nonclassical Bluman-Cole method.

Example 3.1 *[111]. Let us consider the reaction-diffusion (RD) equation*

$$u_t = (e^u u_x)_x + e^{-u} \tag{3.14}$$

and the additional condition of the first-order (i.e., $m = 1$)

$$L_Q \equiv e^u u_x^2 + e^{-u} - 2u_t = 0. \tag{3.15}$$

It can be found using the classical Lie algorithm that system (3.14) and (3.15) admits the following Lie symmetries

$$J_{01} = x\partial_t + (t - e^u)\partial_x,$$
$$D = t\partial_t + x\partial_x + \partial_u,$$
$$K_0 = 2(t - e^u)D - (t^2 - x^2 - 2te^u)\partial_t,$$
$$K_1 = 2xD + (t^2 - x^2 - 2te^u)\partial_x.$$

According to the above definition each of the operators J_{01}, D, K_0 and K_1 is a conditional symmetry of the RD equation (3.14). On the other hand, the operator D only is a Lie symmetry among them (see case 14 in Table 2.7). Thus, J_{01}, K_0 and K_1 are the non-Lie operators. Moreover, these operators are not nonclassical ones [16] (it follows also from Theorem 3.8).

Notably, other highly nontrivial examples of conditional symmetries involving *multidimensional PDEs* are presented in [63, 64].

To conclude this short description of the conditional symmetry concept (this terminology was introduced by Fushchych [114, Section 5.7]), it should be noted that the main problem of the concept is how to define the suitable condition(s), i.e., Eq. (3.15). Similarly to the method of differential constraints [232, 249], any constructive algorithm does not exist and the additional conditions are usually obtained using an ad hoc approach. In the case of Q-conditional symmetry this problem is solved from the very beginning because $L_Q = Q(u)$.

3.2 Q–conditional symmetry of the nonlinear heat equation

Here we consider nonlinear heat (diffusion) equations of the form

$$u_t = [d(u)u_x]_x, \tag{3.16}$$

which were already studied in Section 2.2 in order to provide the Lie symmetry classification.

First of all, it should be noted that the determining equations (DEs) for searching Q-conditional (nonclassical) symmetries of the linear heat equation, i.e., (3.16) with $d(u) = \text{constant}$, were constructed but not solved in the seminal work [26]. Many authors tried to build the general solution of those equations [88, 217, 247]. The most general results were obtained in [15, 115] and [176]. In the papers [115] and [176], it was proved that the general solution is expressed in terms of three solutions of the linear heat equation, while the authors of [15] have shown how the general solution is also obtainable via the matrix Cole–Hopf transformation.

It turns out that the problem becomes much more complicated if one looks for Q-conditional symmetries of Eq. (3.16) with a nonconstant $d(u)$. To the best of our knowledge, there are only particular examples of such symmetries at the present time [16, 120, 216]. In order to obtain more general results, we apply the Kirchhoff transformation [154]

$$v = \int d(u)\mathrm{d}u \equiv A_0(u) \tag{3.17}$$

to Eq. (3.16) (we remind the reader that $d(u) \geq 0$ so that we have bijective transformation), therefore the equation takes the form

$$F(v)v_t = v_{xx}, \tag{3.18}$$

where $F(v)\big|_{v=A_0(u)} = \frac{1}{d(u)}$. Simultaneously operators (3.11) and (3.12) take the forms

$$Q = \partial_t + \xi^{1*}(t,x,v)\partial_x + \eta^*(t,x,v)\partial_v \tag{3.19}$$

and

$$Q = \partial_x + \eta^*(t,x,v)\partial_v. \tag{3.20}$$

Here ξ^{1*} and η^* are related with ξ^1 and η via the the Kirchhoff transformation (the star * is omitted in what follows).

Now we apply Definition 3.2 to Eq. (3.18) and obtain the system of determining equations (DEs)

$$\begin{gathered}
\xi^1_{vv} = 0, \quad \eta_{vv} = -2\xi^1\xi^1_v F + 2\xi^1_{xv}, \\
\xi^1 \eta F_v + (\xi^1_t + 2\xi^1\xi^1_x - 2\xi^1_v\eta)F + 2\eta_{xv} - \xi^1_{xx} = 0, \tag{3.21} \\
\eta^2 F_v + (2\xi^1_x\eta + \eta_t)F - \eta_{xx} = 0;
\end{gathered}$$

if the operator of Q-conditional symmetry has the form (3.11). The single equation

$$\eta(\eta_x + \eta\eta_v)F_v = (\eta_{xx} + 2\eta\eta_{xv} + \eta^2\eta_{vv})F + \eta_t F^2$$

is obtained if the operator in question has the form (3.12) (so that Definition 3.1 should be applied).

Here we do not present the relevant details of how the above system of DEs and the nonlinear PDE were derived because it will be done for the general class of RDC equations in Section 3.3.

Solving the system of DEs presented above for an arbitrary given function F is a difficult task. However, some interesting cases leading to Q-conditional symmetries, which does not coincide with the Lie symmetries presented in Section 2.2, can be identified.

Example 3.2 Setting $d(u) = e^u$ in Eq. (3.16), one easily derives that the Kirchhoff substitution leads to the equation

$$v^{-1}v_t = v_{xx}. \tag{3.22}$$

The system of DEs (3.21) with $F(v) = v^{-1}$ takes the form

$$\begin{aligned}
&\xi^1_{vv} = 0, \quad \eta_{vv} = -\frac{2\xi^1 \xi^1_v}{v} + 2\xi^1_{xv}, \\
&-\frac{\xi^1_t \eta}{v^2} + \frac{\xi^1_t + 2\xi^1 \xi^1_x - 2\xi^1_v \eta}{v} + 2\eta_{xv} - \xi^1_{xx} = 0, \\
&-\frac{\eta^2}{v^2} + \frac{2\xi^1_x \eta + \eta_t}{v} - \eta_{xx} = 0.
\end{aligned} \tag{3.23}$$

The first two equations of system (3.23) are integrable, so that one obtains

$$\begin{aligned}
\xi^1 &= a(t,x)v + b(t,x), \\
\eta &= (a_x - a^2)v^2 - 2abv(\ln|v| - 1) + \alpha(t,x)v + \beta(t,x),
\end{aligned} \tag{3.24}$$

where a, b, α and β are arbitrary (at the moment) functions. Substituting the above functions ξ^1 and η into the third and fourth equations from (3.23), we can reduce the expressions obtained to a set of simple equations because the functions a, b, α and β do not depend on the dependent variable v. As a result, one easily identifies that $b = \beta = 0$, while the other two functions have correctly-specified forms. Substituting the derived functions into (3.24), we obtain the following Q-conditional symmetry operators of Eq. (3.22):

$$\begin{aligned}
Q_1 &= \partial_t - \frac{v}{Ct+x}\partial_x - \frac{Cv}{Ct+x}\partial_v, \\
Q_2 &= \partial_t - \frac{2xv}{x^2+Ct+C_0}\partial_x - \frac{2v^2+Cv}{x^2+Ct+C_0}\partial_v,
\end{aligned} \tag{3.25}$$

where C, C_0 are arbitrary constants. Finally, using Property 1 and the Kirchhoff transformation (3.17), operators (3.25) can be transformed to the form

$$\begin{aligned}
Q_1 &= \partial_t - \frac{e^u}{Ct+x}\partial_x - \frac{C}{Ct+x}\partial_u, \\
Q_2 &= \partial_t - \frac{2xe^u}{x^2+Ct+C_0}\partial_x - \frac{2e^u+C}{x^2+Ct+C_0}\partial_u.
\end{aligned}$$

Thus, Eq. (3.22) admits two operators of Q-conditional symmetry (formally speaking, it is a set consisting of an infinite number of operators because of the parameters C and C_0).

Remark 3.2 The above operators can be derived by solving the system of PDEs presented in case 7 of Table 1 [16] under the restriction $p = q = r = 0$.

Example 3.3 As it was noted above, the general solution of the first equation of (3.21) has the form $\xi^1 = a(t,x)v + b(t,x)$. Assuming that the diffusivity $d(u)$ is an arbitrary smooth function and setting $a(t,x) = 0$, one immediately obtains that $\eta = \alpha(t,x)v + \beta(t,x)$. Thus, the system of DEs (3.21) reduces to the form

$$b(\alpha v + \beta)F_v + (b_t + 2bb_x)F + 2\alpha_x - b_{xx} = 0,$$
$$(\alpha v + \beta)^2 F_v + [(2\alpha b_x + \alpha_t)v + 2\beta b_x + \beta_t]F - \alpha_{xx}v - \beta_{xx} = 0. \quad (3.26)$$

A complete analysis of system (3.26) is still a nontrivial task, hence we present here two interesting cases only.

Assuming that $F(v)$ is an arbitrary function (i.e., the diffusivity $d(u)$ is arbitrary), all the equations from (3.26) are satisfied provided $\alpha = \beta = 0$ and

$$b_t + 2bb_x = 0, \quad b_{xx} = 0 \qquad (3.27)$$

(otherwise the function $F(v)$ should be correctly defined). System (3.27) is integrable, hence we find its general solution

$$b = \frac{x + C_1}{2t + C_0}. \qquad (3.28)$$

So, we obtain the coefficients $\xi^1 = \frac{x+C_1}{2t+C_0}$ and $\eta = 0$ of operator (3.19), i.e.,

$$Q = \partial_t + \frac{x + C_1}{2t + C_0}\partial_x. \qquad (3.29)$$

Multiplying operator (3.29) by $2t + C_0$ (see Property 1), we arrive at the operator

$$X = (2t + C_0)\partial_t + (x + C_1)\partial_x,$$

which is nothing else but a linear combination of Lie symmetry operators (2.26).

Thus, the system of DEs (3.26) may lead to pure conditional symmetries for the correctly-specified functions F only. Let us consider $F = \lambda + v^2$. In this case, the system takes the form

$$2b(\alpha v + \beta)v + (b_t + 2bb_x)(\lambda + v^2) + 2\alpha_x - b_{xx} = 0,$$
$$2(\alpha v + \beta)^2 v + [(2\alpha b_x + \alpha_t)v + 2\beta b_x + \beta_t](\lambda + v^2) - \alpha_{xx}v - \beta_{xx} = 0. \quad (3.30)$$

Splitting Eqs. (3.30) w.r.t. the different exponents of v, we obtain the condition

$$\beta b = 0,$$

and the PDE system

$$\begin{aligned}
b_t + 2bb_x + 2\alpha b &= 0, \\
2\alpha_x - b_{xx} + \lambda(b_t + 2bb_x) &= 0, \\
2\alpha b_x + \alpha_t + 2\alpha^2 &= 0, \\
2\beta b_x + \beta_t + 4\alpha\beta &= 0, \\
\lambda(2\alpha b_x + \alpha_t) - \alpha_{xx} + 2\beta^2 &= 0, \\
\lambda(2\beta b_x + \beta_t) - \beta_{xx} &= 0.
\end{aligned} \qquad (3.31)$$

Determining equations for finding Q-conditional symmetry of RDC   87

In the case $b = 0$, the Lie symmetry operator $2t\partial_t + 3u\partial_u$ is derived only provided $\lambda = 0$. In the case $\beta = 0$, two equations vanish in system (3.31), while the four remaining equations have the solution $b = -\frac{3}{\lambda x}$, $\alpha = -\frac{3}{\lambda x^2}$. Substituting this solution into (3.19) and multiplying by λx^2, we obtain the Q-conditional symmetry operator

$$Q = \lambda x^2 \partial_t - 3x\partial_x - 3v\partial_v, \quad (3.32)$$

of the nonlinear equation

$$(v^2 + \lambda)v_t = v_{xx}, \quad \lambda \neq 0, \quad (3.33)$$

Applying the Kirchhoff transformation (3.17) to Eq. (3.33) and operator (3.32), we derive the equation

$$u_t = [D_u(u)u_x]_x,$$

and the operator

$$Q = \lambda x^2 \partial_t - 3x\partial_x - [9u - 6\lambda D(u)]\partial_u,$$

where the diffusivity is determined by the cubic equation

$$D^3 + 3\lambda D = 3u.$$

The operator obtained is a pure conditional symmetry in contrast to (3.29) because all possible Lie symmetry operators of nonlinear diffusion equations are listed in Table 2.1.

Remark 3.3 Setting $d(u) = u^{-1}$ in Eq. (3.16), one obtains the well-known equation of fast diffusion, which was studied in some papers using symmetry-based methods [120, 216, 221]. In particular, some conditional symmetries of the form (3.12) for the fast diffusion equation were found in [120].

3.3 Determining equations for finding Q-conditional symmetry of reaction-diffusion-convection equations

Here we examine the class of reaction-diffusion-convection (RDC) equations

$$u_t = [A(u)u_x]_x + B(u)u_x + C(u), \quad (3.34)$$

where $u = u(t,x)$ is the unknown function and $A(u), B(u), C(u)$ are arbitrary smooth functions. In order to construct operators of Q-conditional symmetry for equations belonging to class (3.34), one should apply Definition 3.2.

However, it can be noted that the Kirchhoff substitution used in the previous section allows us to simplify the relevant calculations.

In fact, one easily checks that the substitution

$$v = \int A(u)du \equiv A_0(u), \tag{3.35}$$

reduces Eq. (3.34) to the form

$$v_{xx} = F_0(v)v_t + F_1(v)v_x + F_2(v) \tag{3.36}$$

where

$$F_0(v) = \frac{1}{A(u)}\bigg|_{u=A_0^{-1}(v)}, \quad F_1(v) = -\frac{B(u)}{A(u)}\bigg|_{u=A_0^{-1}(v)},$$
$$F_2(v) = -C(u)\bigg|_{u=A_0^{-1}(v)} \tag{3.37}$$

(here A_0^{-1} is the inverse function to $A_0(u)$). Because the diffusivity $A(u)$ must be nonnegative, Eqs. (3.34) and (3.36) are locally equivalent for any sufficiently smooth function $A(u)$.

Theorem 3.1 *[80] An equation of the form (3.36) is Q-conditionally invariant under the operator*

$$Q = \xi^0(t,x,v)\partial_t + \xi^1(t,x,v)\partial_x + \eta(t,x,v)\partial_v, \tag{3.38}$$

if and only if the functions ξ^0, ξ^1, η satisfy the DEs:

Case 1 $\xi^0 = 1$:

$$\xi^1_{vv} = 0, \quad \eta_{vv} = 2\xi^1_v(F_1 - \xi^1 F_0) + 2\xi^1_{xv},$$
$$\eta(F_1 - \xi^1 F_0)_v - (\xi^1_t + 2\xi^1\xi^1_x - 3\xi^1_v\eta)F_0 + \xi^1_x F_1 + 3\xi^1_v F_2 - 2\eta_{xv} + \xi^1_{xx} = 0; \tag{3.39}$$
$$\eta(\eta F_0 + F_2)_v + (2\xi^1_x - \eta_v)(\eta F_0 + F_2) + \eta_t F_0 + \eta_x F_1 - \eta_{xx} = 0;$$

Case 2 $\xi^0 = 0$; $\xi^1 = 1$:

$$\eta(\eta_x + \eta\eta_v - \eta F_1 - F_2)\frac{dF_0}{dv} - \eta_t F_0^2 =$$
$$= (\eta_{xx} + 2\eta\eta_{xv} + \eta^2\eta_{vv} - \eta^2\frac{dF_1}{dv} - \eta_x F_1 - \eta\frac{dF_2}{dv} + \eta_v F_2) F_0. \tag{3.40}$$

Proof First of all, we note that two essentially different cases should be examined (see Section 3.1), i.e., Q-conditional symmetries of the form (3.11) and (3.12) should be found separately.

Let us consider the case $\xi^0 = 1$ corresponding to symmetries of the form (3.19). In order to construct the relevant system of DEs, we apply Definition 3.2 to Eq. (3.36) with an arbitrary given triplet (F_0, F_1, F_2). The manifold $\mathcal{M}$ consists of two equations, hence two derivatives, say, v_t and v_{xx} can be expressed via other variables (we remind the reader that the invariance criteria

Determining equations for finding Q-conditional symmetry of RDC 89

acts in the prolonged space of variables up to the second-order derivatives). So, using the equation $Q(v) = 0$ we obtain

$$v_t = -\xi^1 v_x + \eta, \tag{3.41}$$

while the second-order derivative

$$v_{xx} = F_0(-\xi^1 v_x + \eta) + F_1 v_x + F_2 \tag{3.42}$$

can be derived from from Eq. (3.36), taking into account Eq. (3.41).

Now we should calculate the second prolongation of (3.19):

$$\underset{2}{Q} = Q + \zeta^0 \partial_{v_t} + \zeta^1 \partial_{v_x} + \sigma^{00} \partial_{v_{tt}} + \sigma^{01} \partial_{v_{tx}} + \sigma^{11} \partial_{v_{xx}}, \tag{3.43}$$

where

$$\zeta^0 = \eta_t + v_t \eta_v - v_x(\xi_t^1 + v_t \xi_v^1),$$
$$\zeta^1 = \eta_x + v_x \eta_v - v_x(\xi_x^1 + v_x \xi_v^1),$$
$$\sigma^{00} = \eta_{tt} + 2v_t \eta_{tv} + v_t^2 \eta_{vv} + v_{tt}\eta_v - v_x(\xi_{tt}^1 + 2v_t \xi_{tv}^1 + v_t^2 \xi_{vv}^1 + v_{tt}\xi_v^1) -$$
$$- 2v_{tx}(\xi_t^1 + v_t \xi_v^1),$$
$$\sigma^{01} = \eta_{tx} + v_x \eta_{tv} + v_t \eta_{vx} + v_t v_x \eta_{vv} + v_{tx}\eta_v - v_x(\xi_{tx}^1 + v_x \xi_{tv}^1 + v_t \xi_{vx}^1 +$$
$$+ v_t v_x \xi_{vv}^1 + v_{tx}\xi_v^1) - v_{tx}(\xi_t^1 + v_x \xi_v^1) - v_{xx}(\xi_t^1 + v_t \xi_v^1),$$
$$\sigma^{11} = \eta_{xx} + 2v_x \eta_{xv} + v_x^2 \eta_{vv} + v_{xx}\eta_v - v_x(\xi_{xx}^1 + 2v_x \xi_{xv}^1 + v_x^2 \xi_{vv}^1 + v_{xx}\xi_v^1) -$$
$$- 2v_{xx}(\xi_x^1 + v_x \xi_v^1)$$

Thus, having operator (3.43) and formulae (3.41) and (3.42) for excluding the derivatives v_t and v_{xx}, we apply Definition 3.2 to Eq. (3.36) and derive the expression

$$v_x^3 \xi_{vv}^1 + v_x^2 \Big[2\xi_v^1(F_1 - \xi^1 F_0) + 2\xi_x^1 v - \eta_{vv}\Big] +$$
$$+ v_x\Big[\eta(F_1 - \xi^1 F_0)_v - (\xi_t^1 + 2\xi^1 \xi_x^1 - 3\xi_v^1 \eta)F_0 + \xi^1 F_1 + 3\xi_v^1 F_2 - 2\eta_{xv} + \xi_{xx}^1\Big] + \tag{3.44}$$
$$+ \eta(\eta F_0 + F_2)_v + (2\xi_x^1 - \eta_v)(\eta F_0 + F_2) + \eta_t F_0 + \eta_x F_1 - \eta_{xx} = 0.$$

Expression (3.44) can be split w.r.t. to v_x^3, v_x^2, v_x^1 and v_x^0 because all the functions arising therein do not depend on the variable v_x. As a result, one arrives exactly at the system of DEs (3.39).

DEs for finding Q-conditional symmetries of the form (3.20) (i.e., $\xi^0 = 0$) must be derived using Definition 3.1. In fact, the manifold $\mathcal{M}$ consists of three equations in this case because the differential consequence of the equation $Q(v) = 0$ w.r.t. the variable x belongs to $\mathcal{M}$. As a result, three derivatives v_t, v_x and v_{xx} can be expressed via other variables. Indeed, the equation $Q(v) = 0$ takes the form

$$v_x = \eta, \tag{3.45}$$

differentiating (3.45) w.r.t. x, we obtain the second equation

$$v_{xx} = \eta_x + \eta_v v_x \equiv \eta_x + \eta\eta_v, \tag{3.46}$$

while the third equation is the given RDC equation from class (3.36). Obviously, having (3.45) and (3.46), we arrive at

$$v_t = \frac{1}{F_0}(\eta_x + \eta\eta_v - \eta F_1 - F_2). \tag{3.47}$$

The second prolongation of (3.20) has the form (3.43) with the coefficients

$$\begin{aligned}
\zeta^0 &= \eta_t + v_t \eta_v, \\
\zeta^1 &= \eta_x + v_x \eta_v, \\
\sigma^{00} &= \eta_{tt} + 2v_t \eta_{tv} + v_t^2 \eta_{vv} + v_{tt} \eta_v, \\
\sigma^{01} &= \eta_{tx} + v_x \eta_{tv} + v_t \eta_{vx} + v_t v_x \eta_{vv} + v_{tx} \eta_v, \\
\sigma^{11} &= \eta_{xx} + 2v_x \eta_{xv} + v_x^2 \eta_{vv} + v_{xx} \eta_v.
\end{aligned} \tag{3.48}$$

Applying operator (3.43) with coefficients (3.48) to Eq. (3.36) and excluding the derivatives v_t, v_x and v_{xx} using formulae (3.45), (3.46) and (3.47), we arrive exactly at Eq. (3.40).

The proof is now completed. □

The system of nonlinear equations (3.39) is very complicated. As a result, it is not plausible that one can obtain a complete classification of Q-conditional symmetries of RDC equations belonging to class (3.34) as it was done for Lie symmetry operators in Chapter 2. However, several subclasses of RDC equations arising in real world applications will be successfully examined in subsequent sections. Here we present a nontrivial particular solution of the system of DEs (3.39), which is interesting from a mathematical point of view because one involves an arbitrary function $F_0(v)$.

Indeed, assuming that [80]

$$\xi^1 = v + \lambda_4, \qquad \eta = \mathcal{P}_3(v) \tag{3.49}$$

one finds the solution of (3.39):

$$F_1 = (v + \lambda_4)F_0 + 3\lambda_3 v + \lambda_2, \quad F_2 = -\mathcal{P}_3(v)(F_0 + \lambda_3), \tag{3.50}$$

where $\mathcal{P}_3(v) = \lambda_0 + \lambda_1 v + \lambda_2 v^2 + \lambda_3 v^3$ is an arbitrary polynomial (λ_i, $i = 0, \cdots, 4$ are arbitrary constants) and $F_0(v)$ is an arbitrary smooth function. So, the equation

$$v_{xx} = F_0(v)[v_t + (v + \lambda_4)v_x - \mathcal{P}_3(v)] + (3\lambda_3 v + \lambda_2)v_x - \lambda_3 \mathcal{P}_3(v) \tag{3.51}$$

admits the Q-conditional symmetry operator

$$Q = \partial_t + (v + \lambda_4)\partial_x + \mathcal{P}_3(v)\partial_v. \tag{3.52}$$

Now one can claim that a RDC equation with an arbitrary given diffusivity is invariant w.r.t. the Q-conditional symmetry provided its reaction and convection terms are defined by formulae (3.50). Moreover, taking into account the structure of (3.52), one notes that it is a pure conditional symmetry. In fact, all Lie symmetry operators of RDC equations have the coefficients ξ^1 (see (3.2)), which does not depend on u. Obviously, the Kirchhoff transformation (3.35) does not affect this property, hence (3.52) is a non-Lie operator.

Consider the nonlinear equation (3.40) arising in case 2 of Theorem 3.1. If one assumes that the triplet (F_0, F_1, F_2) is given then it is a nonlinear $(1+2)$-dimensional PDE to find the function $\eta(t, x, v)$. Obviously this equation is more complicated than Eq. (3.36) with the same triplet (F_0, F_1, F_2). It turns out that Eq. (3.40) can be reduced to the later equation by a set of nonlocal substitutions (see for details [252]). Hence the search of Q-conditional symmetries of the form (3.12) for any RDC equation is equivalent (up to nonlocal transformations) to solving of the equation in question.

On the other hand, each *particular solution* of DE in case 2 leads to a Q-conditional symmetry, which does not coincide with one from case 1, hence new exact solutions can be found (see, e.g., [120, 192]). Let us consider an interesting example.

In [80], the following partial solution of Eq. (3.40) was noted:

$$\eta = \frac{H(v)}{t}, \quad F_0 = \lambda_0 \dot{H},$$
$$F_1 = \lambda_1 \dot{H} + \lambda_0 \dot{H} \int \frac{dv}{H(v)}, \quad F_2 = \lambda_2 H \dot{H}, \tag{3.53}$$

where $H(v)$ is an arbitrary nonconstant smooth function. So, the equation

$$v_{xx} = \dot{H}(v)\left[\lambda_0 v_t + \left(\lambda_1 + \lambda_0 \int \frac{dv}{H(v)}\right) v_x + \lambda_2 H(v)\right] \tag{3.54}$$

is invariant under the Q-conditional operator

$$Q = \partial_x + \frac{H(v)}{t}\partial_v,$$

which is equivalent to the Galilei-type operator

$$Q = t\partial_x + H(v)\partial_v. \tag{3.55}$$

In the simplest case $H = v$, $\lambda_0 = 1$, $\lambda_1 = 0$, the equation

$$v_t = v_{xx} - \ln v \, v_x - \lambda_2 v. \tag{3.56}$$

is obtained and operator (3.55) with $H = v$ is nothing else but the Galilei operator

$$G = t\partial_x + v\partial_v. \tag{3.57}$$

Obviously, operator (3.57) is Lie's operator because according to Table 2.5 (see case 21) Eq. (3.56) is invariant with respect to the Galilei algebra. However, operator (3.55) with more complicated forms of the function H is a pure conditional symmetry.

The next three sections are devoted to a complete classification of Q-conditional symmetries of some subclasses of RDC equations belonging to class (3.34). The subclasses, which we are going to examine, include several important equations arising in real world applications. Before proceeding to their examination, we need to formulate an algorithm. That can be formulated on the basis of the algorithm used for Lie symmetry classification (see Section 2.3). In the general case, the algorithm for solving Q-conditional symmetry classification (QSC) problem for a given class of PDEs is as follows.

1. Construction of the group of equivalence transformations (ETs) $\mathcal{E}$ for the class of PDEs in question.

2. Construction of form-preserving transformations (FPTs) for the same class.

3. Deriving the systems of determining equations (DEs) (two different cases should be examined).

4. Deriving QSC for the class of PDEs using the group of ETs $\mathcal{E}$.

5. Deriving QSC for class of PDEs using FPTs.

6. Identifying pure Q-conditional symmetry operators among those obtained at the previous step.

One easily notes that the first three steps of the algorithm are already realized. However, it is not plausible that the next steps can be successfully done because the systems of DEs (3.39) and Eq. (3.40) are very complicated. Hence, our purpose is to derive QSC for some subclasses of RDC equations belonging to class (3.34). Moreover, we search for the symmetries of the form (3.11) only (see the motivation presented in the end of Section 3.1).

3.4 Q–conditional symmetry of reaction-diffusion-convection equations with constant diffusivity

Here we examine reaction-diffusion-convection (RDC) equations of the form
$$u_t = u_{xx} + \lambda u u_x + C(u), \qquad (3.58)$$
with an arbitrary parameter λ, which generate a special subclass of the class of RDC equations (3.34). Our aim is to derive QSC for this subclass.

Q-conditional symmetry of RDC equations with constant diffusivity

It is worth presenting the following observation from the very beginning. Formally speaking, one should construct the group of ETs $\mathcal{E}$ of subclass (3.58) according to the algorithm formulated in the previous section. Obviously this group is the same or smaller comparing with that of a wider class of PDEs. Because we know $\mathcal{E}$ of class (3.34), it is enough to check simply whether this group suits to subclass (3.58). Simple calculations show that it is true provided $e_0 = e_1^2$ if $\lambda \ne 0$ (see (2.36)) and the additional restriction $g = 0$ arises if $\lambda = 0$. Moreover, there is a single FPT (see case 7 in Table 2.6), which occurs for nonlinear equations of the form (3.58). Thus, we combine the above ETs and FPT in order to unite the fourth and fifth steps of the algorithm.

The main result of this section is presented in the form of the theorem, which gives a complete information on Q-conditional symmetries of the form (3.11), i.e.,

$$Q = \partial_t + \xi^1(t, x, u)\partial_x + \eta(t, x, u)\partial_u,$$

for the above subclass of RDC equations.

Theorem 3.2 *[53] An equation of the form (3.58) is Q-conditionally invariant under the operator (3.11) up to equivalent representations generated by transformations of the form*

$$\bar{t} = e_0 t, \quad \bar{x} = e_1 x + g_1 t + g_2 t^2,$$
$$\bar{u} = e_2 u + g_3 t + u_0$$
(3.59)

(here the coefficients are arbitrary parameters obeying the restrictions $e_0 = e_1^2$, $e_1 e_2 \ne 0$) if and only if the equation has one of the following forms.

Case 1. *The Burgers equation*

$$u_t = u_{xx} + \lambda u u_x$$

is Q-conditionally invariant under the operator

$$Q = \partial_t + \left(\frac{\lambda}{2}u + b\right)\partial_x + \left(\alpha + \beta u - \frac{\lambda b}{2}u^2 - \frac{\lambda^2}{4}u^3\right)\partial_u,$$
(3.60)

where triplet of the functions (α, β, b) is the general solution of the system

$$\alpha_t = \alpha_{xx} - 2\alpha b_x,$$
$$\beta_t = \beta_{xx} - 2\beta b_x + \lambda \alpha_x,$$
$$b_t = b_{xx} - 2bb_x - 2\beta_x.$$
(3.61)

Case 2. *The equation*

$$u_t = u_{xx} + \lambda u u_x + \lambda_0 + \lambda_2 u^2$$
(3.62)

is Q-conditionally invariant under the operator

$$Q = \partial_t + \left(\frac{\lambda_2}{\lambda} - \lambda u\right)\partial_x + (\lambda_0 + \lambda_2 u^2)\partial_u,$$
(3.63)

where λ_0, λ and λ_2 are arbitrary constants and $\lambda\lambda_2 \neq 0$.
Case 3. The equation

$$u_t = u_{xx} + \lambda u u_x + \lambda_0 + \lambda_1 u + \lambda_3 u^3 \tag{3.64}$$

is Q-conditionally invariant under the operators

$$Q_i = \partial_t + p_i u \partial_x + \frac{3p_i}{2p_i - \lambda}(\lambda_0 + \lambda_1 u + \lambda_3 u^3)\partial_u, \quad i = 1, 2, \tag{3.65}$$

where p_i are the roots of the quadratic equation

$$2p^2 + \lambda p + 9\lambda_3 - \lambda^2 = 0, \tag{3.66}$$

(λ_0, λ_1 and $\lambda_3 \neq 0$ are arbitrary constants) and the operators

$$Q = \partial_t + b\partial_x + (\gamma\, b_{xx} - b_x u)\partial_u, \tag{3.67}$$

where the function $b(t,x)$ is an arbitrary solution of the overdetermined system

$$\begin{aligned}(\lambda\gamma - 3)b_{xx} + 2bb_x + b_t &= 0, \\ b_{xxx} - bb_{xx} + \lambda_1 b_x &= 0, \\ \lambda\gamma^2\, b_{xxx} + \gamma b_x^2 + 3\gamma\lambda_1 b_x + 3\lambda_0 b &= C_0,\end{aligned} \tag{3.68}$$

where $\gamma = \frac{\lambda}{3\lambda_3}$, $\lambda_0\lambda_1\lambda_3 \neq 0$ and $C_0 \in \mathbb{R}$.

Proof of the theorem is based on solving the system of DEs (3.39), which can be essentially simplified if one examines subclass (3.58) instead of the general class (3.34). In fact, setting $(F_0, F_1, F_2) = (1, -\lambda u, -C)$ and renaming $v \to u$ in (3.39), we obtain the following system for the functions $C(u)$, ξ^1 and η:

$$\xi^1_{uu} = 0, \quad \eta_{uu} = -2\xi^1_u(\lambda u + \xi^1) + 2\xi^1_{xu}, \tag{3.69}$$

$$\lambda\eta + \xi^1_t + 2\xi^1\xi^1_x - 2\xi^1_u\eta + \lambda u\xi^1_x + 2\eta_{xu} - \xi^1_{xx} + 3\xi^1_u C(u) = 0, \tag{3.70}$$

$$\eta\,[\eta - C(u)]_u + (2\xi^1_x - \eta_u)[\eta - C(u)] + \eta_t - \lambda u\eta_x - \eta_{xx} = 0. \tag{3.71}$$

Subsystem (3.69) is easily integrated and its general solution has the form

$$\xi^1 = a(t,x)u + b(t,x), \quad \eta = -\frac{a}{3}(a+\lambda)u^3 + (a_x - ab)u^2 + \alpha(t,x)u + \beta(t,x), \tag{3.72}$$

where a, b, α and β are arbitrary smooth functions at the moment. Obviously, one needs to consider two different cases, namely: **(I)** $a \neq 0$ and **(II)** $a = 0$.

Consider case **(I)**. Substituting (3.72) into (3.70), we immediately establish that the function $C(u)$ can be at maximum a cubic polynomial w.r.t. the variable u:

$$C(u) = \lambda_0 + \lambda_1 u + \lambda_2 u^2 + \lambda_3 u^3, \tag{3.73}$$

where the constants λ_i ($i = 0, ..., 3$) are determined by the functions a, b, α and β. The relevant formulae have the form

$$\begin{aligned} 2a^2 + a\lambda + 9\lambda_3 - \lambda^2 &= 0, \\ b(\lambda - 2a) &= 3\lambda_2, \\ \alpha(\lambda - 2a) &= -3a\lambda_1, \\ \beta(\lambda - 2a) &= -3a\lambda_0, \end{aligned} \qquad (3.74)$$

provided $\lambda_3 \neq 0$. If $\lambda_3 = 0$ then the first equation (with $\lambda_3 = 0$) of (3.74) is again obtained and then $a = -\lambda$ or $a = \frac{\lambda}{2}$. The value $a = -\lambda$ is examined below. The value $a = \frac{\lambda}{2}$ leads to the requirement $\lambda_2 = 0$; therefore the function $C(u)$ (see (3.73)) is linear. In this case the equation can be reduced to the form either

$$u_t = u_{xx} + \lambda u u_x \qquad (3.75)$$

or

$$u_t = u_{xx} + \lambda u u_x + \lambda_1 u. \qquad (3.76)$$

In the case $C(u) = \lambda_0$, the known FPT (2.49)(see also case 7 of Table 2.6) should be applied, while the relevant ET from (2.36) works in the case $C(u) = \lambda_1 u + \lambda_0$. Both transformations are included in set (3.59).

Eq. (3.75) is, of course, the well-known Burgers equation and its Q-conditional symmetry is easily obtained by substitution $C(u) = 0$, $a = \frac{\lambda}{2}$ and (3.72) into the DEs (3.70)–(3.71). After the relevant calculations, we obtain their general solution leading to operator (3.60) with the coefficients satisfying (3.61). The analogous procedure was realized for Eq. (3.76), however, only such Q-conditional symmetries were found, which are equivalent to Lie symmetry operators listed in case 8 of Table 2.7.

Now we consider the case $\lambda_3 \neq 0$ and the subcase $\lambda_3 = 0$ and $a = -\lambda$. Solving the system of the algebraic equation (3.74) w.r.t. the a, b, α, β and substituting the expressions obtained into DE (3.71), we arrive at the general solution (3.73) and

$$\begin{aligned} \xi^1 &= -\lambda u + \tfrac{\lambda_2}{\lambda}, \\ \eta &= \lambda_2 u^2 + \lambda_1 u + \lambda_0, \end{aligned} \qquad (3.77)$$

if $\lambda_3 = 0$ and $a = -\lambda$, and

$$\begin{aligned} \xi^1 &= pu - \tfrac{3\lambda_2}{2p-\lambda}, \\ \eta &= \tfrac{3p}{2p-\lambda}\left(\lambda_3 u^3 + \lambda_2 u^2 + \lambda_1 u + \lambda_0\right), \end{aligned} \qquad (3.78)$$

if $\lambda_3 \neq 0$. Here the constant p is the solution of the quadratic equation (3.66). So, operator (3.11) with coefficients (3.77) is Q-conditional symmetry of the RDC equation

$$u_t = u_{xx} + \lambda u u_x + \lambda_0 + \lambda_1 u + \lambda_2 u^2, \qquad (3.79)$$

while this operator with coefficients (3.78) generates two Q-conditional symmetries of the RDC equation

$$u_t = u_{xx} + \lambda u u_x + \lambda_0 + \lambda_1 u + \lambda_2 u^2 + \lambda_3 u^3. \qquad (3.80)$$

It should be noted that there is only a single Q-conditional symmetry, if $8\lambda_3 = \lambda^2$.

Finally, we note that Eqs. (3.79) and (3.80) are reduced to the same equations with $\lambda_1 = 0$ and $\lambda_2 = 0$, respectively, using a transformation of the form (3.59) with $g_2 = g_3 = 0$. Thus, the examination of case *(I)* is now completed and cases 1, 2 and 3 (excepting operator (3.67)) of the theorem are obtained.

The examination of case *(II)* is simpler and leads to Q-conditional symmetry operator (3.67) only, in which the function $b(t, x)$ is the general solution of the nonlinear overdetermined system (3.68). It should be noted that this system is compatible (for example, $b = constant$ is a solution, if $C_0 = 3\lambda_0 b$). However, finding its general solution is a nontrivial task in the case $\lambda_0\lambda_1 \neq 0$ (we discus this issue below). In the case $\lambda_0\lambda_1 = 0$, the general solution can be found in a straightforward way, however the operators obtained coincide with the Lie symmetry operators listed in case 23 of Table 2.5.

Thus, the proof is now completed. □

Remark 3.4 *Particular cases of Theorem 3.2 were derived a long time ago in papers [6, 13] (case 1) and [49] (subcases of cases 2 and 3).*

Remark 3.5 *All the symmetry operators arising in Theorem 3.2 are non-Lie operators (excepting (3.67)) because the coefficient ξ depends on u.*

As it was already noted the overdetermined system (3.68) with $\lambda_0\lambda_1 \neq 0$ is compatible, however its integration is a nontrivial task. The theoretical background of this difficulty follows from paper [137]. Indeed, one easily checks that both third-order ODEs arising in (3.68) cannot be linearized by point and/or contact transformations (see Theorems 2.1 and 5.1 in [137]). Thus, the system cannot be integrated in a straightforward way.

On the other hand, one may look for techniques used for solving overdetermined systems of PDEs. An overview of possible approaches with attempt to create general algorithm of integrating overdetermined systems is presented in [232] (see also discussion in [41]). However, to the best of our knowledge, there is no general algorithm of integration of such systems at the present time. In order to solve a given nonlinear overdetermined system, one should develop a special algorithm, adopted to the system in question. Here we apply the technique, which is a modification of that used for integration of the overdetermined system (2.28) from [17]. It should be stressed that system (2.28) [17] follows from system (3.68) as a particular case by setting $\lambda = 0$ in the first two equations. However, the case $\lambda \neq 0$ is much more complicated.

Theorem 3.3 *The overdetermined system (3.68) with $\lambda\lambda_0\lambda_1 \neq 0$ possesses the trivial solution $b = const$ only.*

Proof Let us rewrite system (3.68) in the form

$$b_t = kb_{xx} - 2bb_x, \qquad (3.81)$$

$$b_{xxx} = bb_{xx} - \lambda_1 b_x, \tag{3.82}$$

$$(3-k)b_{xxx} + b_x^2 + 3\lambda_1 b_x + \frac{3\lambda\lambda_0}{3-k}b = C_0, \tag{3.83}$$

where $k = 3 - \lambda\gamma \neq 3$, $C_0 = \frac{c_0}{\gamma}$.

The main idea of proof consists of constructing an algebraic equation for the function $b(t,x)$ using equations from (3.68) and their differential consequences. Because such approach involves extensive calculations, the program package Mathematica computer algebra was partly used.

The first step is to derive a second-order ODE for b using Eqs. (3.81) and (3.82). Let us take differential consequences of (3.81) and (3.82) of the first and second orders w.r.t. x. As a result, the following equations are obtained

$$\begin{aligned} b_{tx} &= kb_{xxx} - 2bb_{xx} - 2b_x^2, \\ b_{txx} &= kb_{xxxx} - 2bb_{xxx} - 6b_x b_{xx}, \\ b_{xxxx} &= bb_{xxx} + b_x b_{xx} - \lambda_1 b_{xx}, \\ b_{xxxxx} &= bb_{xxxx} + 2b_x b_{xxx} - \lambda_1 b_{xxx} + b_{xx}^2. \end{aligned} \tag{3.84}$$

In Eqs. (3.84), the terms with b_{xxx} can be excluded using (3.82), hence we obtain

$$\begin{aligned} b_{tx} &= (k-2)bb_{xx} - 2b_x^2 - k\lambda_1 b_x, \\ b_{txx} &= \left[(k-6)b_x + (k-2)b^2 - k\lambda_1\right]b_{xx} - (k-2)\lambda_1 bb_x, \\ b_{xxxx} &= (b_x + b^2 - \lambda_1)b_{xx} - \lambda_1 bb_x, \\ b_{xxxxx} &= b_{xx}^2 + (3b_x + b^2 - 2\lambda_1)bb_{xx} - \lambda_1(2b_x + b^2 - \lambda_1)b_x. \end{aligned} \tag{3.85}$$

Now we take the differential consequence of (3.81) w.r.t. t and that of the third order of (3.82) w.r.t. x. Having done this, the fifth-order PDE is easily derived:

$$kb_{xxxxx} - 2bb_{xxxx} - 8b_x b_{xxx} - 6b_{xx}^2 - b_x b_{xx} - bb_{txx} + \lambda_1 b_{tx} = 0. \tag{3.86}$$

Using (3.85), the time derivatives and higher-order space derivatives can be excluded in (3.86), hence we obtain the nonlinear second-order ODE

$$3b_{xx}^2 - (k-1)bb_x b_{xx} + (k-3)\lambda_1 b_x^2 = 0. \tag{3.87}$$

Differentiating (3.87) w.r.t. x and eliminating b_{xxx} from (3.82), we derive the second ODE

$$(k-7)bb_{xx}^2 + \left[(k-1)b_x + (k-1)b^2 - 2(k-6)\lambda_1\right]b_x b_{xx} - (k-1)\lambda_1 bb_x^2 = 0. \tag{3.88}$$

Now the term with b_{xx}^2 in Eq. (3.88) can be eliminated using (3.87), so that the semilinear ODE

$$\left[3(k-1)b_x + (k-4)(k-1)b^2 - 6(k-6)\lambda_1\right]b_{xx} - (k^2 - 7k + 18)\lambda_1 bb_x = 0 \tag{3.89}$$

is obtained. Note that b_{xx} can be found from (3.89), i.e., $3(k-1)b_x + (k-4)(k-1)b^2 - 6(k-6)\lambda_1 \neq 0$ (otherwise $(k^2 - 7k + 18)\lambda_1 b b_x = 0$ so that $b = const$).

The second step is to derive a first-order ODE for b.

Differentiating (3.89) w.r.t. x, one eliminates b_{xxx} and b_{xx}^2 using Eqs. (3.82) and (3.87). As a result, the equation

$$\left[3(k-1)(k-2)b_x + (k-1)(k-4)b^2 - (k^2 - k - 18)\lambda_1\right] bb_{xx} + \\ +\lambda_1 \left[-2(k^2 - 4k + 9)b_x - (k-1)(k-4)b^2 + 6(k-6)\lambda_1\right] b_x = 0 \quad (3.90)$$

is obtained. Eq. (3.90) reduces to the nonlinear first-order ODE

$$6(k-1)\left(k^2 - 4k + 9\right) b_x^2 - \left[(k-1)^2 \left(k^2 - 13k + 48\right) b^2 + \\ + 6(k-6)\left(2k^2 - 5k + 15\right)\lambda_1\right] b_x + 2(k-1)(k-4)(k-7)b^4 + \quad (3.91) \\ + \left(k^4 - 20k^3 + 139k^2 - 300k - 36\right)\lambda_1 b^2 + 36(k-6)^2\lambda_1^2 = 0,$$

because b_{xx} can be found from Eq.(3.89).

Another first-order ODE is obtained from the differential consequence of (3.91) w.r.t. x. By excluding b_{xx} from the consequence obtained, we obtain the nonlinear first-order ODE

$$6(k-1)^3 \left(k^2 - 13k + 48\right) b_x^2 + \left[2(k-4)(k-1)^2 \left(36 + (k-7)^2 k\right) b^2 - \\ - 6(k-1)\left(5k^4 - 82k^3 + 539k^2 - 1398k + 864\right)\lambda_1\right] b_x - \\ - 8(k-7)\left[(k-1)(k-4)\right]^2 b^4 - \quad (3.92) \\ - (k-1)\left(k^5 - 27k^4 + 213k^3 - 169k^2 - 3618k + 9216\right)\lambda_1 b^2 + \\ + 6(k-6)\left(4k^4 - 59k^3 + 364k^2 - 795k + 198\right)\lambda_1^2 = 0.$$

Solving (3.91) w.r.t. b_x^2 (if $k=1$ then the contradiction $\lambda_1 = 0$ is obtained) and substituting into (3.92), we arrive at the semilinear first-order ODE

$$\left(A_1 b^2 + A_0\right) b_x = A_4 b^4 + A_3 b^2 + A_2, \quad (3.93)$$

where

$$A_0 = 18(1-k)\left(k^6 - 19k^5 + 169k^4 - 813k^3 + 2162k^2 - 2796k + 1152\right)\lambda_1,$$
$$A_1 = 3(k-1)^2\left(k^6 - 24k^5 + 230k^4 - 1096k^3 + 2649k^2 - 2528k - 96\right),$$
$$A_2 = 6(6-k)\left(4k^6 - 81k^5 + 762k^4 - 3772k^3 + 10008k^2 - 12159k + 3510\right)\lambda_1^2,$$
$$A_3 = (k-1)(2k^7 - 65k^6 + 810k^5 - 4778k^4 + 12578k^3 - 2301k^2 - 57222k + 84672)\lambda_1,$$
$$A_4 = 2(k-7)(k-4)(k-1)^2\left(5k^3 - 46k^2 + 161k - 192\right).$$

Notably $|A_1| + |A_0| \neq 0$ in (3.93) (otherwise either $b = const$ or $\lambda_1 = 0$).

Using (3.93) and its differential consequences w.r.t. x, the derivatives b_x, b_{xx}, and b_{xxx} can be expressed as rational functions of b. Substituting b_x, b_{xx}, and b_{xxx} obtained into Eq. (3.82), we obtain the tenth-order polynomial to find the function b with the coefficients, which themselves are polynomials of $A_0, \ldots, A_4$. The simple analysis shows that all the coefficients of the tenth-order polynomial vanish provided either

$$A_1 = A_4 = 0, \quad A_0 = 3A_3, \quad A_2 = -\frac{3}{2}\lambda_1 A_0, \tag{3.94}$$

or

$$A_1 = 3A_4, \quad A_0 = \frac{1}{2}(6A_3 + 27A_4\lambda_1), \quad A_2 = -\frac{9}{4}\lambda_1(2A_3 + 9A_4\lambda_1). \tag{3.95}$$

So, we obtain two overdetermined systems of the algebraic equation for finding k. If each system has no real roots then we conclude that $b = const$.

If there exists $k = k^*$ satisfying system (3.94) (or (3.95)), then we can simplify Eq. (3.93) using (3.94)(or (3.95)). The resulting equation in both cases is

$$b_x = \frac{1}{3}b^2 - \frac{3}{2}\lambda_1. \tag{3.96}$$

Thus, b_{xxx} can be immediately found using differential consequences of the above ODE. Finally, substituting b_x from (3.96) and b_{xxx} into (3.83), the algebraic equation

$$\frac{1}{9}(2k-7)b^4 - \frac{4}{3}(k-3)\lambda_1 b^2 + \frac{3\lambda\lambda_0}{k-3}b + \frac{3}{4}(2k-3)\lambda_1^2 + C_0 = 0 \tag{3.97}$$

is obtained. Because $\lambda\lambda_0\lambda_1 \neq 0$, Eq. (3.96) does not possess nonconstant solutions, so that $b = const$.

The proof is now completed. □

Setting $\lambda = 0$ in (3.58), one arrives at the standard class of reaction-diffusion (RD) equations

$$u_t = u_{xx} + C(u), \tag{3.98}$$

which contains several important (from applicability point of view) equations. It turns out that an equation of the form (3.98) can admit a pure Q-conditional symmetry only under the condition

$$C(u) = \lambda_3 u^3 + \lambda_2 u^2 + \lambda_1 u + \lambda_0, \quad \lambda_3 \neq 0 \tag{3.99}$$

(we remind the reader that the term $\lambda_2 u^2$ in an arbitrary cubic polynomial is removable by means of the transformation $\bar{u} = u + \frac{\lambda_2}{3\lambda_3}$ belonging to set (3.59)).

Theorem 3.4 *The RD equation (3.98) is Q-conditionally invariant under operator (3.11) if and only if $C(u)$ has the form (3.99) and the following operators occur.*

Case 1. $\xi_u^1 \neq 0$, $\lambda_3 \neq 0$,

$$Q = \partial_t + \frac{3}{2}\sqrt{-2\lambda_3}u\partial_x + \frac{3}{2}\left(\lambda_3 u^3 + \lambda_1 u + \lambda_0\right)\partial_u.$$

Case 2. $\xi_u^1 = 0$, $\lambda_0 = 0$, $\lambda_3 \neq 0$,
$$Q = \partial_t + b\partial_x - b_x u \partial_u,$$
where the function $b(t,x)$ is an arbitrary solution of the nonlinear system
$$\begin{aligned}b_t - 3b_{xx} + 2bb_x &= 0,\\ b_{xxx} - bb_{xx} + \lambda_1 b_x &= 0.\end{aligned} \quad (3.100)$$

Case 3. $\xi_u^1 = 0$, $\lambda_0 = \lambda_3 = 0$,
$$Q = \partial_t + b\partial_x + (\alpha + \beta u)\partial_u,$$
where triplet of the functions (α, β, b) is the general solution of the system
$$\begin{aligned}\alpha_t &= \alpha_{xx} - 2\alpha b_x + \lambda_1 \alpha,\\ \beta_t &= \beta_{xx} - 2\beta b_x + 2\lambda_1 \alpha_x,\\ b_t &= b_{xx} - 2bb_x - 2\beta_x.\end{aligned} \quad (3.101)$$

Remark 3.6 *If one compares Theorems 3.2 and 3.4 and takes into account that system (3.68) with $\lambda = 0$ is equivalent to the two-component system (3.100) then one realizes that cases 1 and 2 of Theorem 3.4 follow from Theorem 3.2 as a consequence by setting $\lambda = 0$.*

Case 1 of Theorem 3.4 was firstly identified by Serov in [226], later that and case 2 were independently derived in [17, 88]. System (3.101) with $\lambda_1 = 0$ arising in case 3 was found for the first time in the seminal work [26].

The general solution of the nonlinear system (3.100) was constructed independently in the 1990s by several authors [17, 88]. The Q-conditional operators have the form (one may set $\lambda_1 = 0; \pm 1$ without losing a generality):
$$\begin{aligned}Q &= \partial_t - \tfrac{3}{x}\partial_x - \tfrac{3}{x^2} u \partial_u, \quad \lambda_1 = 0,\\ Q &= \partial_t + \tfrac{3\sqrt{2}}{2}\tan\left(\tfrac{\sqrt{2}}{2}x\right)\partial_x - \tfrac{3}{2}\cos^{-2}\left(\tfrac{\sqrt{2}}{2}x\right) u\partial_u, \quad \lambda_1 = -1,\\ Q &= \partial_t - \tfrac{3\sqrt{2}}{2}\tanh\left(\tfrac{\sqrt{2}}{2}x\right)\partial_x + \tfrac{3}{2}\cosh^{-2}\left(\tfrac{\sqrt{2}}{2}x\right) u\partial_u, \quad \lambda_1 = 1,\\ Q &= \partial_t - \tfrac{3\sqrt{2}}{2}\coth\left(\tfrac{\sqrt{2}}{2}x\right)\partial_x - \tfrac{3}{2}\sinh^{-2}\left(\tfrac{\sqrt{2}}{2}x\right) u\partial_u, \quad \lambda_1 = 1.\end{aligned} \quad (3.102)$$

There is also the time-dependent solution $b = \frac{x+C_1}{2t+C_0}$ of system (3.100) with $\lambda_1 = 0$, however one leads to an operator Q, which is equivalent to the known Lie symmetry operator of the RD equation with the cubic source/sink $\lambda_3 u^3$.

3.5 Q–conditional symmetry of reaction-diffusion-convection equations with power-law diffusivity

Among reaction-diffusion-convection (RDC) equations with variable coefficients, the equations with power-law diffusion and convective terms are the most common equations in real world applications, especially in mathematical biology [181, 182, 194]. In this section, we examine two subclasses of the general class of RDC equations, namely

$$u_t = (u^m u_x)_x + \lambda u^m u_x + C(u) \qquad (3.103)$$

and

$$u_t = (u^m u_x)_x + \lambda u^{m+1} u_x + C(u), \qquad (3.104)$$

where λ and $m \neq 0$ are arbitrary constants, while $C(u)$ is an arbitrary function. Notably, both subclasses with $m = 0$ are reducible to that (or to a particular case) investigated in Section 3.4. In the case $\lambda \neq 0$, equations belonging to the above classes are essentially different because class (3.103) contains the RDC equations with linearly dependent diffusion and convection coefficients, while (3.104) consists of those with different exponents in diffusion and convection coefficients. It should be also noted that we may skip examination of class (3.104) with $m = -1$ because such convection term is removable (see Remark 2.1 in Chapter 2).

In order to construct the group of ETs of the above classes of RDC equations, we again use the group $\mathcal{E}$ (2.36) of the general class. Using this group, one can easily derive the ET groups of (3.103) and (3.104). It turns out that both classes have the same group of ETs:

$$\bar{t} = e_0 t + t_0, \quad \bar{x} = e_1 x + x_0, \quad \bar{u} = e_2 u, \qquad (3.105)$$

where arbitrary parameters obey the restrictions $e_0 = e_1^2 e_2^{-m}$, $e_1 e_2 \neq 0$. Obviously the set of transformations 3.105 is nothing else but a four-parameter group of scale transformations and translations w.r.t. t and x. Using Table 2.6 (see cases 8–10), one can also identify three FPTs occurring for the RDC equations belonging to (3.103), however these transformations do not play any role in solving the QSC problem for (3.103). Setting $\lambda = 0$ in (3.103) and (3.104), one notes that there are three more FPTs occurring for the RD equations obtained (see cases 3–5 in Table 2.6), but again they cannot be applied for simplification of the equations constructed in the next two subsections. Thus, the first two steps of the algorithm for solving the QSC problem for classes (3.103) and (3.104) are completed.

3.5.1 The case of proportional diffusion and convection coefficients

Now we formulate the theorem, which presents a complete QSC for RDC equations of the form (3.103). We remind the reader that the pure conditional symmetry operators are only constructed, i.e., those, which are not reducible to the Lie symmetry operators.

Theorem 3.5 *[75] An equation of the form (3.103) is Q-conditionally invariant under operator (3.11) if and only if the equation (up to transformations (3.105)) and the relevant operator (up to equivalent representations generated by multiplying on the arbitrary smooth function $M(t,x,u)$) have the following forms:*

Case 1.

$$u_t = (u^m u_x)_x + \lambda u^m u_x + (\lambda_1 u^{m+1} + \lambda_2)(u^{-m} - \lambda_3), \tag{3.106}$$

$$Q = \partial_t + (\lambda_1 u + \lambda_2 u^{-m})\partial_u, \quad m \neq -1, \lambda_2 \neq 0; \tag{3.107}$$

Case 2.

$$u_t = (u^{-1} u_x)_x + \lambda u^{-1} u_x + (\lambda_1 \ln u + \lambda_2)(u - \lambda_3), \tag{3.108}$$

$$Q = \partial_t + (\lambda_1 \ln u + \lambda_2) u \partial_u, \quad \lambda_1 \neq 0; \tag{3.109}$$

Case 3.

$$u_t = (u^{-\frac{1}{2}} u_x)_x + \lambda u^{-\frac{1}{2}} u_x + \lambda_1 u + \lambda_2 u^{\frac{1}{2}} + \lambda_3, \tag{3.110}$$

$$Q = \partial_t + b(t,x)\partial_x + 2[\alpha(t,x)u + \beta(t,x)u^{\frac{1}{2}}]\partial_u, \quad \beta \neq 0, \tag{3.111}$$

where the function triplet (b, α, β) is the general solution of the system

$$\begin{aligned} 2bb_x + b_t + b\alpha &= 0, \\ b_{xx} - \lambda b_x - 2\alpha_x - b\beta &= 0, \\ (\alpha - \tfrac{\lambda_1}{2})(\alpha + 2b_x) + \alpha_t &= 0, \\ 2\alpha\beta - \lambda_1\beta + 2b_x\beta - \lambda_2 b_x + \beta_t - \lambda\alpha_x - \alpha_{xx} &= 0, \\ \beta^2 - \tfrac{\lambda_2}{2}\beta - \lambda_3 b_x + \tfrac{\lambda_3}{2}\alpha - \lambda\beta_x - \beta_{xx} &= 0. \end{aligned} \tag{3.112}$$

(λ_1, λ_2 and λ_3 are arbitrary constants).

Proof of Theorem 3.5 is based on the algorithm presented in Section 3.3. The first two steps of the algorithm are already discussed above. In order to derive the system of DEs (the third step), we use the system of DEs (3.39) derived for the general class of RDC equations. Indeed, applying the Kirchhoff substitution (3.35) for $A(u) = u^m$, we obtain

$$v = \begin{cases} u^{m+1}, & m \neq -1, \\ \ln u, & m = -1. \end{cases} \tag{3.113}$$

In the cases $m \neq -1$ and $m = -1$, substitution (3.113) reduces Eq. (3.103) to the forms

$$v_{xx} = v^n v_t - \lambda v_x + F(v), \tag{3.114}$$

(here $n = -\frac{m}{m+1} \neq 0$, $F(v) = -(m+1)C(v^{\frac{1}{m+1}})$, $\lambda \neq 0$) and
$$v_{xx} = e^v v_t - \lambda v_x + F(v), \quad F(v) = C(e^v), \quad (3.115)$$
respectively.

Thus, the system of DEs (3.39) for the triplet $(F_0, F_1, F_2) = (v^n, -\lambda, F(v))$ takes the form
$$\xi^1_{vv} = 0, \quad \eta_{vv} = 2\xi^1_v(-\lambda - \xi^1 v^n) + 2\xi^1_{xv},$$
$$(2\xi^1_v \eta - 2\xi^1 \xi^1_x - \xi^1_t)v^n - n\xi^1 \eta v^{n-1} - \lambda \xi^1_v + 3\xi^1_v F - 2\eta_{xv} + \xi^1_{xx} = 0, \quad (3.116)$$
$$\eta F_v + (2\xi^1_x - \eta_v)F + n\eta^2 v^{n-1} + 2\xi^1_x \eta v^n + \eta_t v^n - \lambda \eta_x - \eta_{xx} = 0.$$

The remaining steps of the algorithm will be realized provided all inequivalent solutions of the nonlinear system (3.116) are found.

Let us analyze (3.116). Solving the first equation of the system, we arrive at the function $\xi^1 = a(t,x)v + b(t,x)$ with $a(t,x)$ and $b(t,x)$ being arbitrary smooth functions at the moment. It turns out that system (3.116) does not possess any Q-conditional symmetry provided $a(t,x) \neq 0$ (the relevant calculations are omitted here because of their awkwardness).

So, we set
$$\xi^1 = b(t,x). \quad (3.117)$$
Solving the second equation of (3.116) under condition (3.117), we arrive at
$$\eta = \alpha(t,x)v + \beta(t,x). \quad (3.118)$$

Taking into account (3.117) and (3.118), the third equation of (3.116) reduces to the form
$$(2bb_x + b_t + nb\alpha)v^n + nb\beta v^{n-1} - b_{xx} + \lambda b_x + 2\alpha_x = 0. \quad (3.119)$$

This equation can be split with respect to (w.r.t.) the different exponents of v. One should consider two cases depending on n:

(a) if $n \neq 1$ then
$$2bb_x + b_t + nb\alpha = 0,$$
$$b\beta = 0, \quad (3.120)$$
$$b_{xx} - \lambda b_x - 2\alpha_x = 0.$$

(b) if $n = 1$ then
$$2bb_x + b_t + b\alpha = 0,$$
$$b_{xx} - \lambda b_x - 2\alpha_x - b\beta = 0.$$

Let us consider case (a). Substituting (3.117) and (3.118) into the fourth equation of (3.116), one arrives at

$$(\alpha v + \beta)F_v + (2b_x - \alpha)F = -nv^{n-1}(\alpha v + \beta)^2 + \beta_{xx} + \lambda\beta_x + \\ +(\alpha_{xx} + \lambda\alpha_x)v - (\alpha_t + 2b_x\alpha)v^{n+1} - (\beta_t + 2b_x\beta)v^n. \quad (3.121)$$

To solve (3.121) and (3.120) one needs to consider two subcases, $b = 0$ and $\beta = 0$ (see the second line in (3.120)).

The subcase $b = 0$ leads to $\alpha = \alpha(t)$ and the equation

$$(\alpha v + \beta)F_v - \alpha F = -nv^{n-1}(\alpha v + \beta)^2 + \beta_{xx} + \lambda\beta_x - \alpha_t v^{n+1} - \beta_t v^n. \quad (3.122)$$

In the simplest case $g = const$ and $\beta = const$, we immediately arrive at

$$b = 0, \ \alpha = \lambda_1^*, \ \beta = \lambda_2^*, \ F = (\lambda_1^* v + \lambda_2^*)(\lambda_3 - v^n),$$

therefore the equation

$$v_{xx} = v^n v_t - \lambda v_x + (\lambda_1^* v + \lambda_2^*)(\lambda_3 - v^n) \quad (3.123)$$

admits the Q-conditional symmetry

$$Q = \partial_t + (\lambda_1^* v + \lambda_2^*)\partial_v. \quad (3.124)$$

Applying substitution (3.113) to Eq. (3.123) and operator (3.124) we obtain case 1 of the theorem (new notations $\lambda_i = \frac{\lambda_i^*}{m+1}$, $i = 1, 2$ should be used).

Now we assume that $\alpha \neq const$, so that the third equation of (3.122) can be reduced to the form

$$\left(v + \frac{\beta}{\alpha}\right)F_v - F = -n\alpha v^{n-1}\left(v + \frac{\beta}{\alpha}\right)^2 + \frac{\beta_{xx} + \lambda\beta_x - \beta_t v^n - \alpha_t v^{n+1}}{\alpha}. \quad (3.125)$$

It turns out that the last equation can be satisfied only under condition $\frac{\beta}{\alpha} = const$ (see the proof below). So, setting $\frac{\beta}{\alpha} = const$ into (3.125) and making the relevant calculations, we obtain the Lie symmetry operators only and a particular case of operator (3.124) for Eq. (3.123).

Let us prove that $\frac{\beta}{\alpha} = const$. By differentiating Eq. (3.125) w.r.t. the variables x and t, one obtains two equations (they are omitted here). Assuming $(\frac{\beta}{\alpha})_t(\frac{\beta}{\alpha})_x = 0$, one easily arrives at the condition $\frac{\beta}{\alpha} = const$.

Assuming $(\frac{\beta}{\alpha})_t(\frac{\beta}{\alpha})_x \neq 0$, we immediately obtain the restriction $\beta_x \neq 0$ because the condition $\alpha_x = 0$ follows from the last equation of (3.120). Now we differentiate Eq. (3.125) w.r.t. the variable x and obtain the equation

$$F_v = -\frac{1}{\beta_x}(2n\alpha\beta + \beta_t)_x v^n - 2n\beta v^{n-1} + \frac{\beta_{xxx} + \lambda\beta_{xx}}{\beta_x}.$$

Since the functions v^n, v^{n-1} and 1 on the right-hand side are functionally independent (we consider the case $n \neq 1$), all the coefficients must be constants.

It means that $\beta = const$, so that $\beta_x = 0$, i.e., we arrive at the contradiction. Thus, the condition $\frac{\beta}{\alpha} = const$ indeed takes place.

Now we turn to the subcase $\beta = 0$, so that (3.120) and (3.121) take the form

$$2bb_x + b_t + nb\alpha = 0,$$
$$b_{xx} - \lambda b_x - 2\alpha_x = 0, \qquad (3.126)$$
$$\alpha v F_v + (2b_x - \alpha)F = -n\alpha^2 v^{n+1} + (\alpha_{xx} + \lambda\alpha_x)v - (\alpha_t + 2b_x\alpha)v^{n+1}.$$

The above system can be easily solved and its general solution has the form

$$b = \frac{C_1 \exp(\lambda_1 nt)}{C_2 \exp(\lambda_1 nt) + 1}, \quad \alpha = -\frac{\lambda_1}{C_2 \exp(\lambda_1 nt) + 1}, \quad F = \lambda_1 v^{n+1} + \lambda_2 v,$$

where $C_k \in \mathbb{R}$, $k = 1, 2$. Thus, we arrive at the RDC equation

$$v_{xx} = v^n v_t - \lambda v_x + \lambda_1 v^{n+1} + \lambda_2 v$$

and the Q-conditional operator

$$Q = \partial_t + \frac{C_1 \exp(\lambda_1 nt)}{C_2 \exp(\lambda_1 nt) + 1}\partial_x - \frac{\lambda_1 v}{C_2 \exp(\lambda_1 nt) + 1}\partial_v. \qquad (3.127)$$

However, one easily proves multiplying (3.127) with the function $M(t, x, v) = C_2 \exp(\lambda_1 nt) + 1$ that this operator is nothing else but a linear combination of the Lie symmetry operators. In fact, the above RDC is equivalent to that listed in case 24 of Table 2.5.

This completes the examination of case (a).

Consider case (b). Substituting (3.117) and (3.118) into the fourth equation of (3.116), we arrive at (3.121) with $n = 1$. Dealing with this equation in the same way as in case (a), we obtain equation

$$v_{xx} = vv_t - \lambda v_x + \lambda_1^* v^2 + \lambda_2^* v + \lambda_3^*, \qquad (3.128)$$

and Q-conditional operators of the form

$$Q = \partial_t + b(t, x)\partial_x + [\alpha(t, x)v + \beta(t, x)]\partial_v, \qquad (3.129)$$

where the triplet (b, α, β) is an arbitrary solution of system (3.112). Applying formula (3.113) with $m = -\frac{1}{2}$ we obtain case 3 of the theorem (note one should use new notations $\lambda_i = -2\lambda_i^*$, $i = 1, 2, 3$).

The overdetermined system (3.112) can be easily solved under the restriction $\beta = 0$ because one reduces to a linear system. However, operator (3.129) with $\beta = 0$ and the functions b and α derived is equivalent to known Lie's operators. Thus, one needs $\beta \neq 0$ in order to guarantee that operator (3.129) leads to pure Q-conditional symmetries. We were unable to integrate system (3.112) with $\beta \neq 0$, however, its particular solutions were identified (see examples below).

Finally, we analyze Eq. (3.115), which is equivalent to Eq. (3.103) with $m = -1$. The system of DEs (3.39) for the triplet $(F_0, F_1, F_2) = (e^v, -\lambda, F(v))$ takes the form

$$\xi^1_{vv} = 0, \quad \eta_{vv} = 2\xi^1_v(-\lambda - \xi^1 e^v) + 2\xi^1_{xv},$$
$$(\xi^1_t + 2\xi^1 \xi^1_x - 2\xi^1_v \eta + \xi^1 \eta)e^v + \lambda \xi^1_x - 3\xi^1_v F + 2\eta_{xv} - \xi^1_{xx} = 0, \quad (3.130)$$
$$\eta F_v + (2\xi^1_x - \eta_v)F + (\eta^2 + 2\xi^1_x \eta + \eta_t)e^v - \lambda \eta_x - \eta_{xx} = 0,$$

where ξ^1, η and F are yet-to-be determined functions. Solving the first and second equations of this system, we establish that the functions ξ^1 and η must be given by formulae (3.117) and (3.118), respectively, otherwise Q-conditional symmetry does not exist. Substituting (3.117) and (3.118) into the third equation of (3.130) we obtain the equation

$$(b\beta + b_t + 2bb_x)e^v + b\alpha v e^v + \lambda b_x + 2\alpha_x - b_{xx} = 0.$$

Since the functions b, α and h do not depend on v, one can split this equation w.r.t. e^v and ve^v, so that the system

$$b_t + 2bb_x + b\beta = 0,$$
$$b\alpha = 0, \quad (3.131)$$
$$\lambda b_x + 2\alpha_x - b_{xx} = 0.$$

is derived.

Substituting (3.117) and (3.118) into the fourth equation of (3.130) we arrive at the equation

$$(\alpha v + \beta)F_v + (2b_x - \alpha)F = -(\alpha v + \beta)^2 e^v + \beta_{xx} + \lambda \beta_x +$$
$$+(\alpha_{xx} + \lambda \alpha_x)v - (\alpha_t + 2b_x \alpha)v e^v - (\beta_t + 2b_x \beta)e^v. \quad (3.132)$$

Now we apply to (3.132) the same approach, which has been used for solving Eq. (3.121). Thus, solving system (3.131)–(3.132), we finally obtain the expressions

$$F = (\lambda_1 v + \lambda_2)(\lambda_3 - e^v), \quad b = 0, \quad \alpha = \lambda_1, \quad \beta = \lambda_2,$$

which lead to the equation

$$v_{xx} = e^v v_t - \lambda v_x + (\lambda_1 v + \lambda_2)(\lambda_3 - e^v) \quad (3.133)$$

and the operator

$$Q = \partial_t + (\lambda_1 v + \lambda_2)\partial_v. \quad (3.134)$$

Applying substitution (3.113) with $m = -1$ to (3.133) and (3.134), one obtains case 2 of the theorem.

The proof is now completed. □

Remark 3.7 *Cases 1 and 2 with $\lambda = 0$ immediately give the RD equations and the relevant symmetries obtained earlier in [16].*

The overdetermined system (3.112) contains five equations on three unknown functions. A natural question arises about its compatibility. The answer is positive, i.e., that is compatible. In fact, the system with $b = \alpha = 0$, $\lambda_1 = 0$ is reduced to the ordinary differential equation (ODE)

$$\beta_{xx} + \lambda \beta_x + \frac{\lambda_2}{2}\beta - \beta^2 = 0, \tag{3.135}$$

therefore

$$Q = \partial_t + 2\beta(x) u^{\frac{1}{2}} \partial_u \tag{3.136}$$

is the Q-conditional symmetry operator for an arbitrary nonzero solution of (3.135). Unfortunately, ODE (3.135) cannot be integrated for the arbitrary coefficients λ and λ_2, however some particular solutions can be easily established. For example, setting $\beta = \frac{\lambda_2}{2}$ case 1 with $m = -\frac{1}{2}$, $\lambda_1 = 0$ is obtained.

Setting $\lambda = \lambda_2 = 0$ in (3.135), we arrive at the known ODE $\beta_{xx} = \beta^2$ with the general solution $\beta = \mathcal{W}(0, C_1, x + C_2)$, where C_1 and C_2 are arbitrary constants, $\mathcal{W}$ is the Weierstrass function with the periods 0 and C_1. The simplest solution takes the form $\beta = 6x^{-2}$ and leads to the known Q-conditional symmetry operator $Q = \partial_t + 12x^{-2} u^{\frac{1}{2}} \partial_u$ of the nonlinear diffusion equation $u_t = (u^{-\frac{1}{2}} u_x)_x$ [16]. However, this result can be generalized as follows. One can easily check that an arbitrary particular solution of (3.112) with $\lambda = 0$ generates the Q-conditional symmetry operator (3.111) of the RD equation

$$u_t = (u^{-\frac{1}{2}} u_x)_x + \lambda_1 u + \lambda_2 u^{\frac{1}{2}} + \lambda_3. \tag{3.137}$$

Obviously, the operator presented in Table 3 of [16] is obtainable from (3.111) and (3.112) by setting $\lambda = \lambda_1 = 0$ and $b = \alpha = 0$ but not vice versa.

Construction of the general solution of the overdetermined system (3.112) is a highly nontrivial task and lies beyond the scope of this research.

3.5.2 The case of different diffusion and convection coefficients

Here we examine the class of RDC equations (3.104) and its natural generalization

$$u_t = (u^m u_x)_x + \lambda u^n u_x + C(u) \tag{3.138}$$

with arbitrary parameters $\lambda \neq 0$ and $n \neq 0, m, m+1$.

Theorem 3.6 *[75] An equation of the form (3.104) is Q-conditionally invariant under the operator (3.11) if and only if the equation (up to transformations (3.105)) and the relevant operator (up to equivalent representations* generated by multiplying on the arbitrary smooth function $M(t,x,u)$) have the following forms:

Case 1.
$$u_t = (u^m u_x)_x + \lambda u^{m+1} u_x + \lambda_1 u + \lambda_2 u^{-m}, \quad m \neq -1, \tag{3.139}$$

$$Q = \partial_t - \lambda u^{m+1}\partial_x + (\lambda_1 u + \lambda_2 u^{-m})\partial_u; \tag{3.140}$$

Case 2.
$$u_t = (u^{-\frac{1}{2}} u_x)_x + \lambda u^{\frac{1}{2}} u_x + (\lambda_1 u^{\frac{3}{2}} + \lambda_2 u^{\frac{1}{2}} + \lambda_3)\left(\frac{\lambda_1}{2\lambda^2} + u^{\frac{1}{2}}\right), \tag{3.141}$$

$$Q = \partial_t + \left(\frac{3\lambda_1}{2\lambda} - \lambda u^{\frac{1}{2}}\right)\partial_x + (\lambda_1 u^{\frac{3}{2}} + \lambda_2 u^{\frac{1}{2}} + \lambda_3)\partial_u. \tag{3.142}$$

Proof Here a sketch of the proof is presented because one is similar to the proof of Theorem 3.5. Let us use again substitution (3.113), which reduces Eq. (3.104) to the form

$$v_{xx} = v^n v_t - \lambda v^{n+1} v_x + F(v) \tag{3.143}$$

(here $n = -\frac{m}{m+1} \neq 0, -1$, $F(v) = -(m+1)C(v^{\frac{1}{m+1}})$).

Let us analyze Eq. (3.143). The system of DEs (3.39) for the triplet $(F_0, F_1, F_2) = (v^n, -\lambda v^{n+1}, F(v))$ takes the form

$$\xi^1_{vv} = 0, \quad \eta_{vv} = 2\xi^1_v(-\lambda v^{n+1} - \xi^1 v^n) + 2\xi^1_{xv},$$

$$\lambda \xi^1_x v^{n+1} + \left[\left(-2\xi^1_v + \lambda(n+1)\right)\eta + 2\xi^1 \xi^1_x + \xi^1_t\right]v^n + \tag{3.144}$$

$$+\xi^1 \eta n v^{n-1} - 3\xi^1_v F + 2\eta_{xv} - \xi^1_{xx} = 0,$$

$$\eta F_v + (2\xi^1_x - \eta_v)F + n\eta^2 v^{n-1} + 2\xi^1_x \eta v^n + \eta_t v^n - \lambda v^{n+1}\eta_x - \eta_{xx} = 0.$$

So, we need to find all possible functions ξ^1, η and F, which are solutions of the overdetermined system (3.144).

Because the first equation in system (3.144) is a trivial linear ODE, while the second equation can be easily integrated using the solution of the latter, one establishes that there are only three possibilities for the functions ξ^1 and η:

$$\begin{aligned}&(a) \; \xi^1 = \lambda_1^* v + \lambda_2^*, \; \eta = \eta(v), \quad \lambda_1^*, \lambda_2^* \in \mathbb{R}, \\ &(b) \; \xi^1 = b(t,x), \; \eta = \alpha(t,x)v + \beta(t,x), \\ &(c) \; \xi^1 = a(t,x)v + b(t,x), \; \eta = \eta(t,x,v), \quad a(t,x) \neq 0.\end{aligned} \tag{3.145}$$

Here a, b, α and β are to-be-determined functions. The function η in case (c) takes the forms

$$\eta = \begin{cases} -\dfrac{2a(a+\lambda)}{(n+2)(n+3)} v^{n+3} - \dfrac{2ab}{(n+1)(n+2)} v^{n+2} + a_x v^2 + \alpha v + \beta, \\ \\ n \neq -2, -3; \\ \\ 2a[b - (a+\lambda)v] \ln v + a_x v^2 + [2a(a+\lambda) + \alpha]v + \beta, \\ \\ n = -2; \\ \\ 2a(a+\lambda) \ln v - abv^{-1} + a_x v^2 + \alpha v + \beta, \\ \\ n = -3, \end{cases} \qquad (3.146)$$

where $\alpha = \alpha(t,x)$, $\beta = \beta(t,x)$.

It turns out that a detailed examination of cases (b) and (c) lead only to Lie symmetry operators.

Let us briefly consider case (a), which is the most complicated. The system of DEs (3.144) should be solved. Since the function ξ^1 is the linear function, the general solution of the second equation of (3.144) is the function

$$\eta = -\frac{2\lambda_1^*(\lambda + \lambda_1^*)}{(n+2)(n+3)} v^{n+3} - \frac{2\lambda_1^* \lambda_2^*}{(n+1)(n+2)} v^{n+2} + \lambda_3^* v + \lambda_4^*, \qquad (3.147)$$

where $\lambda_3^*, \lambda_4^* \in \mathbb{R}$, $n \neq -2, -3$. Substituting (3.147) into the third equation of (3.144), one obtains

$$F = \frac{\eta}{3\lambda_1^*} \left[\left(\lambda_1^*(n-2) + \lambda(n+1) \right) v^n + n\lambda_2^* v^{n-1} \right], \qquad (3.148)$$

if $\lambda_1^* \neq 0$ (the case $\lambda_1^* = 0$ leads only to the Lie symmetry operators). Substituting (3.148) into the fourth equation of system (3.144) one arrives at the expression

$$\eta \left[\left(1 + \frac{1}{3\lambda_1^*} \left(\lambda_1^*(n-2) + \lambda(n+1) \right) \right) v^{n-1} + (n-1)\lambda_2^* v^{n-2} \right] = 0. \qquad (3.149)$$

The first possibility is

$$\left[1 + \frac{1}{3\lambda_1^*} \left(\lambda_1^*(n-2) + \lambda(n+1) \right) \right] v^{n-1} + (n-1)\lambda_2^* v^{n-2} = 0 \qquad (3.150)$$

while the second one is $\eta = 0$.

Splitting (3.150) with respect to (w.r.t.) the different exponents of v, one obtains the system of algebraic conditions

$$\begin{aligned} (\lambda_1^* + \lambda)(n+1) &= 0, \\ (n-1)\lambda_2^* &= 0. \end{aligned} \qquad (3.151)$$

The first condition in (3.151) immediately gives $\lambda_1^* = -\lambda$ because $n \neq -1$. The second one is fulfilled provided $\lambda_2^* = 0$ and/or $n = 1$. Hence we obtain the equation

$$v_{xx} = v^n v_t - \lambda v^{n+1} v_x - (\lambda_3^* v + \lambda_4^*) v^n, \qquad (3.152)$$

and the Q-conditional symmetry operator

$$Q = \partial_t - \lambda v \partial_x + (\lambda_3^* v + \lambda_4^*) \partial_v \qquad (3.153)$$

if $\lambda_2^* = 0$, and the equation

$$v_{xx} = vv_t - \lambda v^2 v_x - \frac{\lambda_2^* + 3\lambda v}{3\lambda} \left(\frac{1}{3} \lambda_2^* \lambda v^3 + \lambda_3^* v + \lambda_4^* \right) \qquad (3.154)$$

and the operator

$$Q = \partial_t + (-\lambda v + \lambda_2^*) \partial_x + \left(\frac{1}{3} \lambda_2^* \lambda v^3 + \lambda_3^* v + \lambda_4^* \right) \partial_v, \qquad (3.155)$$

if $n = 1$. Finally, applying substitution (3.113) with $m \neq -1$ to (3.152)–(3.155) and introducing new notations $\lambda_3^* = \lambda_1(m+1)$, $\lambda_4^* = \lambda_2(m+1)$ and $\lambda_2^* = \frac{3\lambda_1}{2\lambda}$, $\lambda_3^* = \frac{\lambda_2}{2}$, $\lambda_4^* = \frac{\lambda_3}{2}$ in Eqs. (3.152) and (3.154), respectively, we arrive exactly at cases 1 and 2 of the theorem.

The case $\eta = 0$ (see (3.149)) leads to Eq. (3.152) and operator (3.153) with $\lambda_3^* = \lambda_4^* = 0$.

We have also proved that the special values $n = -2$ and $n = -3$ (see (3.147)) lead to particular cases of Eq. (3.152) and operator (3.153)only. The sketch of the proof is now completed. □

Finally we present the result of QSC for the class of RDC equations (3.138).

Theorem 3.7 *If an RDC equation belonging to class (3.138) with $\lambda n \neq 0$ and $n \neq m, m+1$ is Q-conditionally invariant w.r.t. an operator of the form (3.11), then this operator is equivalent to a linear combination of the Lie symmetry operators of the equation in question.*

Proof is very cumbersome and can be found in [76]. Here we present the proof under the assumption that the coefficients of operator (3.11) do not depend on the variables t and x, but depend on u. In this case the system of DEs (3.39) is reduced to a much simpler one. It turns out that it is possible to construct the general solution provided $F_0(v)$ is a given function. In fact, the last equation of (3.11) takes the form

$$\eta(\eta F_0 + F_2)_v - \eta_v(\eta F_0 + F_2) = 0, \qquad (3.156)$$

which is integrable and has the solution

$$F_2 = (C - F_0)\eta, \qquad (3.157)$$

being C an arbitrary constant. Note that the special case $\eta = 0$ when Eq. (3.156) vanishes immediately leads to the condition $n = m + 1$.

Taking into account (3.157), the third equation of system (3.39) has the form
$$\eta(\eta F_1 - \xi^1 F_0)_v - 3C\xi_v^1 \eta = 0. \tag{3.158}$$

Because $\xi^1 = av + b$ is the linear function (see the first equation in (3.39)), Eq. (3.158) is integrable, so that we obtain
$$F_1 = (av + b)F_0 - 3aCv + C_0. \tag{3.159}$$

Now we turn to the given class of equations (3.138). Obviously, $F_0 = v^p$, $p = -\frac{m}{m+1}$ if $m \neq -1$ and $F_0 = e^v$ if $m = -1$ (see (3.113)). Because the convective term in (3.138) is λu^n, we obtain $F_1 = -\lambda v^k$, $k = \frac{n-m}{m+1}$ if $m \neq -1$ and $F_1 = -\lambda e^{(n+1)v}$ if $m = -1$. Thus, the following equalities should take place
$$-\lambda v^k = (av + b)v^p - 3aCv + C_0 \tag{3.160}$$
and
$$-\lambda e^{(n+1)v} = (av + b)e^v - 3aCv + C_0, \tag{3.161}$$
respectively. Eq. (3.161) leads immediately to the contradiction $\lambda = 0$ because $n \neq 0$.

Consider Eq. (3.160). Because $p \neq k$, we conclude that $b = 0$. Thus, depending on the parameters a, C and C_0, there are only two possibilities, which lead either to the condition $n = m$ or the condition $n = m+1$. However, the class of RDC equations (3.138) with such exponents was already studied and we assume in this theorem that $n \neq m, m+1$.

The proof is now completed. $\square$

Remark 3.8 *Formulae (3.157) and (3.159) allow us to construct the general solution of the system of DEs (3.39) provided the function F_0 is given and the coefficients of operator (3.11) do not depend on the variables t and x.*

3.6 Q–conditional symmetry of reaction-diffusion-convection equations with exponential diffusivity

In this section, we deal with the reaction-diffusion-convection (RDC) equations of the form
$$u_t = (e^{nu} u_x)_x + \lambda e^{mu} u_x + C(u), \tag{3.162}$$
which is a special subclass of (3.34). Hereafter n and m are arbitrary constants with the restriction $m \neq 0$ (otherwise the convective term is removable). A

significant feature of this section is the study of nonlinear equations involving three transport mechanisms (diffusion, convection and reaction), which can describe real world processes with very fast diffusion/heat transfer and convection/advection ($n > 0$, $m > 0$). Thus, we consider the cases when these equations model such complicated processes that all the transport mechanisms must be taken into account.

We aim to solve the QSC problem for class (3.162). According to the algorithm derived in Section 3.3, equivalence transformations (ETs) and form-preserving transformations (FPTs) should be identified from the very beginning. In order to construct the ET group of the above class of RDC equations, we again use the group $\mathcal{E}$ (2.36) of the general class (3.34). Using this group, one can easily derive the ET group of (3.162), which is the five-parameter group of scale transformations and translations w.r.t. t, x and u:

$$\bar{t} = e_0 t + t_0, \quad \bar{x} = e_1 x + x_0, \quad \bar{u} = e_2 u + u_0, \qquad (3.163)$$

where arbitrary parameters obey the restrictions $e_0 = e_1^2 \exp(-\frac{n u_0}{e_2})$, $e_1 e_2 \neq 0$.

Using Table 2.6 (see Cases 6, 12–14), one can also identify four FPTs occurring for the RDC equations belonging to class (3.162), however these transformations do not play any role in solving the QSC problem for (3.162). In fact, the transformations are applicable only to particular cases of the RDC equations arising in the theorem presented below. Thus, the first two steps of the algorithm for solving the QSC problem for class (3.162) are completed. The third step is rather trivial because we have already derived the system of DEs for the general class of RDC equations (see Theorem 3.1). In order to complete the next steps of the algorithm for the class of RDC equations (3.162), we start from the theorem.

Theorem 3.8 *[78] A RDC equation from class (3.162) is Q-conditionally invariant under operator (3.11) if and only if the equation (up to ETs (3.163)) and the relevant operator (up to ETs generated by multiplying on an arbitrary smooth function $M(t, x, u)$) have the following forms.*

Case 1.
$$u_t = (e^u u_x)_x + \lambda e^u u_x + \lambda_0 + \lambda_1 e^u + \lambda_2 e^{-u}, \qquad (3.164)$$

$$Q = \partial_t + a e^u \partial_x + \left[\alpha + (a_x - a^2 - \lambda a) e^u + \lambda_2 e^{-u}\right] \partial_u, \qquad (3.165)$$

where the functions a and α satisfy the overdetermined system

$$\begin{aligned}
& a_t + 2\alpha_x - 3 a \alpha + 3 \lambda_0 a = 0, \\
& a_{xx} - 3 a a_x + a^3 + \lambda a^2 - \lambda a_x + \lambda_1 a = 0, \\
& \alpha_t + \lambda_2 a_x - \alpha^2 - \lambda_2 a^2 + \lambda_0 \alpha - \lambda \lambda_2 a - \lambda_1 \lambda_2 = 0, \\
& \alpha_{xx} + \lambda \alpha_x - a_{tx} + 2 a a_t + \lambda a_t - 2 a(a + \lambda)(\alpha - \lambda_0) = 0.
\end{aligned} \qquad (3.166)$$

Case 2
$$u_t = (e^u u_x)_x + \lambda e^{2u} u_x + \frac{1}{9} \lambda^2 e^{3u} + \lambda_0 + \lambda_1 e^u + \lambda_2 e^{-u}, \qquad (3.167)$$

$$Q = \partial_t + ae^u\partial_x + \left[\alpha - \frac{\lambda a}{3}e^{2u} + (a_x - a^2)e^u + \lambda_2 e^{-u}\right]\partial_u, \quad (3.168)$$

where the functions a and α satisfy the overdetermined system

$$\begin{aligned}
& a_t + 2\alpha_x - 3a\alpha + 3\lambda_0 a + \lambda\lambda_2 = 0, \\
& a_{xx} - 3aa_x + a^3 + \lambda_1 a + \tfrac{1}{3}\lambda\alpha = 0, \\
& \alpha_t + \lambda_2 a_x - \alpha^2 - \lambda_2 a^2 + \lambda_0 \alpha - \lambda_1 \lambda_2 = 0, \\
& \alpha_{xx} - a_{tx} + 2aa_t - 2a^2\alpha + 2\lambda_0 a^2 + \tfrac{2}{3}\lambda\lambda_2 a = 0.
\end{aligned} \quad (3.169)$$

Proof The proof is based on the extensive analysis of the system of DEs. From the very beginning, we apply the Kirchhoff substitution in order to transform the given class of equations (3.162) to the form (3.36). In the case $n = 0$, the substitution is rather trivial

$$u = \frac{1}{m}v, \quad (3.170)$$

and transforms (3.162) to the form

$$v_{xx} = v_t - \lambda e^v v_x + F(v), \quad (3.171)$$

where $F(v) = -mC(\frac{1}{m}u)$. In the case $n \neq 0$, the substitution

$$u = \frac{1}{n}\ln v, \quad (3.172)$$

reduces the class of equations to the form

$$v_{xx} = v^{-1}v_t - \lambda v^q v_x + F(v), \quad (3.173)$$

where $q = \frac{m}{n} - 1 \neq -1$, $F(v) = -nC(u)$.

Thus, the system of DEs (3.39) for the triplet $(F_0, F_1, F_2) = (1, -\lambda e^v, F(v))$ takes the form

$$\begin{aligned}
& \xi^1_{vv} = 0, \\
& \eta_{vv} = -2\xi^1_v(\lambda e^v + \xi^1) + 2\xi^1_{xv}, \\
& 3\xi^1_v F - \lambda(\eta + \xi^1_x)e^v + \xi^1_x \eta - \xi^1_t - 2\xi^1 \xi^1_x + 3\xi^1_v \eta - 2\eta_{xv} + \xi^1_{xx} = 0, \\
& \eta F_v + (2\xi^1_x - \eta_v)F - \lambda\eta_x e^v - \eta_{xx} + \eta_t + 2\xi^1_x \eta = 0,
\end{aligned} \quad (3.174)$$

In the case of Eq. (3.173), the triplet $(F_0, F_1, F_2) = (v^{-1}, -\lambda v^q, F(v))$, hence (3.39) takes the form

$$\begin{aligned}
& \xi^1_{vv} = 0, \\
& \eta_{vv} = -2\xi^1_v(\lambda v^q + \xi^1 v^{-1}) + 2\xi^1_{xv}, \\
& 3\xi^1_v F - \lambda\xi^1_x v^q - \lambda q\eta v^{q-1} - (\xi^1_t + 2\xi^1 \xi^1_x - 2\xi^1_v \eta)v^{-1} + \xi^1 \eta v^{-2} - 2\eta_{xv} + \xi^1_{xx} = 0, \\
& \eta F_v + (2\xi^1_x - \eta_v)F - \lambda\eta_x v^q - \eta^2 v^{-2} + (\eta_t + 2\xi^1_x \eta)v^{-1} - \eta_{xx} = 0.
\end{aligned}$$
$$(3.175)$$

It can be noted that systems (3.174) and (3.175) are nonlinear and their general solutions cannot be derived in a simple way.

The first equation in both systems is linear and has the general solution

$$\xi^1 = a(t,x)v + b(t,x), \qquad (3.176)$$

where the functions $a(t,x)$ and $b(t,x)$ are arbitrary at the moment.

Let us use formula (3.176) for integration of system (3.174). Substituting the function ξ obtained into the second equation of (3.174), one derives the equation

$$\eta_{vv} = -2\lambda a e^v - 2a^2 v + 2a_x - 2ab,$$

which immediately produces

$$\eta = -2\lambda a e^v - \frac{1}{3}a^2 v^3 + (a_x - ab)v^2 + \alpha(t,x)v + \beta(t,x), \qquad (3.177)$$

where the functions $\alpha(t,x)$ and $\beta(t,x)$ are arbitrary at the moment.

Taking into account formulae (3.176) and (3.177), the third equation of system (3.174) can be solved with respect to (w.r.t.) the function F:

$$F = \frac{\lambda}{3a}\left[-\frac{a^2}{3}v^3 + (a_x - ab)v^2 + (\alpha + a_x)v + b_x + 4a^2 + \beta - 4a_x\right]e^v -$$
$$-\frac{2}{3}e^{2v}\lambda^2 + \frac{2}{9}v^3 a^2 + \frac{1}{3a}\left[2a(ab - 2a_x)v^2 + (3a_{xx} + a_t - \right. \qquad (3.178)$$
$$\left. -2a\alpha - 2(ab)_x)v + 2bb_x - 2a\beta + 2\alpha_x - b_{xx} + b_t\right], \quad a \neq 0.$$

It should be noted that a special case $a(t,x) = 0$ leads only to Lie symmetries.

Since the function F depends only on the variable v, we conclude that all the coefficients next to the linearly independent terms e^{2v}, $v^3 e^v$, $v^2 e^v$, $v e^v$, v^3, v^2, v and v^0 arising in right-hand side of (3.178) must be constants. Among others the condition $-\frac{1}{9}\lambda a = const$ arises (see the term with $e^v v^3$), so that $a = const \neq 0$. Now one sees that the requirement $a = const$ immediately leads to $b = const$ (see the term with $e^v v^2$). Finally, we obtain that $\alpha = const$, $\beta = const$, hence all the functions in right-hand side of (3.176) and (3.177) are constants. In this case, straightforward calculations show that one needs $a = 0$ to satisfy the fourth equation of system (3.174), i.e., the contradiction is obtained. Thus, the integration of system (3.174) does not lead to any Q-conditional symmetry operators.

The examination of system (3.175) gives positive results. In fact, substituting the function ξ from (3.176) into the second equation of (3.175), one obtains the linear equation

$$\eta_{vv} = -2\lambda a v^q - 2a^2 - 2abv^{-1} + 2a_x.$$

In the special case $q = -2$, i.e., $m = -n$ in (3.162), the general solution takes the form

$$\eta = 2\lambda a \ln v - 2abv(\ln v - 1) + (a_x - a^2)v^2 + \alpha(t,x)v + \beta(t,x) \qquad (3.179)$$

Q-conditional symmetry of RDC equations with exponential diffusivity 115

and then the straightforward calculations show that the Lie symmetries derived in Chapter 2 can be only obtained.

In the case $q \neq -2$, one obtains

$$\eta = -\frac{2\lambda a}{(q+1)(q+2)} v^{q+2} - 2abv(\ln v - 1) + (a_x - a^2)v^2 + \alpha(t,x)v + \beta(t,x). \tag{3.180}$$

Setting $a = 0$ in (3.180), one sees that η is a linear function and straightforward calculations lead to the operator

$$Q = \partial_t + \left[\alpha(t) + \lambda_2 e^{-u}\right] \partial_u, \tag{3.181}$$

where the function α depends on $D = \lambda_0^2 - 4\lambda_1\lambda_2$ and takes the forms

$$\alpha(t) = \begin{cases} \frac{\lambda_0 \pm \sqrt{D}}{2}, \\ -\frac{1}{t} + \frac{\lambda_0}{2}, \quad D = 0, \\ \frac{\sqrt{-D}}{2} \tan\left(\frac{\sqrt{-D}}{2}t\right) + \frac{\lambda_0}{2}, \quad D < 0, \\ -\frac{\sqrt{D}}{2} \coth(\frac{\sqrt{D}}{2}t) + \frac{\lambda_0}{2}, \quad (2\alpha - \lambda_0)^2 > D > 0, \\ -\frac{\sqrt{D}}{2} \tanh(\frac{\sqrt{D}}{2}t) + \frac{\lambda_0}{2}, \quad D > (2\alpha - \lambda_0)^2 > 0. \end{cases} \tag{3.182}$$

On the other hand, one may check that (3.182) is nothing else but the solution of (3.166) with $a(t,x) = 0$ (see case 1 of Table 3.1). Moreover, the corresponding RDC equation is exactly (3.164). Thus, a subcase of case 1 of the theorem is derived.

Now we assume that $a(t,x) \neq 0$. Using formulae (3.176) and (3.180), the third equation of system (3.175) can be solved w.r.t. the function F. Omitting the relevant expression for F because of its awkwardness, one may note that the expression obtained contains the term $\frac{2}{3}b^2v^{-1}\ln v$, which is linearly independent (w.r.t. the variable v) from other terms. Since the function F depends only on the variable v we conclude that $b = const$. Thus, b is a constant in what follows.

Now the above mentioned expression for F simplifies to the form

$$F = \tfrac{2}{3}b\left(-\lambda qv^q + bv^{-1} + 3a - 2\tfrac{a_x}{a}\right)\ln v + \tfrac{\lambda}{3a}\left[\tfrac{(q-1)(q+3)}{q+1}a_x + \right.$$
$$\left. + \tfrac{6-q(q+1)(q+2)}{(q+1)(q+2)}a^2\right]v^{q+1} + \tfrac{\lambda}{3a}\left[qa + \left(\tfrac{2}{(q+1)(q+2)} + 2q\right)ab\right]v^q + $$
$$+ \tfrac{\lambda q\beta}{3a}v^{q-1} + \tfrac{1}{a}(a_{xx} - 3aa_x + a^3)v + \tag{3.183}$$
$$+ \tfrac{1}{3a}(2\alpha_x + a_xb - 5a^2b + a_t - 3a\alpha) - \tfrac{1}{3}(2b^2 + \tfrac{b\alpha}{a} + 3\beta)v^{-1} - $$
$$- \tfrac{b\beta}{3a}v^{-2} - \tfrac{2\lambda^2 q}{3(q+1)(q+2)}v^{2q+1}.$$

Obviously, the coefficient in first term of (3.183) must be a constant because other terms do not contain $\ln v$, hence

$$b\left(3a - 2\frac{a_x}{a}\right) = A \tag{3.184}$$

where A is an arbitrary constant.

Since the function F depends only on the variable v, all coefficients next to different exponents of this variable must be constants. If q is arbitrary then one arrives at 8 equations (including Eq. (3.184)) to find the functions a, b, α, and β, which can be satisfied if and only if all these functions are constants. The further calculations lead to the conclusion that Lie symmetry operators are obtained only.

The number of equations for the functions a, b, α, and β may be shortened for certain values of q, when some exponents take equal values. The simple analysis shows that there are only four possibilities:

$$1)\ q = -3,\ 2)\ q = 2,\ 3)\ q = 0,\ 4)\ q = 1.$$

The straightforward calculations show that the exponents arising in 1) and 2) do not produce any Q-conditional symmetry operators.

Consider the third possibility $q = 0$. In this case, F takes the form

$$F = \tfrac{2}{3}b\left[bv^{-1} + (3a - 2\tfrac{a_x}{a})\right]\ln v + \tfrac{1}{a}\left[\lambda(a^2 - a_x) + a_{xx} - 3aa_x + a^3\right]v +$$
$$+ \tfrac{1}{3a}(2\alpha_x + a_x b - 5a^2 b + a_t - 3a\alpha + \lambda ab) + \quad (3.185)$$
$$+ \tfrac{1}{3}(-2b^2 - \tfrac{b\alpha}{a} - 3\beta)v^{-1} - \tfrac{b\beta}{3a}v^{-2}.$$

If $b \neq 0$ then the equation

$$a_{xx} = 3aa_x + A_1 a_x,\ A_1 = const \quad (3.186)$$

is obtained by differentiating (3.184) multiplied by a. On the other hand, the linear term in the right-hand side of (3.185) produces the second equation for a:

$$\frac{\lambda}{a}(a^2 - a_x) + \frac{1}{a}(a_{xx} - 3aa_x + a^3) = A_2$$

with another arbitrary constant A_2. The last equation reduces to the algebraic equation

$$a^3 + \frac{1}{2}(3A_1 - \lambda)a^2 + (A_1 - \lambda)A_1 a - A_2 a = 0 \quad (3.187)$$

by using (3.184) and (3.186). Obviously Eq. (3.187) implies that $a = const$. However, other terms in (3.185) with $a = const \neq 0$ immediately produce $\alpha = const$, $\beta = const$. Finally, we arrive at the contradiction $b = 0$ in order to satisfy the last equation of system (3.175).

If $b = 0$ then the function F takes the form

$$F = \frac{1}{a}\left(a_{xx} - 3aa_x - \lambda a_x + a^3 + \lambda a^2\right)v + \frac{1}{3a}(2\alpha_x + a_t - 3a\alpha) - \beta v^{-1}, \quad (3.188)$$

which leads to the RDC equation (see (3.173) with $q = 0$)

$$v_{xx} = v^{-1}v_t - \lambda v_x - \lambda_2 v^{-1} - \lambda_1 v - \lambda_0.$$

The corresponding Q-conditional symmetry operator is (see (3.180) with $b = q = 0$)
$$Q = \partial_t + av\partial_x + \Big[(a_x - a^2 - \lambda a)v^2 + \alpha v + \lambda_2\Big]\partial_v.$$

Substitution (3.172) reduces the above equation and operator to those (3.164) and (3.165), respectively.

Simultaneously, the overdetermined system (3.166) is obtained. Indeed, the function F in (3.188) depends only on v, hence $\beta = \lambda_2$ and

$$\begin{aligned}\tfrac{1}{3a}(2\alpha_x + a_t - 3a\alpha) &= -\lambda_0,\\ \tfrac{1}{a}\left(a_{xx} - 3aa_x - \lambda a_x + a^3 + \lambda a^2\right) &= -\lambda_1,\end{aligned} \quad (3.189)$$

where $\lambda_1, \lambda_2, \lambda_3$ are arbitrary constants. Obviously, it is nothing else but the first and second equations from system (3.166). Substituting F (3.188) with coefficients (3.189) and $\beta = \lambda_2$ into the fourth equation of (3.175) and splitting the expression obtained w.r.t. the different exponents of the variable v, one arrives at the equations

$$\begin{aligned}&\alpha_t + \lambda_2 a_x - \alpha^2 - \lambda_2 a^2 + \lambda_0 \alpha - \lambda\lambda_2 a - \lambda_1\lambda_2 = 0,\\ &\alpha_{xx} + \lambda\alpha_x - a_{tx} + 2aa_t + \lambda a_t + 2a(\lambda_0 a - a\alpha - \lambda\alpha + \lambda\lambda_0) = 0,\\ &(a + \lambda)\Big[a_{xx} - 3aa_x + a^3 + \lambda a^2 - \lambda a_x + \lambda_1 a\Big] +\\ &+\partial_x\Big[a_{xx} - 3aa_x + a^3 + \lambda a^2 - \lambda a_x + \lambda_1 a\Big] = 0.\end{aligned} \quad (3.190)$$

Now one notes that the first and the second equations coincide with the third and fourth ones from system (3.166), while the last one in (3.190) vanishes because of the second equation from (3.166). Thus, Case 1 is identified.

The last possibility 4) $q = 1$ has been examined in a quite similar way. If $b \neq 0$ then we have again shown that any Q-conditional symmetry operators cannot be derived. If $b = 0$ then the nonlinear RDC

$$v_{xx} = v^{-1}v_t - \lambda v v_x - \frac{1}{9}\lambda^2 v^3 - \lambda_2 v^{-1} - \lambda_1 v - \lambda_0 \quad (3.191)$$

was derived, which is Q-conditionally invariant w.r.t. the operator

$$Q = \partial_t + av\partial_x + \Big[-\frac{1}{3}\lambda av^3 + (a_x - a^2)v^2 + \alpha v + \lambda_2\Big]\partial_v, \quad (3.192)$$

where the functions $a(t,x)$ and $\alpha(t,x)$ satisfy the nonlinear system (3.169). One easily checks that substitution (3.172) reduces Eq. (3.191) and operator (3.192) to those (3.167) and (3.168), respectively.

The proof is now completed. □

Remark 3.9 *System (3.166) with $\lambda = 0$ can be easily integrated and its solutions lead to the Q-conditional symmetries found earlier in paper [16].*

It should be noted that symmetries derived therein in cases 3 and 6 of Table 1 [16], which are treated as new nonclassical symmetries, are equivalent to the Lie symmetries presented in Cases 27 and 28 of Table 2.5 (actually the corresponding equations are equivalent if one applies FPT from Case 14 of Table 2.6).

The theorem proved above identifies all the RDC equations belonging to class (3.162), which admit Q-conditional symmetries. However, the symmetries are not presented in explicit form, hence QSC of the class in question is incomplete. In order to obtain a complete QSC, the nonlinear systems (3.166) and (3.169) should be solved. Notably, both systems (3.166) and (3.169) are compatible because they possess constant solutions, which where found earlier in [77].

Systems (3.166) and (3.169) are overdetermined systems of nonlinear PDEs. In the next subsections we present the algorithm for solving both systems. Because both systems with $\lambda = 0$ coincide and the system obtained is easily integrable, it is assumed $\lambda \neq 0$ in what follows.

3.6.1 Solving the nonlinear system (3.166)

Now our aim is to integrate the overdetermined nonlinear system (3.166), i.e., to find all possible solutions. As it was stressed in Section 3.4, there are no general methods of integrating overdetermined systems at the present time. Integration of such systems is a highly nontrivial task. We were able to derive a constructive way for integration of (3.166). As a result, the following theorem was proved.

Theorem 3.9 *[78] The RDC equation (3.164) is Q-conditionally invariant under operator (3.11) if and only if the operator has the form (3.165), i.e.,*

$$Q = \partial_t + ae^u \partial_x + \left[\alpha + (a_x - a^2 - \lambda a)e^u + \lambda_2 e^{-u}\right]\partial_u,$$

where the functions $a(t,x)$ and $\alpha(t,x)$ are listed in Table 3.1.

Proof Taking into account Theorem 3.8, the proof of the theorem is equivalent to solving the overdetermined system (3.166). Let us rewrite system (3.166) in the form

$$\begin{aligned} a_{xx} - (3a + \lambda)a_x + a^3 + \lambda a^2 + \lambda_1 a &= 0, \\ a_t + 2\alpha_x - 3a(\alpha - \lambda_0) &= 0, \\ \alpha_{xx} + \lambda \alpha_x - (a_x - a^2 - \lambda a)_t - 2a(\lambda + a)(\alpha - \lambda_0) &= 0, \\ \alpha_t + \lambda_2(a_x - a^2 - \lambda a) - \alpha(\alpha - \lambda_0) - \lambda_1 \lambda_2 &= 0, \end{aligned} \quad (3.193)$$

where $a(t,x)$ and $\alpha(t,x)$ are to-be-determined functions.

In the very beginning, we examine the simplest case when a and α are some constants, hence system (3.193) is reduced to the system of algebraic

equations. Making straightforward calculations (see some details in [77]), we obtain
$$a = \frac{-\lambda \pm \sqrt{\lambda^2 - 4\lambda_1}}{2}, \quad \alpha = \lambda_0. \tag{3.194}$$

Thus, the Q-conditional symmetry operator
$$Q = \partial_t + \frac{-\lambda \pm \sqrt{\lambda^2 - 4\lambda_1}}{2} e^u \partial_x + \left(\lambda_0 + \lambda_1 e^u + \lambda_2 e^{-u}\right) \partial_u \tag{3.195}$$

of Eq. (3.164) is derived. On the other hand, (3.194) is nothing else but Case 1 of Table 3.1. Moreover, the corresponding RDC equation is exactly (3.164).

It can be also noted that this system is easily integrated provided $a(t, x) = 0$. In fact, the first equation vanishes, the second one immediately gives $\alpha_x = 0$ and then the third equation vanishes, hence, the general solution of the fourth equation takes the form presented in case 2 of Table 3.1.

The algorithm of integrating system (3.193) with a nonconstant $a(t, x) \neq 0$ is based on the following steps. The first step is to examine the first and third equations of the system. Because the first equation is autonomous and contains the time variable as a parameter, we can integrate one in a quite similar way as it was done in [53] for the similar equation (see Eq. (4.47) in Section 4.3.2). Having the function $a(t, x)$, the unknown function $\alpha(t, x)$ can be found as follows. The second equation of system (3.193) is equivalent to
$$\alpha_x = \frac{1}{2} \left[3a \left(\alpha - \lambda_0\right) - a_t\right]. \tag{3.196}$$

One also obtains α_{xx} by differentiating (3.196) w.r.t. x. Having the expressions for α_x and α_{xx}, we substitute them into the third equation of system (3.193) and solve the equation obtained w.r.t. the function $\alpha(t, x)$:
$$\alpha = \frac{\left(2\lambda a + \frac{5}{2}a^2 - 6a_x\right)_t}{2\lambda a - a^2 - 6a_x} + \lambda_0. \tag{3.197}$$

The special case $2\lambda a - a^2 - 6a_x = 0$ leads to the requirement $a = const$ and $\alpha = const$. As a result, a particular case of the Q-conditional symmetry operator (3.195) is obtained.

The second step of the algorithm is to find all possible conditions when the function $a(t, x)$ and $\alpha(t, x)$ derived from the first and third equation will satisfy second and fourth equations of system (3.193).

Nevertheless the main idea is rather simple, the corresponding calculations are nontrivial and cumbersome. The result of its application is presented in cases 3–6 of Table 3.1. The most important details are as follows.

Thus, applying the substitution
$$a(t, y) = \frac{1}{3} \left(\frac{w_y}{w} - \lambda\right), \quad x = -3y, \tag{3.198}$$

where $w = w(t, y)$ is new unknown function to the first equation of (3.193), we arrive at the linear ODE

$$w_{yyy} + 3\left(3\lambda_1 - \lambda^2\right) w_y + \lambda\left(2\lambda^2 - 9\lambda_1\right) w = 0. \qquad (3.199)$$

The characteristic equation of (3.199) can be written as follows

$$(k - \lambda)(k^2 + \lambda k + 9\lambda_1 - 2\lambda^2) = 0. \qquad (3.200)$$

Now one notes that the form of the general solution of (3.199) depends on the value $P = \lambda^2 - 4\lambda_1$, hence, three cases occur

$$\begin{aligned}&\text{(i) } P > 0;\\ &\text{(ii) } P < 0;\\ &\text{(iii) } P = 0, \; i.e., \; \lambda_1 = \tfrac{\lambda^2}{4}.\end{aligned} \qquad (3.201)$$

Case (i) leads to the roots of (3.200):

$$k_1 = \lambda, \; k_2 = \frac{-\lambda + 3\sqrt{P}}{2}, \; k_3 = \frac{-\lambda - 3\sqrt{P}}{2}.$$

If $\sqrt{P} = \pm\lambda$, i.e., $\lambda_1 = 0$ then two roots are equal and we may write $k_2 = k_1 = \lambda$, $k_3 = -2\lambda$. Thus, the general solution of ODE (3.199) takes the form

$$w = \left(\frac{1}{3}\hat{\theta} + \hat{\rho}y\right)e^{\lambda y} + \frac{1}{3}\hat{\gamma}e^{-2\lambda y}, \qquad (3.202)$$

where $\hat{\theta}(t)$, $\hat{\rho}(t)$ and $\hat{\gamma}(t)$ are arbitrary functions at the moment. Taking into account the substitution used above, we arrive at the solution of the first equation in system (3.193)

$$a = -\partial_x\left(\ln\left|\hat{\gamma}e^{\lambda x} - \hat{\rho}x + \hat{\theta}\right|\right). \qquad (3.203)$$

There are two possibilities, $\hat{\gamma} \neq 0$ and $\hat{\gamma} = 0, \hat{\rho} \neq 0$, leading to different results. If $\hat{\gamma} \neq 0$ then solution (3.203) is equivalent to

$$a = -\partial_x(\ln|e^{\lambda x} - \rho x + \theta|), \qquad (3.204)$$

where $\theta(t) = \tfrac{\hat{\theta}}{\hat{\gamma}}, \rho(t) = \tfrac{\hat{\rho}}{\hat{\gamma}}$. Having the function $a(t, x)$, we immediately calculate the function $g(t, x)$ via formula (3.197) and insert both functions into the second equation of (3.193):

$$\rho_t\left[6\lambda(\lambda\rho x - \lambda\theta - 3\rho)e^{x\lambda} + \lambda^2\rho^2 x^2 + 2\lambda\rho(2\rho - \lambda\theta)x + \right.$$
$$\left. +\lambda^2\theta^2 - 4\lambda\theta\rho + 7\rho^2\right] = 0. \qquad (3.205)$$

Equation (3.205) reduces to equality if and only if $\rho = A_1$ provided A_1 is an arbitrary constant. So, the formulae (3.204) and (3.197) take the form

$$a = -\partial_x\left(\ln\left|e^{\lambda x} - A_1 x + \theta\right|\right), \quad \alpha = \lambda_0 - \partial_t\left(\ln\left|e^{\lambda x} - A_1 x + \theta\right|\right). \tag{3.206}$$

Now we substitute the functions $a(t,x)$ and $\alpha(t,x)$ derived into the last equation of (3.193). Making the relevant calculations, one obtains the linear ODE

$$\theta_{tt} - \lambda_0 \theta_t + A_1 \lambda \lambda_2 = 0 \tag{3.207}$$

with the general solution

$$\theta = A_2 e^{\lambda_0 t} + \frac{A_1 \lambda \lambda_2}{\lambda_0} t + A_3, \tag{3.208}$$

if $\lambda_0 \neq 0$ and

$$\theta = -\frac{1}{2} A_1 \lambda \lambda_2 t^2 + A_2 t + A_3, \tag{3.209}$$

if $\lambda_0 = 0$. Thus, substituting (3.208) and (3.209) into (3.206), one arrives exactly at case 3 (with $A_0 = 1$) of Table 3.1. Notably, A_0 can be reduced either to 1, or to 0 without losing a generality.

If $\hat{\gamma} = 0$ then formulae (3.203) and (3.197) simplifies to the form

$$a = -\partial_x\left(\ln|\theta - x|\right), \quad \alpha = \lambda_0 - \partial_t\left(\ln|\theta - x|\right). \tag{3.210}$$

Dealing with formulae (3.210) in the same way as above, we conclude that the last equation of (3.193) again takes the form (3.207) but with the fixed value $A_1 = 1$. Thus, we obtain the function θ of the form (3.208) or (3.209) with $A_1 = 1$. Finally, case 3 of Table 3.1 under restriction $A_0 = 0$, $A_1 = 1$ was derived (A_1 can be reduced to 1 without losing a generality provided $A_0 = 0$).

If $\lambda_1 \neq 0$ then three roots of (3.200) are real and different, so that the general solution of ODE (3.199) takes the form

$$w = \left(\hat{\rho} e^{-\frac{3}{2}\sqrt{P}y} + \hat{\gamma} e^{\frac{3}{2}\sqrt{P}y}\right) e^{-\frac{\lambda}{2}y} + \hat{\theta} e^{\lambda y}.$$

Taking into account substitution (3.198), we find the general solution of the first equation of system (3.193):

$$a = -\partial_x\left[\ln\left|\hat{\theta} + e^{\frac{\lambda}{2}x}(\hat{\rho} e^{\frac{1}{2}\sqrt{P}x} + \hat{\gamma} e^{-\frac{1}{2}\sqrt{P}x})\right|\right]. \tag{3.211}$$

Now we need to consider two possibilities, $\hat{\rho} \neq 0$ and $\hat{\rho} = 0, \hat{\gamma} \neq 0$.

Assuming $\hat{\rho} \neq 0$, solution (3.211) can be rewritten in the form

$$a = -\partial_x\left[\ln\left|\theta + e^{\frac{\lambda}{2}x}(e^{\frac{1}{2}\sqrt{P}x} + \gamma e^{-\frac{1}{2}\sqrt{P}x})\right|\right], \tag{3.212}$$

where $\theta(t) = \frac{\hat{\theta}}{\hat{\rho}}, \gamma(t) = \frac{\hat{\gamma}}{\hat{\rho}}$. Having the function $a(t,x)$, we immediately calculate the function $\alpha(t,x)$ via formula (3.197) and insert both functions into the second equation of (3.193). Omitting the corresponding calculations, we present the final expression:

$$\gamma_t \left[\left(\lambda^2 - 9P\right)\theta^2 e^{\sqrt{P}x} + 2\left(3P + 16\sqrt{P}\lambda + 5\lambda^2\right)\theta e^{\frac{1}{2}(3\sqrt{P}+\lambda)x} + \right.$$
$$+ 2\left(3P - 16\sqrt{P}\lambda + 5\lambda^2\right)\theta\gamma e^{\frac{1}{2}(\sqrt{P}+\lambda)x} + \left(P + 4\sqrt{P}\lambda - 5\lambda^2\right)e^{(2\sqrt{P}+\lambda)x} +$$
$$\left. + 2\left(29P - 5\lambda^2\right)\gamma e^{(\sqrt{P}+\lambda)x} + \left(P - 4\sqrt{P}\lambda - 5\lambda^2\right)\gamma^2 e^{\lambda x}\right] = 0.$$

Because x is the independent variable this expression reduces to equality if and only if $\gamma = A_1$ (A_1 is an arbitrary constant). So, the formulae (3.212) and (3.197) take the forms

$$a = -\partial_x \left[\ln\left|\theta + e^{\frac{\lambda}{2}x}\left(e^{\frac{1}{2}\sqrt{P}x} + A_1 e^{-\frac{1}{2}\sqrt{P}x}\right)\right|\right],$$
$$\alpha = \lambda_0 - \partial_t \left[\ln\left|\theta + e^{\frac{\lambda}{2}x}\left(e^{\frac{1}{2}\sqrt{P}x} + A_1 e^{-\frac{1}{2}\sqrt{P}x}\right)\right|\right]. \quad (3.213)$$

Now we substitute the derived functions $a(t,x)$ and $\alpha(t,x)$ into the fourth equation of (3.193) and obtain the linear ODE

$$\theta_{tt} - \lambda_0 \theta_t + \lambda_1 \lambda_2 \theta = 0. \quad (3.214)$$

Depending on the value $D = \lambda_0^2 - 4\lambda_1\lambda_2$, three types of the general solutions of Eq. (3.214) are constructed:

1. $D = 0$, $\theta = e^{\frac{\lambda_0}{2}t}(A_2 t + A_3)$; $\quad (3.215)$
2. $D > 0$, $\theta = e^{\frac{\lambda_0}{2}t}\left(A_2 e^{-\frac{\sqrt{D}}{2}t} + A_3 e^{\frac{\sqrt{D}}{2}t}\right)$; $\quad (3.216)$
3. $D < 0$, $\theta = e^{\frac{\lambda_0}{2}t}\left[A_2 \cos\left(\frac{\sqrt{-D}}{2}t\right) + A_3 \sin\left(\frac{\sqrt{-D}}{2}t\right)\right]$. $\quad (3.217)$

Substituting (3.215)–(3.217) into (3.213), one arrives exactly at case 4 (with $A_0 = 1$) of Table 3.1.

Assuming $\hat{\rho} = 0$, formulae (3.211) and (3.197) give the functions

$$a = -\partial_x\left(\ln\left|\theta + e^{\frac{1}{2}(\lambda - \sqrt{P})x}\right|\right), \quad \alpha = \lambda_0 - \partial_t\left(\ln\left|\theta + e^{\frac{1}{2}(\lambda - \sqrt{P})x}\right|\right). \quad (3.218)$$

Substituting (3.218) into the last equation of (3.193), we again obtain ODE (3.214). Finally, case 4 of Table 3.1 under restriction $A_0 = 0$, $A_1 = 1$ was derived.

Case (ii) leads to the roots of (3.200):

$$k_1 = \lambda, \quad k_{2,3} = -\frac{\lambda}{2} \pm \frac{3}{2}i\sqrt{-P}, \quad i = \sqrt{-1}.$$

Thus, the general solution of (3.199) takes the form

$$w(t,y) = \left[\hat{\rho}\cos\left(\frac{3}{2}\sqrt{-P}y\right) + \hat{\gamma}\sin\left(\frac{3}{2}\sqrt{-P}y\right)\right]e^{-\frac{\lambda}{2}y} + \hat{\theta}e^{y\lambda},$$

Q-conditional symmetry of RDC equations with exponential diffusivity 123

hence, the first equation in (3.193) possesses the solution

$$a = -\partial_x \left[\ln \left| \hat{\theta} + e^{\frac{\lambda}{2}x} \left(\hat{\rho} \cos\left(\frac{1}{2}\sqrt{-P}x\right) - \hat{\gamma} \sin\left(\frac{1}{2}\sqrt{-P}x\right) \right) \right| \right]. \qquad (3.219)$$

The further analysis can be done in the quite similar way as for case (i). So, we present the final form of the functions $a(t,x)$ and $\alpha(t,x)$:

$$a = -\partial_x \left[\ln \left| \theta + e^{\frac{\lambda}{2}x} \left(A_0 \cos(\tfrac{1}{2}\sqrt{-P}x) - A_1 \sin(\tfrac{1}{2}\sqrt{-P}x) \right) \right| \right],$$

$$\alpha = \lambda_0 - \partial_t \left[\ln \left| \theta + e^{\frac{\lambda}{2}x} \left(A_0 \cos(\tfrac{1}{2}\sqrt{-P}x) - A_1 \sin(\tfrac{1}{2}\sqrt{-P}x) \right) \right| \right],$$

where $A_0^2 + A_1^2 \neq 0$ and θ is determined in (3.215)–(3.217).

Thus, case 5 of Table 3.1 has been derived.

Case (iii) leads to the roots of (3.200): $k_1 = \lambda$, $k_2 = -\frac{\lambda}{2}$, $k_3 = -\frac{\lambda}{2}$. Thus, the general solution of (3.199) is

$$w = \frac{1}{6}\hat{\theta} e^{\lambda y} + \left(\frac{1}{3}\hat{\rho} + \hat{\gamma} y\right) e^{-\frac{\lambda}{2}y},$$

and one generates the solution

$$a = -\partial_x \left(\ln \left| \hat{\theta} - 2(\hat{\gamma} x - \hat{\rho}) e^{\frac{\lambda}{2}x} \right| \right) \qquad (3.220)$$

of the first equation in (3.193).

We need to analyze again two possibilities, $\hat{\gamma} \neq 0$ and $\hat{\gamma} = 0, \hat{\rho} \neq 0$.

Let us assume $\hat{\gamma} \neq 0$. Then the second equation of (3.193) can be satisfied if and only if

$$a = -\partial_x \left(\ln \left| \theta - 2(x - A_1) e^{\frac{\lambda}{2}x} \right| \right) \qquad (3.221)$$

and

$$\alpha = \lambda_0 - \partial_t \left(\ln \left| \theta - 2(x - A_1) e^{\frac{\lambda}{2}x} \right| \right). \qquad (3.222)$$

Substituting (3.221) and (3.222) into the fourth equation of (3.193), one derives the linear ODE

$$\theta_{tt} - \lambda_0 \theta_t + \frac{\lambda^2 \lambda_2}{4}\theta = 0. \qquad (3.223)$$

Depending on the value $R = \lambda_0^2 - \lambda^2 \lambda_2$, three types of the general solutions of ODE (3.223) occur:

1. $R = 0$, $\theta = e^{\frac{\lambda_0}{2}t}(A_2 t + A_3)$; $\qquad (3.224)$
2. $R > 0$, $\theta = e^{\frac{\lambda_0}{2}t}(A_2 e^{-\frac{1}{2}\sqrt{R}t} + A_3 e^{\frac{1}{2}\sqrt{R}t})$; $\qquad (3.225)$
3. $R < 0$, $\theta = e^{\frac{\lambda_0}{2}t}\left[A_2 \sin\left(\frac{1}{2}\sqrt{-R}t\right) + A_3 \cos\left(\frac{1}{2}\sqrt{-R}t\right)\right]$. (3.226)

TABLE 3.1: A complete list of exact solutions of the nonlinear system (3.166)

	The explicit forms of $a(t,x)$ and $\alpha(t,x)$
1.	$a = \dfrac{-\lambda \pm \sqrt{\lambda^2 - 4\lambda_1}}{2}$, $\alpha = \lambda_0$
2.	$a = 0$, $\alpha = \begin{cases} \dfrac{\lambda_0 \pm \sqrt{D}}{2}, & D \equiv \lambda_0^2 - 4\lambda_1\lambda_2, \\ -\dfrac{1}{t} + \dfrac{\lambda_0}{2}, & D = 0, \\ \dfrac{\sqrt{-D}}{2} \tan\left(\dfrac{\sqrt{-D}}{2} t\right) + \dfrac{\lambda_0}{2}, & D < 0, \\ -\dfrac{\sqrt{D}}{2} \coth(\dfrac{\sqrt{D}}{2} t) + \dfrac{\lambda_0}{2}, & (2\alpha - \lambda_0)^2 > D > 0, \\ -\dfrac{\sqrt{D}}{2} \tanh(\dfrac{\sqrt{D}}{2} t) + \dfrac{\lambda_0}{2}, & D > (2\alpha - \lambda_0)^2 > 0 \end{cases}$
3.	$a = -\partial_x \left(\ln \left| A_0 e^{\lambda x} - A_1 x + \theta(t) \right| \right)$, $\alpha = \lambda_0 - \partial_t \left(\ln \left| A_0 e^{\lambda x} - A_1 x + \theta(t) \right| \right)$
4.	$a = -\partial_x \left(\ln \left| \theta(t) + e^{\frac{\lambda}{2} x} (A_0 e^{\frac{1}{2}\sqrt{P}x} + A_1 e^{-\frac{1}{2}\sqrt{P}x}) \right| \right)$, $\alpha = \lambda_0 - \partial_t \left(\ln \left| \theta(t) + e^{\frac{\lambda}{2} x} (A_0 e^{\frac{1}{2}\sqrt{P}x} + A_1 e^{-\frac{1}{2}\sqrt{P}x}) \right| \right)$
5.	$a = -\partial_x \left(\ln \left| \theta(t) + e^{\frac{\lambda}{2} x} \left(A_0 \cos(\dfrac{\sqrt{-P}}{2} x) - A_1 \sin(\dfrac{\sqrt{-P}}{2} x) \right) \right| \right)$, $\alpha = \lambda_0 - \partial_t \left(\ln \left| \theta(t) + e^{\frac{\lambda}{2} x} \left(A_0 \cos(\dfrac{\sqrt{-P}}{2} x) - A_1 \sin(\dfrac{\sqrt{-P}}{2} x) \right) \right| \right)$
6.	$a = -\partial_x \left(\ln \left| \theta(t) - (A_0 x - A_1) e^{\frac{\lambda}{2} x} \right| \right)$, $\alpha = \lambda_0 - \partial_t \left(\ln \left| \theta(t) - (A_0 x - A_1) e^{\frac{\lambda}{2} x} \right| \right)$

Finally, we can remove the multiplier 2 in formulae (3.221) and (3.222) because the RDC equation (3.164) is invariant under x-translations. Thus, case 6 of Table 3.1 is derived.

The assumption $\hat{\gamma} = 0, \hat{\rho} \neq 0$ leads also to case 6 of Table 3.1 but under the restriction $A_0 = 0$.

Thus, all possible solutions of the overdetermined system (3.193) are constructed and they are listed in Table 3.1.

The proof is now completed. $\square$

Remark 3.10 *In Table 3.1, two arbitrary constants must satisfy the condition $A_0^2 + A_1^2 \neq 0$ (otherwise $a(t,x) = 0$). However, one may assume (without losing a generality) that either the constant $A_0 = 1$ and $A_1 \in \mathbb{R}$, or $A_0 = 0$ and $A_1 \in \mathbb{R} \setminus 0$. Moreover, the following restrictions take place in cases 3–6:*

Case 3. $P \equiv \lambda^2 - 4\lambda_1 > 0,\ \lambda_1 = 0,\ \theta(t) = \begin{cases} (3.208), & if\ \lambda_0 \neq 0, \\ (3.209), & if\ \lambda_0 = 0 \end{cases}$

Case 4. $P > 0,\ \lambda_1 \neq 0,\ \theta(t) = \begin{cases} (3.215), & if\ D = 0, \\ (3.216), & if\ D > 0, \\ (3.217), & if\ D < 0 \end{cases}$

Case 5. $P < 0,\ \theta(t) = \begin{cases} (3.215), & if\ D = 0, \\ (3.216), & if\ D > 0, \\ (3.217), & if\ D < 0 \end{cases}$

Case 6. $P = 0,\ R \equiv \lambda_0^2 - \lambda^2 \lambda_2,\ \theta(t) = \begin{cases} (3.224), & if\ R = 0, \\ (3.225), & if\ R > 0, \\ (3.226), & if\ R < 0 \end{cases}$

3.6.2 Solving the nonlinear system (3.169)

The algorithm applied to integrate the overdetermined system (3.166) does not work in the case of system (3.169) because the latter contains no autonomous equation w.r.t. either $a(t,x)$ or $\alpha(t,x)$. Thus, another algorithm for its solving was developed. The main idea of the algorithm is based on the construction of integrable ODEs for the function $a(t,x)$ and $\alpha(t,x)$ using appropriate differential consequences of PDEs from system (3.169). Having the integrable ODEs, we construct their general solution and substitute that into system (3.169). Finally, the necessary restrictions of the system coefficients should be found in order to obtain all possible solutions. We remind the reader that system (3.169) with $\lambda \neq 0$ is under study.

Theorem 3.10 *The RDC equation (3.167) is Q-conditionally invariant under operator (3.11) if and only if the operator has the form (3.168), i.e.,*

$$Q = \partial_t + ae^u \partial_x + \left[\alpha - \frac{\lambda a}{3}e^{2u} + (a_x - a^2)e^u + \lambda_2 e^{-u}\right]\partial_u,$$

where the functions $a(t,x)$ and $\alpha(t,x)$ have either the form

$$a(t,x) = a,\ \alpha(t,x) = \frac{\lambda \lambda_2}{3a} + \lambda_0, \tag{3.227}$$

where a is a root of the algebraic equation $9a^4 + 9\lambda_1 a^2 + 3\lambda\lambda_0 a + \lambda^2 \lambda_2 = 0$, or

$$a(t,x) = -\partial_x \ln|\gamma(t,x)|,\ \alpha(t,x) = -\partial_t \ln|\gamma(t,x)|,\quad \gamma(t,x) \neq 0$$

with the function γ, having one of the forms presented in Table 3.2.

Proof Taking into account Theorem 3.8, the proof of the theorem is equivalent to solving the overdetermined system (3.169), i.e.,

$$a_t + 2\alpha_x - 3a\alpha + 3\lambda_0 a + \lambda\lambda_2 = 0,$$
$$a_{xx} - 3aa_x + a^3 + \lambda_1 a + \tfrac{1}{3}\lambda\alpha = 0,$$
$$\alpha_t + \lambda_2 a_x - \alpha^2 - \lambda_2 a^2 + \lambda_0 \alpha - \lambda_1\lambda_2 = 0,$$
$$\alpha_{xx} - a_{tx} + 2aa_t - 2a^2\alpha + 2\lambda_0 a^2 + \tfrac{2}{3}\lambda\lambda_2 a = 0.$$

Among four equations of system (3.169), the second is the second-order ODE (with the variable t as a parameter), while other equations are PDEs. The crucial step is to reduce this equation to the first-order ODE using differential consequences of other equations. Let us differentiate the second equation of (3.169) w.r.t. the variable t:

$$a_{txx} - 3aa_{tx} - 3a_x a_t + 3a^2 a_t + \lambda_1 a_t + \tfrac{1}{3}\lambda\alpha_t = 0. \tag{3.228}$$

All the time derivatives in Eq. (3.228) can be excluded using the third equation of (3.169), the first and second differential consequences (w.r.t. x) of the first equation and the first differential consequence (w.r.t. x) of the fourth equation. As a result, we obtain the equation

$$(24a_x - 10a^2 - 9\lambda_1)(\alpha_x - a\alpha + \lambda_0 a + \tfrac{1}{3}\lambda\lambda_2) = 0.$$

Thus, two possibilities should be examined. Assuming $24a_x - 10a^2 - 9\lambda_1 = 0$, one obtains by straightforward calculations that the functions a and α must be constants. So, system (3.169) reduces to that of algebraic equations with the solution (3.227).

Let us assume that the second possibility, i.e., the equality

$$\alpha_x - a(\alpha - \lambda_0)a + \tfrac{1}{3}\lambda\lambda_2 = 0$$

takes place. In this case, system (3.169) can be simplified to the form

$$\begin{aligned}
\alpha_x &= a(\alpha - \lambda_0) - \tfrac{1}{3}\lambda\lambda_2, \\
a_t &= a(\alpha - \lambda_0) - \tfrac{1}{3}\lambda\lambda_2, \\
a_{xx} &- 3aa_x + a^3 + \lambda_1 a + \tfrac{1}{3}\lambda\alpha = 0, \\
\alpha_t &+ \lambda_2 a_x - \alpha^2 - \lambda_2 a^2 + \lambda_0 \alpha - \lambda_1\lambda_2 = 0.
\end{aligned} \tag{3.229}$$

Since $a_t = \alpha_x$, we may introduce the so-called stream function $\gamma(t, x)$ as follows

$$a = -\partial_x \ln\gamma(t,x), \quad \alpha = -\partial_t \ln\gamma(t,x), \quad \gamma(t,x) \neq 0 \tag{3.230}$$

(generally speaking, one should write $|\gamma(t,x)|$ in the above formulae, however the results are the same because of the differentiation operation). Using

Eq. (3.230), the first and the third equations of the system (3.229) are transformed to the form

$$\gamma_{tx} + \lambda_0 \gamma_x - \frac{1}{3}\lambda\lambda_2 \gamma = 0, \quad \gamma_{xxx} + \lambda_1 \gamma_x + \frac{1}{3}\lambda\gamma_t = 0. \tag{3.231}$$

Having the system of linear PDEs (3.231), one easily derives the fourth-order ODE (with t as a parameter)

$$\gamma_{xxxx} + \lambda_1 \gamma_{xx} - \frac{1}{3}\lambda\lambda_0 \gamma_x + \frac{1}{9}\lambda^2 \lambda_2 \gamma = 0. \tag{3.232}$$

Now we should construct all possible solutions of Eq. (3.232) depending on three arbitrary parameters $\lambda \neq 0$, λ_0 and λ_1. Because the corresponding characteristic equation is

$$k^4 + \lambda_1 k^2 - \frac{1}{3}\lambda\lambda_0 k + \frac{1}{9}\lambda^2 \lambda_2 = 0, \tag{3.233}$$

it is a standard routine to establish that nine different cases occur and each of them lead to the general solution of Eq. (3.232). Now we present the list of these solutions together with the relevant restrictions on the coefficients.

1. If four different real roots:
$(k - k_1)(k - k_2)(k - k_3)(k - k_4) = 0$, $k_1 + k_2 + k_3 + k_4 = 0$, $\lambda\lambda_0 = -3(k_1 + k_2)(k_1 + k_3)(k_2 + k_3)$, $\lambda_1 = -k_1^2 - k_2^2 - k_3^2 - k_1 k_2 - k_1 k_3 - k_2 k_3$, $\lambda^2 \lambda_2 = -9 k_1 k_2 k_3 (k_1 + k_2 + k_3)$
then
$$\gamma = A_1(t) e^{k_1 x} + A_2(t) e^{k_2 x} + A_3(t) e^{k_3 x} + A_4(t) e^{k_4 x}. \tag{3.234}$$

2. If three different real roots and one of them occurs twice:
$(k - k_1)^2 (k - k_3)(k - k_4) = 0$, $2k_1 + k_3 + k_4 = 0$, $\lambda\lambda_0 = -6k_1(k_1 + k_3)^2$, $\lambda_1 = -2k_1^2 - (k_1 + k_3)^2$, $\lambda^2 \lambda_2 = -9 k_1^2 k_3 (2k_1 + k_3)$
then
$$\gamma = [A_1(t) + A_2(t)x] e^{k_1 x} + A_3(t) e^{k_3 x} + A_4(t) e^{k_4 x}.$$

3. If two different real roots and one of them occurs three times:
$(k - k_1)^3 (k - k_4) = 0$, $k_4 = -3k_1$, $\lambda\lambda_0 = -24 k_1^3$, $\lambda_1 = -6k_1^2$, $\lambda^2 \lambda_2 = -27 k_1^4$
then
$$\gamma = [A_1(t) + A_2(t)x + A_3(t)x^2] e^{k_1 x} + A_4(t) e^{k_4 x}.$$

4. If a single real root occurring four times:
$(k - k_1)^4 = 0$, $k_1 = 0$, $\lambda_0 = \lambda_1 = \lambda_2 = 0$
then
$$\gamma = A_4(t) x^3 + A_3(t) x^2 + A_2(t) x + A_1(t).$$

5. If two different real roots and each of them occurs twice:
$(k - k_1)^2 (k - k_3)^2 = 0$, $k_1 + k_3 = 0$, $\lambda_0 = 0$, $\lambda_1 = -2k_1^2$, $\lambda^2 \lambda_2 = 9 k_1^4$
then
$$\gamma = [A_1(t) + A_2(t)x] e^{k_1 x} + [A_3(t) + A_4(t)x] e^{k_3 x}.$$

6. If two different real roots and two complex roots:
$(k - k_1)(k - k_2)(k - \alpha - \imath\beta)(k - \alpha + \imath\beta) = 0$, $\imath^2 = -1$, $k_1 + k_2 + 2\alpha = 0$,
$\lambda\lambda_0 = -6\alpha[(\alpha + k_1)^2 + \beta^2]$, $\lambda_1 = -3\alpha^2 + \beta^2 - k_1(2\alpha + k_1)$,
$\lambda^2\lambda_2 = -9(\alpha^2 + \beta^2)k_1(2\alpha + k_1)$
then

$$\gamma = A_1(t)e^{k_1 x} + A_2(t)e^{k_2 x} + e^{\alpha x}[A_3(t)\sin(\beta x) + A_4(t)\cos(\beta x)].$$

7. If two complex roots and a single real root occurring twice:
$(k - k_1)^2(k - \alpha - \imath\beta)(k - \alpha + \imath\beta) = 0$, $\alpha + k_1 = 0$, $\lambda\lambda_0 = 6k_1\beta^2$,
$\lambda_1 = \beta^2 - 2k_1^2$, $\lambda^2\lambda_2 = 9k_1^2(\beta^2 + k_1^2)$
then

$$\gamma = [A_1(t) + A_2(t)x]e^{k_1 x} + e^{\alpha x}[A_3(t)\sin(\beta x) + A_4(t)\cos(\beta x)].$$

8. If four different complex roots:
$(k - \alpha_1 - \imath\beta_1)(k - \alpha_1 + \imath\beta_1)(k - \alpha_2 - \imath\beta_2)(k - \alpha_2 + \imath\beta_2) = 0$, $\alpha_2 = -\alpha_1$,
$\lambda\lambda_0 = 6\alpha_1(\beta_2^2 - \beta_1^2)$, $\lambda_1 = \beta_1^2 + \beta_2^2 - 2\alpha_1^2$, $\lambda^2\lambda_2 = 9(\alpha_1^2 + \beta_1^2)(\alpha_1^2 + \beta_2^2)$
then

$$\gamma = e^{\alpha_1 x}[A_1(t)\sin(\beta x) + A_2(t)\cos(\beta x)] + e^{\alpha_2 x}[A_3(t)\sin(\beta x) + A_4(t)\cos(\beta x)].$$

9. If two different complex roots and each of them occurs twice:
$(k - \alpha - \imath\beta)^2(k - \alpha + \imath\beta)^2 = 0$, $\alpha = 0$, $\lambda_0 = 0$, $\lambda_1 = 2\beta^2$, $\lambda^2\lambda_2 = 9\beta^4$
then

$$\gamma = [A_1(t) + A_2(t)x]\sin(\beta x) + [A_3(t) + A_4(t)x]\cos(\beta x).$$

In the above formulae, $A_1, \ldots, A_4$ are arbitrary smooth functions at the moment.

In order to solve the linear system (3.231), one substitutes the functions $\gamma(t, x)$ obtained above into the second equation of the system. The equations obtained can be split w.r.t. the relevant functionally independent functions of the variable x. As a result, a four-dimensional system of linear first-order ODEs will be obtained for each form of $\gamma(t, x)$. For example, if one takes $\gamma(t, x)$ of the form (3.234) then the corresponding ODE system has the form

$$\frac{\lambda}{3}\frac{dA_i}{dt} + k_i^3 A_i(t) + k_i\lambda_1 A_i(t) = 0, \ i = 1, 2, 3, 4. \tag{3.235}$$

Obviously, the general solution of system (3.235) can be easily constructed:

$$A_i(t) = C_i \exp\left[-\frac{3}{\lambda}k_i(k_i^2 + \lambda_1)t\right], \ i = 1, 2, 3, 4, \tag{3.236}$$

where $C_1, \ldots C_4$ are arbitrary constants. Finally, we substitute (3.236) into (3.234) and obtain the function $\gamma(t, x)$ presented in case 1 of Table 3.2. Examination of the other 8 forms of the function $\gamma(t, x)$ leads exactly to cases 2–9 of Table 3.2.

It turns out that the fourth equation of (3.229) is a differential consequence of other equations, therefore there is no need to examine that. In fact, substituting (3.230) into the fourth equation of (3.229), we obtain

$$\gamma_{tt} + \lambda_0 \gamma_t + \lambda_2 \gamma_{xx} + \lambda_1 \lambda_2 \gamma = 0. \tag{3.237}$$

The derivatives γ_t and γ_{tt} can be eliminated in (3.237), using the second equation of (3.231) and its differential consequence w.r.t. t:

$$\gamma_t = -\frac{3}{\lambda}(\gamma_{xxx} + \lambda_1 \gamma_x), \quad \gamma_{tt} = -\frac{3}{\lambda}(\gamma_{txxx} + \lambda_1 \gamma_{tx}).$$

Substituting the above expressions into (3.237) we derive

$$\gamma_{txxx} + \lambda_0 \gamma_{xxx} - \frac{1}{3}\lambda\lambda_2 \gamma_{xx} - \lambda_1 \left(\frac{1}{3}\lambda\lambda_2 \gamma - \lambda_0 \gamma_x - \gamma_{tx}\right) = 0. \tag{3.238}$$

Finally, the time derivatives γ_{tx} and γ_{txxx} can be eliminated using the first equation of (3.231) and its second-order differential consequence w.r.t. x. The simple calculations lead to

$$\gamma_{tx} = \frac{1}{3}\lambda\lambda_2 \gamma - \lambda_0 \gamma_x, \quad \gamma_{txxx} = \frac{1}{3}\lambda\lambda_2 \gamma_{xx} - \lambda_0 \gamma_{xxx}.$$

Now one realizes that (3.238) vanishes.

The proof is now completed. □

Remark 3.11 *In Table 3.2, the coefficients k_i, $i = 1, \ldots, 4$ are the roots of polynomial (3.233) and related with the coefficients $\lambda, \lambda_0, \lambda_1$ and λ_2 (see items 1,..., 9 after (3.233)).*

3.7 Nonlinear equations arising in applications and their conditional symmetry

In Sections 3.4–3.6, several RDC equations admitting Q-conditional symmetries were found. Here we identify some equations among them, which are used in real world applications, in order to establish their symmetry. We restrict ourselves on some well-known examples only and the reader may examine other nonlinear RDC arising in applications.

The most interesting results from the applicability point of view are presented in Theorem 3.2. The general form of the RDC equations listed in the theorem can be written as Eq. (3.80), i.e.,

$$u_t = u_{xx} + \lambda u u_x + \lambda_0 + \lambda_1 u + \lambda_2 u^2 + \lambda_3 u^3.$$

TABLE 3.2: A complete list of the solutions $a = -\partial_x \ln|\gamma|$, $\alpha = -\partial_t \ln|\gamma|$ of system (3.169)

	The explicit forms of $\gamma(t,x)$
1.	$\gamma = \sum_{i=1}^{4} C_i \exp\left[k_i\left(x - \frac{3(k_i^2 + \lambda_1)}{\lambda}t\right)\right]$
2.	$\gamma = C_3 \exp\left[k_3\left(x - \frac{3(k_3^2 + \lambda_1)}{\lambda}t\right)\right] + C_4 \exp\left[k_4\left(x - \frac{3(k_4^2 + \lambda_1)}{\lambda}t\right)\right] +$ $+ \left[C_2\left(x - (3k_1^2 + \lambda_1)t\right) + C_1\right] \exp\left[k_1\left(x - \frac{3(k_1^2 + \lambda_1)}{\lambda}t\right)\right]$
3.	$\gamma = C_4 \exp\left[-3k_1\left(x - \frac{9}{\lambda}k_1^2 t\right)\right] +$ $+ \left[C_3\left(\frac{\lambda}{3}x + 3k_1^2 t\right)^2 + C_2\left(\frac{\lambda}{3}x + 3k_1^2 t\right) - 2C_3 k_1 \lambda t + C_1\right] \exp[k_1\left(x + \frac{15}{\lambda}k_1^2 t\right)]$
4.	$\gamma = C_4 \lambda x^3 + C_3 x^2 + C_2 x + C_1 - 18 C_4 t$
5.	$\gamma = \left[C_2\left(\frac{\lambda}{3}x - k_1^2 t\right) + C_1\right] \exp[k_1\left(x + \frac{3}{\lambda}k_1^2 t\right)] +$ $+ \left[C_4\left(\frac{\lambda}{3}x - k_1^2 t\right) + C_3\right] \exp[-k_1\left(x + \frac{3}{\lambda}k_1^2 t\right)]$
6.	$\gamma = C_1 \exp\left[k_1\left(x - \frac{3(k_1^2 + \lambda_1)}{\lambda}t\right)\right] + C_2 \exp\left[k_2\left(x - \frac{3(k_2^2 + \lambda_1)}{\lambda}t\right)\right] +$ $+ C_3 \sin\left[\beta\left(x - \frac{3(3\alpha^2 - \beta^2 + \lambda_1)}{\lambda}t\right) - C_4\right] \exp\left[\alpha\left(x - \frac{3(\alpha^2 - 3\beta^2 + \lambda_1)}{\lambda}t\right)\right]$
7.	$\gamma = C_3 \sin\left[\beta\left(x - k_1^2 t\right) + C_4\right] \exp\left[-k_1\left(x + \frac{3(2\beta^2 + k_1^2)}{\lambda}t\right)\right] +$ $+ \left[C_2\left(\frac{\lambda}{3}x - (\beta^2 + k_1^2)t\right) + C_1\right] \exp\left[k_1\left(x + \frac{3(k_1^2 - \beta^2)}{\lambda}t\right)\right]$
8.	$\gamma = C_1 \exp\left[\alpha_1\left(x + \frac{3(\alpha_1^2 + 2\beta_1^2 - \beta_2^2)}{\lambda}t\right)\right] \sin\left[\beta_1\left(x - \frac{3(\alpha_1^2 + \beta_2^2)}{\lambda}t\right) + C_2\right] +$ $+ C_3 \exp\left[-\alpha_1\left(x + \frac{3(\alpha_1^2 - \beta_1^2 + 2\beta_2^2)}{\lambda}t\right)\right] \sin\left[\beta_2\left(x - \frac{3(\alpha_1^2 + \beta_1^2)}{\lambda}t\right) + C_4\right]$
9.	$\gamma = C_1 \sin\left[\beta\left(x - \frac{3\beta^2}{\lambda}t\right) + C_2\right] + C_3 \left(\frac{\lambda}{3}x + \beta^2 t\right) \sin\left[\beta\left(x - \frac{3\beta^2}{\lambda}t\right) + C_4\right]$

Setting $\lambda_i = 0, i = 0, \ldots, 3$, one obtains the Burgers equation arising in several applications (see Section 2.6). As it follows from Theorem 3.2, the Burgers equation admits an infinite-dimensional set of Q-conditional symmetries of the form (3.60). On the other hand, the Burgers equation is linearizable by the Cole–Hopf transformation, therefore one may claim that its reach Q-conditional symmetry (Lie symmetry as well) is a consequence of the integrability.

Setting $\lambda_0 = \lambda_3 = 0$ and $\lambda_2 \to -\lambda_2$, one obtains the equation

$$u_t = u_{xx} + \lambda u u_x + u(\lambda_1 - \lambda_2 u). \tag{3.239}$$

Eq. (3.239) with $\lambda_1 > 0, \lambda_2 > 0$ was introduced in [179] as a generalization of the famous Fisher equation

$$u_t = u_{xx} + u(\lambda_1 - \lambda_2 u), \tag{3.240}$$

by adding the simplest convection term therefore it can be named the Murray equation. Eq. (3.239) is a typical equation in mathematical biology involving three transport mechanisms and was extensively studied in [179, 180]. As it was already noted in Section 2.6, this equation is invariant w.r.t. the trivial Lie algebra spanned by the time and space translation operators P_t and P_x. On the other hand, according to Theorem 3.2 (ETs from set (3.59) should be used), this equation admits the Q-conditional symmetry operator

$$Q = \partial_t - (\lambda u + \frac{\lambda_2}{\lambda})\partial_x + (\lambda_1 u - \lambda_2 u^2)\partial_u. \tag{3.241}$$

This operator was identified for the first time in [80]. Notably, operator (3.241) is not valid for the Fisher equation because $\lambda \neq 0$ in the above formula. Moreover, the Fisher equation does not admit any operator of Q-conditional symmetry of the form (3.11).

Theorem 3.5 (see Eq. (3.106) with $m = 1$) says that a natural generalization of the Murray equation

$$u_t = (u u_x)_x + \lambda u u_x + u(\lambda_1 - \lambda_2 u),$$

which is called the Murray equation with porous diffusivity, is conditionally invariant. However, the relevant operator is equivalent to the Lie symmetry discussed in Section 2.6 (it can be easily shown using Property 3).

The Murray equation with the fast diffusion (i.e., power-law diffusivity with a negative exponent) can be also obtained from Eq. (3.106). Indeed, setting $m = -2, \lambda_3 = 0, \lambda_1 > 0$ and $\lambda_2 \to -\lambda_2 < 0$, we arrive at the equation

$$u_t = \left(u^{-2} u_x\right)_x + \lambda u^{-2} u_x + u(\lambda_1 - \lambda_2 u),$$

which admits the operator

$$Q = \partial_t + (\lambda_1 u + \lambda_2 u^2)\partial_u,$$

which is a pure Q-conditional symmetry. It should be noted that the above equation with $\lambda_1 = \lambda_2 = 0$ arises in applications and is linearizable by the known integral substitution [102].

Let us consider the Fitzhugh–Nagumo (FN) equation

$$u_t = u_{xx} + u(u - \delta)(1 - u), \ 0 < \delta < 1, \tag{3.242}$$

which is a simplification of the classical model describing nerve impulse propagation [100, 182, 185]. This nonlinear RD equation was extensively studied in [17, 88, 191] using the nonclassical method. The Q-conditional symmetry operator

$$Q = 2\partial_t + \sqrt{2}\left(3u - \delta - 1\right)\partial_x + 3u(u - \delta)(1 - u)\partial_u$$

of the FN equation was found and applied for constructing exact solutions.

In the case $\delta = 0$, Eq. (3.242) becomes the known Huxley equation [43, 88] (terminology "Zeldovich equation" [92] is also used)

$$u_t = u_{xx} + u^2(1 - u). \tag{3.243}$$

Equation (3.243) is reduced to the equation

$$u_t = u_{xx} + u(1 - u)^2 \tag{3.244}$$

by the transformation (here $i^2 = -1$)

$$t \to -t, \ x \to ix, \ u \to 1 - u.$$

This equation was introduced in [156] (see formulae (13) and (16) therein) in order to generalize Fisher's model [98, 99] describing the spread in space of a favored gene in a population, hence (3.244) can be called Kolmogorov–Petrovskii–Piskunov (KPP) equation. Notably, some authors wrongly refer to the Fisher equation as Fisher–KPP equation although Eq. (3.240) does not occur in [156] (see also English translation in [195]).

Finally, we point out that Eq. (3.80) contains as a particular case the equation

$$u_t = u_{xx} + \lambda_1 u + \lambda_3 u^3.$$

In the case $\lambda_1 > 0$ and $\lambda_3 < 0$, this equation is nothing else but the Newell–Whitehead equation occurring in fluid dynamics [187] (u is a complex function in the original equation).

All the above RD equations with cubic nonlinearities can be generalized to the relevant RDC equations by adding the Burgers term $\lambda u u_x$ in order to take into account possible convection/advection. According to Theorem 3.2, the equations obtained are still Q-conditionally invariant. In fact, let us consider the equations

$$u_t = u_{xx} + \lambda u u_x + \lambda_3 u(u - \delta)(1 - u), \ \lambda_3 > 0, \ 0 < \delta < 1 \tag{3.245}$$

and
$$u_t = u_{xx} + \lambda u u_x + \lambda_3 u^2(1-u), \quad \lambda_3 > 0, \qquad (3.246)$$
which can be refereed as the generalized Fitzhugh–Nagumo equation, and the generalized Huxley equation, respectively.

Both equations (3.245) and (3.246) can be reduced to the form ($\delta = 0$ for Eq. (3.246))
$$v_t = v_{yy} + \lambda v v_y + \lambda_0 + \lambda_1 v - \lambda_3 v^3, \qquad (3.247)$$
where
$$\lambda_1 = \lambda_3 \left[\frac{1}{3}(\delta+1)^2 - \delta\right], \quad \lambda_0 = \frac{\lambda_3}{3}(\delta+1)\left[\frac{2}{9}(\delta+1)^2 - \delta\right], \qquad (3.248)$$
by the transformation
$$v(t,y) = u - \frac{\delta+1}{3}, \quad y = x + \frac{\lambda}{3}(\delta+1)t, \qquad (3.249)$$
belonging to set (3.59). Thus, each of Eqs. (3.245) and (3.246) admit two Q-conditional symmetries of the form (3.65).

Chapter 4

Exact solutions of reaction-diffusion-convection equations and their applications

4.1	Classification of exact solutions from the symmetry point of view	135
4.2	Examples of exact solutions for some well-known nonlinear equations	138
4.3	Solutions of some reaction-diffusion-convection equations arising in biomedical applications	143
	4.3.1 The Fisher and Murray equations	143
	4.3.2 The Fitzhugh–Nagumo equation and its generalizations	146
4.4	Solutions of reaction-diffusion-convection equations with power-law diffusivity	156
	4.4.1 Lie's solutions of an equation with power-law diffusion and convection	156
	4.4.2 Non-Lie solutions of some equations with power-law diffusion and convection	159
4.5	Solutions of reaction-diffusion-convection equations with exponential diffusivity	176
	4.5.1 Lie's solutions of an equation with exponential diffusion and convection	176
	4.5.2 Non-Lie solutions of an equation with exponential diffusion and convection	180
	4.5.3 Application of the solutions obtained for population dynamics	188

4.1 Classification of exact solutions from the symmetry point of view

This chapter is devoted to construction of exact solutions of several nonlinear reaction-diffusion-convection (RDC) equations belonging to class (2.35). We remind the reader that there is no general theory for integrating nonlinear PDEs at the present time. The theoretical obstacle for developing the general theory follows from the well-known fact that the principle of linear superposition of solutions cannot be applied to generate new exact solutions for a given *nonlinear PDE*. Thus, the classical methods for solving linear PDEs are not applicable for solving nonlinear ones.

Of course, a change of variables can sometimes be found that transforms the given nonlinear PDE into a linear equation. There are two generic equations, the fast diffusion equation with the exponent -2 and the Burgers equation, which are linearizable and they were discussed in Section 2.6. To the best of our knowledge, any other RDC equation of the form (2.35), which can be linearized, is related to one of these equations.

Thus, construction of particular exact solutions for nonlinear RDC equations of the form (2.35) is *a nontrivial and important problem*. Finding exact solutions that have a physical, chemical or biological interpretation is of *fundamental importance*.

The plane wave solutions (in particular, traveling fronts[1]) form the most important class of the exact solutions of two-dimensional PDEs because such solutions are important from an applicability point of view. Properties of such solutions in the case of the RDC equations were extensively studied during the last decades using different mathematical techniques (see, e.g., monographs [123, 244] and references cited therein).

The corresponding ansatz is given by formula (1.29), i.e.,

$$u = \varphi(\omega), \ \omega = x - vt.$$

Of course, it is Lie's ansatz and one reduces RDC equations to the second-order ODEs of the form

$$[A(\varphi)\varphi_\omega]_\omega + [v + B(\varphi)]\varphi_\omega + C(\varphi) = 0. \tag{4.1}$$

Although this ODE cannot be solved for an arbitrary triplet (A, B, C), there are many cases when the exact solutions can be constructed in an explicit form and some of them are presented in subsequent sections. In the case of power-law triplets (A, B, C), almost all plane wave solutions (up to date 2004) are summarized in book [123](see Chapter 13 therein). In the case of the correctly-specified coefficients with a more complicated structure, integration of ODE (4.1) is a nontrivial task. Notably, the handbooks like [142, 212] can be useful.

Classification of exact solutions of nonlinear PDEs from the symmetry point of view is based on the type of symmetry allowing to construct the solution in question. All invariant solutions, i.e., those obtainable by the Lie method (see the algorithm in Section 1.3), can be called *Lie's solutions*. Exact solutions, which cannot be constructed using Lie symmetries are called *non-Lie solutions*. It is worth noting that a non-Lie ansatz does not guarantee

---

[1] A plane wave solution, which is nonnegative, bounded and satisfies the zero Neumann conditions at infinity is usually called a traveling front.

construction of non-Lie exact solutions. In other words, the non-Lie ansatz can lead only to invariant solutions, especially it happens when the nonlinear PDE in question has a reach Lie symmetry (see the examples in the next section). Notably, the researchers did not pay attention to the above "contradiction" for a long time. Probably this issue was extensively discussed for the first time in paper [48] and we present some interesting examples in this chapter.

At the present time, several types of non-Lie symmetries exist (see Section 3.1 and references therein); therefore different types of non-Lie ansätze can be constructed. Again the question arises: Do two different non-Lie ansätze lead always to different exact solutions? The answer is again negative. For example, the exact solutions of the nonlinear RDC equation (3.64) constructed in [101] via generalized conditional symmetries, are obtainable via Q-conditional symmetries (3.65) (see Subsection 4.3.2). We point out also that it was shown in [192] how to derive via symmetries of the form (3.12) the exact solutions obtained in [251] via the generalized conditional symmetry.

On the other hand, there are examples of exact solutions constructed by application of an non-Lie ansatz and it is unclear how to find these solutions using Q-conditional (nonclassical) symmetry. For example, the RDC equation

$$u_t = (u^{-\frac{3}{2}} u_x)_x + \lambda u^{-\frac{3}{2}} u_x + 4\lambda^2 u^{-\frac{1}{2}} + r u^{\frac{5}{2}}, \quad r\lambda \neq 0 \qquad (4.2)$$

possesses the exact solution [141](see example 8)

$$u = (\varphi(t) + Ce^{3\lambda x})^{-\frac{2}{3}}, \qquad (4.3)$$

where $\varphi(t)$ is a nonconstant solution of the Riccati equation $\dot\varphi + 6\lambda^2 \varphi^2 + \frac{3}{2} r = 0$. Solution (4.3) was found using the corresponding generalized conditional symmetry and is a non-Lie solution because Eq. (4.2) admits the trivial Lie algebra of invariance leading only to the plane wave anzatz (1.29). Moreover, Theorem 3.5 says that Eq. (4.2) does not admit any Q-conditional symmetry of the form (3.11). Thus, solution (4.3) is not obtainable via Q-conditional symmetries derived in Chapter 3. Notably, Eq. (4.2) is a particular case of a more general equation derived firstly in [48], which possesses a reduction leading to non-Lie solutions (see for details Section 5.2).

In our opinion, the situation is still unclear. In fact, all possible Q-conditional symmetries of a nonlinear evolution equation usually cannot be found in so-called no-go case (see operator (3.12)). Thus, one may assume that there exists a Q-conditional symmetry of the form (3.12), which produces the non-Lie solution in question. The proof of the above hypothesis lies beyond the scope of this monograph.

Thus, we distinguish two types of exact solutions, Lie's and non-Lie ones, in what follows. Moreover, it is assumed that the second type solutions can be derived by application of Q-conditional symmetries only.

4.2 Examples of exact solutions for some well-known nonlinear equations

In this section, we present some well-known exact solutions of the nonlinear equations arising in real world applications. We start from the nonlinear heat(diffusion) equation

$$u_t = [d(u)u_x]_x, \qquad (4.4)$$

which was studied in Sections 2.2 and 3.2. The principal algebra of invariance of this equation is the three-dimensional Lie algebra (2.26). Two essentially different ansätze can be constructed using the Lie symmetry operators belonging to this algebra. The first one is ansatz (1.29) for finding plane wave solutions. The second is generated by the operator $2t\partial_t + x\partial_x$, so that the invariance surface condition is $2tu_t + xu_x = 0$. Thus, one obtains the ansatz

$$u = \varphi(\omega), \quad \omega = \frac{x}{\sqrt{t}}. \qquad (4.5)$$

Both ansätze are rather trivial and known since 19th century (notably, ansatz (4.5) was firstly found by L. Boltzmann [31]). Any other linear combination of the basic operators (2.26) produces ansätze, which are reducible to (1.29) or (4.5) by the invariance transformations of Eq. (4.4) (algebra (2.26) generates the three-parameter Lie group of invariance). Formally speaking, the operator ∂_x generates the third inequivalent ansatz $u = \varphi(t)$, however one produces the constant solutions only and is not taken into account in what follows.

In the case of ansatz (1.29), the reduction equation is

$$v\varphi_\omega + [d(\varphi)\varphi_\omega]_\omega = 0, \qquad (4.6)$$

which is easily integrated and the general solution can be presented in an implicit form. As a result, the exact solutions of Eq. (4.4) have the form:

$$\int \frac{d(u)}{u+u_0} du = -v(x-vt), \ v \neq 0, \quad \int d(u)du = Cx, \ v = 0, \qquad (4.7)$$

where u_0 and C are arbitrary constants. Because the restriction $d(u) \geq 0$ takes place, unique inverse functions exist for those in the left-hand side of (4.7) with $u_0 \geq 0$. Notably, the above solutions can be rewritten in explicit forms provided $d(u)$ is correctly-specified (see examples in [73, 55] used for solving the Stefan problem).

In the case of ansatz (4.5), the reduction equation is

$$\omega\varphi_\omega + 2\left[d(\varphi)\varphi_\omega\right]_\omega = 0, \qquad (4.8)$$

which is not integrable provided the diffusivity is an arbitrary smooth function. However, this equation has known exact solutions if the function d has specified forms. Probably the first nontrivial example for $d \neq u^k$ (the case of the power-law function will be examined below), was presented in [104]–[106] for the diffusivity $(u^2 + \lambda_1 u + \lambda_0)^{-1}$. More examples can be found, e.g., in [38, 208].

The most typical form of the coefficient $d(u)$ arising in real world applications is the power-law function u^k, i.e., the equation

$$u_t = \left(u^k u_x\right)_x \tag{4.9}$$

is obtained. Its importance was already discussed in Section 2.6. As it follows from Theorem 2.2, MAI of Eq. (4.9) with $k \neq -4/3$ is the four-dimensional algebra

$$AG^4 = \langle \partial_t, \partial_x, 2t\partial_t + x\partial_x, kx\partial_x + 2u\partial_u \rangle.$$

The corresponding MGI reads as

$$\bar{t} = e^{2v_0} t + t_0, \quad \bar{x} = e^{v_0 + kv_1} x + x_0, \quad \bar{u} = e^{2v_1} u, \tag{4.10}$$

where v_0, v_1, t_0 and x_0 are arbitrary group parameters. MGI immediately leads to the formula of multiplication of the solutions

$$u = e^{-2v_1} u^* [e^{2v_0} t + t_0, e^{v_0 + kv_1} x + x_0], \tag{4.11}$$

where $u^*(t,x)$ is a given solution of Eq. (4.9). Thus, each exact solution of Eq. (4.9) can be multiplied to the four-parameter set using the above formula. In particular, formula (4.11) allows us to simplify exact solutions of the form $u(t+t_0, x+x_0)$ by removing constants t_0 and x_0 in what follows.

In order to construct a set of inequivalent ansätze, we consider the most general form of the operator belonging to AG^4, i.e., the linear combination of the basic operators

$$X = (\alpha_0 + 2\alpha_2 t)\partial_t + (\alpha_1 + \alpha_2 x + \alpha_3 kx)\partial_x + 2\alpha_3 u \partial_u,$$

where α_i, $i = 0, \ldots, 3$ are arbitrary parameters. We also assume that $|\alpha_2| + |\alpha_3| \neq 0$, otherwise ansatz (1.29) leading to the integrable reduction equation is obtained, so that the exact solutions (4.7) with $d = u^k$ are found.

The corresponding invariance surface condition takes the form

$$(\alpha_0 + 2\alpha_2 t)u_t + (\alpha_1 + \alpha_2 x + \alpha_3 kx)u_x = 2\alpha_3 u. \tag{4.12}$$

Now we can identify that there are four essentially different cases: I $\alpha_2 \neq 0, \alpha_3 = 0$, II $\alpha_2 + \alpha_3 k = 0$, III $\alpha_2 = 0$, $\alpha_3 \neq 0$, IV $\alpha_2 + \alpha_3 k \neq 0$, $\alpha_2 \alpha_3 \neq 0$.

In case I, taking into account the time and space invariance transformations from (4.10), we may set $\alpha_0 = \alpha_1 = 0$ without loosing a generality, so that ansatz (4.5) is obtained, which gives the reduction equation

$$\omega \varphi_\omega + 2\left(\varphi^k \varphi_\omega\right)_\omega = 0. \tag{4.13}$$

TABLE 4.1: Lie's ansätze and reduction equations for Eq. (4.9)

	Ansatz	Reduction equation
1	$u = \varphi(\omega),$ $\omega = x - vt$	$\left(\varphi^k \varphi_\omega\right)_\omega = -v\varphi_\omega$
2	$u = \varphi(\omega),$ $\omega = \frac{x}{\sqrt{t}}$	$\left(\varphi^k \varphi_\omega\right)_\omega = -\frac{1}{2}\omega\varphi_\omega$
3	$u = \varphi(\omega)t^{-\frac{1}{k}},$ $\omega = \alpha \ln t + x$	$(\varphi^k \varphi_\omega)_\omega = \alpha\varphi_\omega - \frac{1}{k}\varphi$
4	$u = \varphi(\omega)x^{\frac{2}{k}},$ $\omega = \alpha \ln x + t$	$\alpha^2(\varphi^k \varphi_\omega)_\omega = [1 - \alpha(3 + \frac{4}{k})\varphi^k]\varphi_\omega - \frac{2(k+2)}{k^2}\varphi^{k+1}$
5	$u = \varphi(\omega)t^\alpha,$ $\omega = xt^{-\frac{1+k\alpha}{2}}$	$(\varphi^k \varphi_\omega)_\omega = -\frac{1+k\alpha}{2}\omega\varphi_\omega + \alpha\varphi$

In case *II*, we may set $\alpha_0 = 0$ without loosing a generality, however α_1 cannot be skipped. In fact, setting $\alpha_1 = 0$ one obtains the ansatz $u = t^{-\frac{1}{k}}\varphi(x)$, which is a particular case of that obtained for $\alpha_1 \neq 0$. Thus, we should keep α_1, so that the corresponding characteristic equation is

$$tu_t - \alpha u_x = -\frac{1}{k}u, \tag{4.14}$$

where the renaming $2\alpha = -\frac{\alpha_1}{\alpha_2}$ is used.

In case *III*, we may set $\alpha_1 = 0$ without loosing a generality, hence the characteristic equation

$$-\alpha k u_t + kx u_x = 2u,$$

is obtained (here the renaming $\alpha = -\frac{\alpha_0}{\alpha_3 k}$ is used).

In case *IV*, we may again set $\alpha_0 = \alpha_1 = 0$ without loosing a generality, so that the characteristic equation

$$2tu_t + (1 + k\alpha)xu_x = 2\alpha u,$$

is obtained (here the renaming $\alpha = \frac{\alpha_3}{\alpha_2}$ is used).

Obviously, the above first-order PDEs can be easily solved. The ansätze and reduction equations obtained in cases *I*–*IV* are summarized in Table 4.1.

Remark 4.1 *Case 2 of Table 4.1 can be united with case 5 provided the restriction $\alpha \neq 0$ is skipped.*

Now we present several well-known exact solutions of Eq. (4.9) in order to show that they are invariant solutions and are obtainable from the ansätze and equations presented in Table 4.1.

To the best of our knowledge, J. Boussinesq was the first to construct the exact solution of Eq. (4.9) [32]. He studied this equation with $k = 1$, which describes the solute (water) filtration in soil and is often called the Boussinesq equation, and found the exact solution

$$u = \frac{C_1 \varphi(x)}{1 + C_0 t}, \quad x = \int_0^\varphi \frac{y \mathrm{d}y}{\sqrt{1-y^3}}, \tag{4.15}$$

where C_0 and C_1 are correctly-specified constants [32](actually, one of them can be arbitrary). Obviously, the exact solution (4.15) can be found using the ansatz (with $k = 1, \alpha = 0$) listed in case 3 of Table 4.1 and the time translation $t \to t + C_0^{-1}$. Notably, J. Boussinesq applied the method of separable variables for constructing (4.15).

Although ODE (4.5) is not integrable, its particular solution

$$u = \left[-\frac{k}{2(k+2)} \frac{x^2}{t} \right]^{\frac{1}{k}} \tag{4.16}$$

(hereafter $k \neq -2$ and we remind the reader that the exponent $k = -2$ corresponds to the case when Eq. (4.9) is linearizable) can be easily found. Solution (4.16) is often called waiting-time solution [162].

A more complicated solution of Eq. (4.9) was found in [19] and independently rediscovered by several other authors [33, 207, 239]. The solution has the form

$$u = \left[Ct^{-\frac{k}{k+2}} - \frac{k}{2(k+2)} \frac{x^2}{t} \right]^{\frac{1}{k}} \tag{4.17}$$

and can be found using the reduction equation (with $\alpha = -\frac{1}{k+2}$) listed in case 5 of the table. It is worth noting that the corresponding ansatz was constructed in [19] without application of the Lie method.

Some other solutions were derived in the 1970s–1980s and almost all exact solutions of Eq. (4.9) with the arbitrary exponent k were summarized in paper [131] and handbook [213]). It is easily seen that each of them is obtainable via the corresponding Lie ansatz from Table 4.1 and invariance transformations, hence those are invariant.

Now we turn to exact solutions of Eq. (4.9), which can be derived using non-Lie ansätze. As it was stressed in Section 4.1, a non-Lie ansatz does not guarantee construction of new non-Lie exact solutions, which are not obtainable by the Lie method. In other words, the non-Lie ansatz can lead only to invariant solutions, especially it happens when the nonlinear PDE in question has a nontrivial Lie symmetry. Because Eq. (4.9) admits the four-dimensional Lie algebra, one should not expect that there are many non-Lie solutions of this equation in literature.

As it was pointed out in Section 3.2, there is not a complete Q-conditional symmetry classification (QSC) of the nonlinear heat equation. In particular, a very few Q-conditional symmetries are known for Eq. (4.9). Consider the Q-conditional symmetry operator [16, 120] $Q = \partial_t + 12x^{-2}u^{\frac{1}{2}}\partial_u$ of the fast diffusion equation

$$u_t = (u^{-\frac{1}{2}}u_x)_x. \qquad (4.18)$$

The corresponding non-Lie ansatz

$$u = \left[\varphi(x) + \frac{6t}{x^2}\right]^2 \qquad (4.19)$$

leads to the reduction equation

$$x^2\varphi_{\omega\omega} = 6\varphi.$$

This linear ODE is integrable [142], so that ansatz (4.19) leads to the exact solution of (4.18)

$$u = \left(C_0 x^3 + \frac{C_1 + 6t}{x^2}\right)^2. \qquad (4.20)$$

We have checked that the exact solution (4.20) cannot be derived using any ansatz from Table 4.1 (the invariance transformations (4.10) were also taken into account), so that it is indeed a non-Lie solution.

It turns out that the exact solution can be essentially generalized if one takes into account formulae (4.115)–(4.117) with $\lambda = \lambda_2 = \lambda_3 = 0$ from Section 4.4. In fact, Eq. (4.116) with $\lambda = \lambda_2 = \lambda_3 = 0$ coincides with Eq. (4.18), while system (4.117) takes the form

$$\beta_{xx} - \beta^2 = 0,$$

$$\varphi_{xx} - \beta\varphi = 0.$$

This ODE system can be integrated. As a result, we obtain the exact solution of Eq. (4.18)

$$u = \left[(C_0 + t)\mathcal{W} + C_3 \int \mathcal{W}^{-2} dx\right]^2, \qquad (4.21)$$

where $\mathcal{W}(0, C_1, x + C_2)$ $\mathcal{W}$ is the Weierstrass function with the periods 0 and C_1 ($C_0, \dots, C_3$ are arbitrary constants). Solution (4.21) contains the exact solution (4.20) as a particular case and was derived firstly in [200] using an ad hoc ansatz. Notably, C_0 and C_2 can be removed from (4.21) because Eq. (4.18) is invariant under the time and space translations.

In paper [120], the fast diffusion equation

$$u_t = (u^{-1}u_x)_x \qquad (4.22)$$

is also studied by the nonclassical method. As a result, the author found four Q-conditional symmetries of the form (3.12), assuming that the coefficient η in the operator Q is quadratic w.r.t. the variable u. The symmetries obtained have been used for constructing exact solutions. In particular, the exact solutions (we have simplified those using the invariance transformations from MGI (4.10)):

$$u = \frac{2t}{x^2 - C_0^2 t^2} \qquad (4.23)$$

and

$$u = \frac{1}{x - t + C_1 t \exp(-\frac{x}{t})}. \qquad (4.24)$$

However, one may note that both solutions (4.23) and (4.24) can be derived using ansatz with $\alpha = k = -1$ listed in case 5 of Table 4.1. Thus, these solutions are invariant, although it is claimed in [120] that all the solutions found therein are unobtainable by Lie symmetries.

On the other hand, the exact solutions of Eq. (4.22)

$$u = \frac{C_0 \cos(C_0 t)}{\sin(C_0 x) - \sin(C_0 t)}$$

and

$$u = \frac{C_1 \cosh(C_1 t)}{\sinh(C_1 x) - \sinh(C_1 t)}$$

are indeed the non-Lie solutions and were derived by application of the corresponding Q-conditional symmetries [120] (notably, four more solutions of this type were identified in [216]).

4.3 Solutions of some reaction-diffusion-convection equations arising in biomedical applications

In this section, we present exact solutions of several reaction-diffusion-convection (RDC) equations arising in mathematical biology. We show how these solutions can be found via the Q-conditional symmetries derived in Section 3.4.

4.3.1 The Fisher and Murray equations

The most famous PDE arising in biomedical applications is the Fisher equation [98]:

$$u_t = u_{xx} + \lambda_1 u(1 - \lambda_2 u), \qquad (4.25)$$

where $\lambda_1 > 0$, $\lambda_2 > 0$ and the unknown function $u(t, x)$ means usually density (of cells, population, chemicals etc.). Eq. (4.25) was suggested by R.A. Fisher in 1937 and is a governing equation for the model describing the spread in space of a favored gene in a population. Because the group of equivalence transformations (ETs) of Eq. (4.25) contains scale transformations, both parameters λ_1 and λ_2 can be reduced to 1, hence the standard form is

$$u_t = u_{xx} + u(1-u). \tag{4.26}$$

In contrast to the classical logistic model, i.e., Eq. (4.26) without the diffusion term, the Fisher equation cannot be exactly solved taking into account any reasonable initial and boundary conditions in a domain (e.g., a finite interval). There is only known the exact solution in the form of traveling front satisfying the zero flux conditions $u_x = 0$ at infinity $x = \pm\infty$, which are typical requirements for possible interpretation of such type solutions. The solution has the form

$$u = \frac{1}{4}\left[1 - \tanh\left(\frac{1}{2\sqrt{6}}(x - \frac{5}{\sqrt{6}}t + x_0)\right)\right]^2, \tag{4.27}$$

and was found for the first time in [2]. This solution can be rewritten in the equivalent form

$$u = \left[1 + C\exp\left(\frac{1}{\sqrt{6}}(x - \frac{5}{\sqrt{6}}t)\right)\right]^{-2} \tag{4.28}$$

where C is an arbitrary positive constant (if $C < 0$ then (4.28) is still a solution but with singularities).

There are some studies devoted to searching for exact solutions of the Fisher equation claiming that new ones have been found. However, to the best of our knowledge, the solutions either can be reduced to (4.28), or do not satisfy any realistic boundary conditions. For example, the plane wave solutions in terms of the Weierstrass function are known [92] but they are not useful for any interpretation.

It can be easily checked using the determining equations from Section 2.4 that the Fisher equation admits only the trivial Lie algebra with the basic operators ∂_t and ∂_x. Moreover, it follows from Section 3.4 and the earlier papers [17, 88] that Eq. (4.26) does not admit any Q-conditional symmetries of the form (3.11) while those of the form (3.12) are unknown.

The situation with finding exact solutions of the well-known generalization of Eq. (4.26), the Murray equation [179, 180]

$$u_t = u_{xx} + \lambda u u_x + u(1-u) \tag{4.29}$$

(here the parameters λ_1 and λ_2 are skipped like in Eq. (4.26)) is more optimistic. The Murray equation has no special status with respect to (w.r.t.) the Lie symmetry, so that one admits only the trivial Lie algebra of the time and space translations. Thus, the plane wave ansatz

$$u = \phi(\omega), \quad \omega = x - v_1 t, \quad v_1 \in \mathbb{R} \tag{4.30}$$

can be only constructed, which reduces Eq. (4.29) to the ordinary differential equation (ODE)

$$\phi_{\omega\omega} + (v_1 + \lambda\phi)\phi_\omega + \phi(1-\phi) = 0. \tag{4.31}$$

Solutions of some RDC equations arising in biomedical applications 145

This nonlinear ODE with the arbitrary coefficients λ and v_1 is not integrable, however, several particular solutions can be found. The simplest solutions can be readily constructed and they lead to the plane wave solutions [53]

$$u = 1 + C \exp\left[\tfrac{1}{\lambda}(x + \tfrac{1}{\lambda}t)\right],$$
$$u = C \exp\left[\tfrac{1}{\lambda}x + (1 + \tfrac{1}{\lambda^2})t\right]. \tag{4.32}$$

Solutions (4.32) are unbounded (grow to infinity if $t \to \infty$), hence one doubts whether they are important from applicability point of view.

We were able to derive plane wave solutions of Eq. (4.29) possessing more attractive properties. Omitting the relevant details, we present here the result. The Murray equation possesses the traveling front [53]

$$u = \left[1 + C\exp\left(-\frac{\lambda}{2}x - \frac{\lambda}{2}\left(\frac{\lambda}{2} + \frac{2}{\lambda}\right)t\right)\right]^{-1}. \tag{4.33}$$

Assuming $\lambda < 0$ and $C > 0$, we note that this traveling front is one with the minimal velocity $|\frac{\lambda}{2} + \frac{2}{\lambda}|$, which was found in [180] (see formula (11.60) therein). In fact, solution (4.33) connects the steady-state points $u = 0$ and $u = 1$ and there is no traveling front with a smaller velocity provided $\lambda \leq -2$. Moreover, one sees that this traveling wave (see Figure 4.1) is very similar to the *numerical* solution pictured in Murray's book [180] (see Figure 11.5,b therein).

Finally, we present the plane wave solution [53]

$$u = \left[C\exp\left(-\frac{1}{\sqrt{6}}(\mp x + \frac{5}{\sqrt{6}}t)\right) - \frac{C^2}{3}\exp\left(-\frac{2}{\sqrt{6}}(\mp x + \frac{5}{\sqrt{6}}t)\right)\right]^{-1} \tag{4.34}$$

of the Murray equation (4.29) with the restriction $\lambda^2 = 6$. This solution with the upper sign "−" is valid if $\lambda < 0$ while one with the lower sign "+" is valid if $\lambda > 0$. Interestingly that solution (4.34) is a wave with the velocity $5/\sqrt{6}$ and this velocity coincides with that of the traveling front (4.28) of the Fisher equation.

In contrast to the Fisher equation, the Murray equation admits a Q-conditional symmetry. In fact, according to case 2 of Theorem 3.2 (see also (3.241)) this equation admits the Q-conditional symmetry operator

$$Q = \partial_t - \left(\lambda u + \frac{1}{\lambda}\right)\partial_x + (u - u^2)\partial_u. \tag{4.35}$$

Using this operator one can construct the ansatz

$$\varphi(\omega) = \int \frac{\lambda u + \lambda^{-1}}{u^2 - u}du - x, \quad \omega = \int \frac{du}{u - u^2} - t. \tag{4.36}$$

Although formulae (4.36) present the non-Lie ansatz in implicit form,

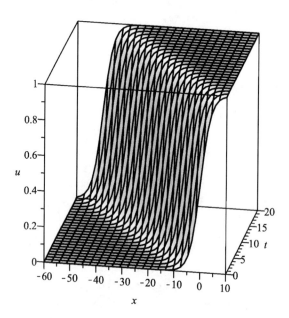

FIGURE 4.1: Exact solution (4.33) with $\lambda = 1$, $C = 1$

Eq. (4.29) successfully reduces to the second-order ODE

$$\varphi_{\omega\omega} = \frac{1}{\lambda}\varphi_\omega^2 + \left(\frac{2}{\lambda^2}+1\right)\varphi_\omega + \frac{1}{\lambda} + \frac{1}{\lambda^3} \qquad (4.37)$$

which is integrable because one reduces to the known first-order ODE [142].

Substituting the general solution of ODE (4.37) into ansatz and making the corresponding calculations, we arrive at the exact solution of the Murray equation[80, 53]

$$u = \frac{1 + C\exp\left[\frac{1}{\lambda}(x + \frac{1}{\lambda}t)\right]}{1 + C_0 e^{-t}}. \qquad (4.38)$$

This solution with $C_0 \neq 0$ is not of the form (4.30), hence it is a non-Lie solution. Notably, solution (4.38) with $C_0 < 0$ possesses a singularity point.

4.3.2 The Fitzhugh–Nagumo equation and its generalizations

Now we turn to Case 3 of Theorem 3.2, which involves as particular cases the well-known Fitzhugh–Nagumo (FN), Huxley and Kolmogorov–Petrovskii–Piskunov (KPP) equations, and their generalizations by adding the convective

Solutions of some RDC equations arising in biomedical applications 147

term $\lambda u u_x$ (see Eqs. (3.242)–(3.246)). As it was noted in Section 3.7, all these equations can be transformed to the form (3.247)

$$v_t = v_{yy} + \lambda v v_y + \lambda_0 + \lambda_1 v - \lambda_3 v^3,$$

which is nothing else but the RDC equation listed in Case 3 of Theorem 3.2 with the correctly-specified coefficients (see 3.248). Thus, we find exact solutions firstly for Eq. (3.247) and afterwards transform the solutions obtained to those of the equations in question [53].

It can be easily identified using the results of Chapter 2 (see Table 2.5) that (3.247) admits a nontrivial Lie symmetry only with $\lambda_0 = \lambda_1 = 0$. However, it is impossible because $\lambda_3 \neq 0$ (see formulae (3.248)), hence Eq. (3.247) is invariant only with respect to (w.r.t.) the trivial Lie algebra $< P_t, P_y >$. The plane wave ansatz

$$v = \phi(\omega), \quad \omega = y - v_1 t, \quad v_1 \in \mathbb{R} \tag{4.39}$$

reduces Eq. (3.247) to the nonlinear ODE

$$\phi_{\omega\omega} + (v_1 + \lambda\phi)\phi_\omega + \lambda_0 + \lambda_1 \phi - \lambda_3 \phi^3 = 0, \tag{4.40}$$

which cannot be integrated for the arbitrary coefficients $\lambda, \lambda_0, \lambda_1, \lambda_3$ and v_1. The known particular solutions of (4.40) generate the plane wave solutions of Eq. (3.247), which will be presented later as particular cases of more general solutions.

Remark 4.2 *Eq. (4.40) with $v_1 = \lambda = 0$ is integrable and its solution can be presented in an implicit form. Hence, the stationary solutions of Eq. (3.247) with $\lambda = 0$ can be completely described.*

Now we construct families of non-Lie solutions of Eq. (3.247). According to Theorem 3.2, Eq. (3.247) admits two Q-conditional symmetries

$$Q_1 = \partial_t + \frac{3\kappa - \lambda}{4} v \partial_y + \frac{3\kappa - \lambda}{2(\kappa - \lambda)}(\lambda_0 + \lambda_1 v - \lambda_3 v^3)\partial_v \tag{4.41}$$

and

$$Q_2 = \partial_t - \frac{3\kappa + \lambda}{4} v \partial_y + \frac{3\kappa + \lambda}{2(\kappa + \lambda)}(\lambda_0 + \lambda_1 v - \lambda_3 v^3)\partial_v \tag{4.42}$$

where $\kappa = \sqrt{\lambda^2 + 8\lambda_3}$ (in the case $\lambda^2 = -8\lambda_3$, a single symmetry is obtained). Using these symmetries, one can construct the non-Lie ansätze

$$\phi(\omega) = \int \frac{v dv}{\lambda_0 + \lambda_1 v - \lambda_3 v^3} - \frac{2}{\kappa - \lambda} y, \quad \omega = \int \frac{dv}{\lambda_0 + \lambda_1 v - \lambda_3 v^3} - \frac{3\kappa - \lambda}{2(\kappa - \lambda)} t \tag{4.43}$$

and

$$\phi(\omega) = \int \frac{v dv}{\lambda_0 + \lambda_1 v - \lambda_3 v^3} + \frac{2}{\kappa + \lambda} y, \quad \omega = \int \frac{dv}{\lambda_0 + \lambda_1 v - \lambda_3 v^3} - \frac{3\kappa + \lambda}{2(\kappa + \lambda)} t \tag{4.44}$$

by solving the invariance surface conditions $Q_1(v) = 0$ and $Q_2(v) = 0$, respectively. Since these ansätze cannot be written in an explicit form (w.r.t. the variable v), we prefer to solve directly the overdetermined systems

$$\begin{aligned} v_t &= v_{yy} + \lambda v v_y + \lambda_0 + \lambda_1 v - \lambda_3 v^3, \\ Q_1(v) &\equiv v_t + \tfrac{3\kappa-\lambda}{4} v v_y - \tfrac{3\kappa-\lambda}{2(\kappa-\lambda)}(\lambda_0 + \lambda_1 v - \lambda_3 v^3) = 0 \end{aligned} \qquad (4.45)$$

and

$$\begin{aligned} v_t &= v_{yy} + \lambda v v_y + \lambda_0 + \lambda_1 v - \lambda_3 v^3, \\ Q_2(v) &\equiv v_t - \tfrac{3\kappa+\lambda}{4} v v_y - \tfrac{3\kappa+\lambda}{2(\kappa+\lambda)}(\lambda_0 + \lambda_1 v - \lambda_3 v^3) = 0, \end{aligned} \qquad (4.46)$$

which is equivalent to the substitution (4.43) and (4.44) into (3.247).

Consider firstly the overdetermined system (4.45). Eliminating v_t in the first equation using the second one, we arrive at the second-order ODE

$$v_{yy} + \frac{3(\kappa+\lambda)}{4} v v_y - \frac{\kappa+\lambda}{2(\kappa-\lambda)}(\lambda_0 + \lambda_1 v - \lambda_3 v^3) = 0, \qquad (4.47)$$

which can be reduced to the form

$$v_{y^*y^*} + 3 v v_{y^*} + v^3 - \frac{\lambda_1}{\lambda_3} v - \frac{\lambda_0}{\lambda_3} = 0 \qquad (4.48)$$

by the local substitution

$$v(t,y) = v(t,y^*), \quad y^* = \frac{\kappa+\lambda}{4} y. \qquad (4.49)$$

As we already know, ODE (4.48) is reduced to the linear third-order ODE

$$w_{y^*y^*y^*} - \frac{\lambda_1}{\lambda_3} w_{y^*} - \frac{\lambda_0}{\lambda_3} w = 0 \qquad (4.50)$$

by the nonlocal substitution [142](see item (6.38)therein)

$$v = \frac{w_{y^*}}{w}. \qquad (4.51)$$

To construct the general solution of (4.50), one needs to solve the cubic equation

$$\lambda_0 + \lambda_1 \alpha - \lambda_3 \alpha^3 = 0, \qquad (4.52)$$

which has three different roots $\alpha_1, \alpha_2, \alpha_3$ in the case of the generalized FN equation and two different roots α_1 and α_2 in the case of the generalized Huxley and KPP equations. The relevant general solutions take the form

$$w = \phi_1(t) \exp(\alpha_1 y^*) + \phi_2(t) \exp(\alpha_2 y^*) + \phi_3(t) \exp(\alpha_3 y^*) \qquad (4.53)$$

ODE system

$$3\alpha_2(\dot\phi_1\phi_2 - \dot\phi_2\phi_1) + \dot\phi_1\phi_3 - \dot\phi_3\phi_1 = \tfrac{\lambda_3(3\kappa-\lambda)}{2(\kappa-\lambda)}(9\alpha_2^3\phi_1\phi_2 - 3\alpha_2^2\phi_1\phi_3),$$

$$\dot\phi_1\phi_3 - \dot\phi_3\phi_1 = 3\alpha_2^2 \tfrac{\lambda_3(3\kappa-\lambda)}{2(\kappa-\lambda)}\phi_1\phi_3, \qquad (4.65)$$

$$\dot\phi_2\phi_3 - \dot\phi_3\phi_2 = 2\alpha_2 \tfrac{\lambda_3(3\kappa-\lambda)}{2(\kappa-\lambda)}\phi_3^2.$$

In the case $\phi_3 = 0$, the solution (4.64) is reduced to a particular case of (4.55), so that we assume $\phi_3 \neq 0$. The ODE system (4.65) is integrable under this restriction and its general has the form

$$\phi_1 = C_1\exp(3\beta_1\alpha_2^2 t)\phi(t), \quad \phi_2 = (C_2 + 2\beta_1\alpha_2 t)\phi(t), \quad \phi_3 = \phi(t). \qquad (4.66)$$

Finally, substituting (4.66) into (4.64), we obtain the two-parameter family of exact solutions

$$v = \frac{1 + \alpha_2 C_2 + \gamma_2 y + 2\beta_1\alpha_2^2 t - 2\alpha_2 C_1 \exp[3(\beta_1\alpha_2^2 t - \gamma_2 y)]}{C_2 + \tfrac{\kappa+\lambda}{4} y + 2\beta_1\alpha_2 t + C_1 \exp[3(\beta_1\alpha_2^2 t - \gamma_2 y)]} \qquad (4.67)$$

of the nonlinear RDC equation (3.247) when this equation takes the form

$$v_t = v_{yy} + \lambda v v_y - \lambda_3(2\alpha_2 + v)(v - \alpha_2)^2. \qquad (4.68)$$

In a quite similar way the overdetermined system (4.46) has been also solved and the following family of exact solutions of the nonlinear RDC equation (4.68)

$$v = \frac{1 + \alpha_2 C_2 + \sigma_2 y + 2\beta_2\alpha_2^2 t - 2\alpha_2 C_1 \exp[3(\beta_2\alpha_2^2 t - \sigma_2 y)]}{C_2 + \tfrac{\lambda-\kappa}{4} y + 2\beta_2\alpha_2 t + C_1 \exp[3(\beta_2\alpha_2^2 t - \sigma_2 y)]}, \qquad (4.69)$$

has been found (the values β_2 and σ_2 are the same as in (4.63)).

Remark 4.3 *Generally speaking, the cubic equation (4.52) can have complex roots, however such possibility is excluded here because the coefficients of Eq. (4.52) are not arbitrary (see (3.248)). The most general case is examined in the next section (see Eq. (4.128)).*

Obviously, each exact solution of the forms (4.59)–(4.62) with $C_1 C_2 C_3 \neq 0$, (4.67) and (4.69) with $C_1 C_2 \neq 0$ cannot be found using the Lie ansatz (4.39), hence those are two-parameter families of non-Lie solutions of the nonlinear RDC equation (3.247).

One observes that the exact solution families constructed above contain several plane wave solutions (traveling fronts) of the form (4.39). Indeed, vanishing one of the constants $C_i, i = 1, 2, 3$ in (4.59) and (4.61), we obtain traveling fronts

$$v = \frac{\alpha_i + \alpha_j C \exp[\tfrac{\lambda+\kappa}{4}(\alpha_j - \alpha_i)(y + \beta_1^{ij} t)]}{1 + C \exp[\tfrac{\lambda+\kappa}{4}(\alpha_j - \alpha_i)(y + \beta_1^{ij} t)]} \qquad (4.70)$$

and
$$v = \frac{\alpha_i + \alpha_j C \exp[\frac{\lambda-\kappa}{4}(\alpha_j - \alpha_i)(y + \beta_2^{ij} t)]}{1 + C \exp[\frac{\lambda-\kappa}{4}(\alpha_j - \alpha_i)(y + \beta_2^{ij} t)]}, \quad (4.71)$$
respectively. Here C is a positive constant and
$$\beta_1^{ij} = \frac{1}{4}(3\kappa - \lambda)(\alpha_i + \alpha_j), \ \beta_2^{ij} = -\frac{1}{4}(3\kappa + \lambda)(\alpha_i + \alpha_j), \ i \neq j, \ i,j = 1,2,3.$$

Obviously, each traveling wave solution of the form (4.70)–(4.71) tends either to α_i or to α_j if $\omega = y + \beta_k^{ij} t \to \pm\infty$, $k = 1, 2$. We remind the reader that the roots $\alpha_1, \alpha_2, \alpha_3$ are the steady-state points of the nonlinear equation (3.247). Thus, the traveling fronts found possess the same properties as those for the Fisher and Murray equations presented in the previous subsection. It should be also noted that each traveling front has the speed β_k^{ij}, therefore, taking into account the identity $\beta_k^{ij} = \beta_k^{ji}$, six different traveling fronts are generated by formulae (4.70) and (4.71).

In the case of Eq. (4.68), two plane wave solutions with different structure follow from (4.67) and (4.69), namely:
$$v = \alpha_2 + \frac{1}{\frac{\kappa+\lambda}{4} y + 2\beta_1 \alpha_2 t}$$
and
$$v = \alpha_2 + \frac{1}{\frac{\lambda-\kappa}{4} y + 2\beta_2 \alpha_2 t}.$$
Both solutions are not continuous in the domain $(t, x) \in \mathbb{R}^+ \times \mathbb{R}$, hence their applicability is not clear. On the other hand, traveling fronts can be obtained from (4.70) and (4.71) under the restriction $\alpha_1 = -2\alpha_2, \alpha_3 = 0$ because the corresponding solution of the ODE system (4.65) with $\phi_3 = 0$ leads to traveling wave solutions.

Now we turn to the generalized FN equation (3.245):
$$u_t = u_{xx} + \lambda u u_x + \lambda_3 u(u - \delta)(1 - u), \quad \lambda_3 > 0, \ 0 < \delta < 1,$$
which can be also called the Fitzhugh–Nagumo–Burgers equation. Exact solutions of the generalized FN equation can be easily constructed using the solutions found above and formulae (3.248) and (3.249). Making simple calculations, we obtain two families of exact solutions
$$u = u_0 + \frac{\sum_{i=1}^{3} \alpha_i C_i \exp\left[\alpha_i \frac{\lambda+\kappa}{4}\left(x + (\lambda u_0 + \frac{3\kappa-\lambda}{4}\alpha_i)t\right)\right]}{\sum_{i=1}^{3} C_i \exp\left[\alpha_i \frac{\lambda+\kappa}{4}\left(x + (\lambda u_0 + \frac{3\kappa-\lambda}{4}\alpha_i)t\right)\right]} \quad (4.72)$$
and
$$u = u_0 + \frac{\sum_{i=1}^{3} \alpha_i C_i \exp\left[\alpha_i \frac{\lambda-\kappa}{4}\left(x + (\lambda u_0 - \frac{3\kappa+\lambda}{4}\alpha_i)t\right)\right]}{\sum_{i=1}^{3} C_i \exp\left[\alpha_i \frac{\lambda-\kappa}{4}\left(x + (\lambda u_0 - \frac{3\kappa+\lambda}{4}\alpha_i)t\right)\right]} \quad (4.73)$$

where
$$u_0 = \frac{1}{3}(\delta+1), \ \alpha_1 = -\frac{1}{3}(\delta+1), \ \alpha_2 = \frac{1}{3}(2\delta-1), \ \alpha_3 = \frac{1}{3}(2-\delta), \ \kappa = \sqrt{\lambda^2 + 8\lambda_3}.$$

In the case $C_i > 0, i = 1, 2, 3$, such solutions are known in applications as two-shock waves. An example of solutions of the form (4.72) is presented in Figure 4.2.

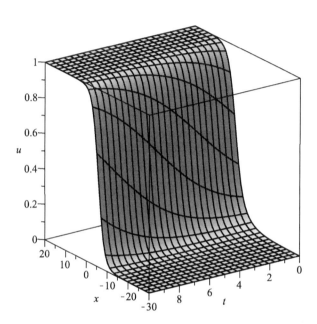

FIGURE 4.2: Exact solution (4.72) with $\lambda = 1, \kappa = 2, \delta = \frac{1}{3}, C_1 = C_2 = C_3 = 1$

Note that the family of solutions (4.73) was found in [101] using a generalized conditional symmetry and both families (4.72) and (4.73) were constructed in [53] using Q-conditional symmetries.

As particular cases, we obtain the traveling wave solutions

$$u = u_0 + \frac{\alpha_i + \alpha_j C \exp\left[\frac{\lambda+\kappa}{4}(\alpha_i - \alpha_j)\left(x + (\lambda u_0 + \beta_1^{ij})t\right)\right]}{1 + C \exp\left[\frac{\lambda+\kappa}{4}(\alpha_i - \alpha_j)\left(x + (\lambda u_0 + \beta_1^{ij})t\right)\right]}, \tag{4.74}$$

and

$$u = u_0 + \frac{\alpha_i + \alpha_j C \exp\left[\frac{\lambda-\kappa}{4}(\alpha_i - \alpha_j)\left(x + (\lambda u_0 + \beta_2^{ij})t\right)\right]}{1 + C \exp\left[\frac{\lambda-\kappa}{4}(\alpha_i - \alpha_j)\left(x + (\lambda u_0 + \beta_2^{ij})t\right)\right]} \tag{4.75}$$

(here $i \neq j$, $i,j = 1, 2, 3$ and a summation over i and/or j is not assumed).

All the solutions of the generalized FN equation constructed above generate the known solutions of the FN equation (3.242). However, one may note that families (4.72) and (4.73) with $\lambda = 0$ can be thought as the single family

$$u = u_0 + \frac{\sum_{i=1}^{3} \alpha_i C_i \exp\left[\alpha_i \frac{\sqrt{2\lambda_3}}{2}\left(x + 3\alpha_i \frac{\sqrt{2\lambda_3}}{2} t\right)\right]}{\sum_{i=1}^{3} C_i \exp\left[\alpha_i \frac{\sqrt{2\lambda_3}}{2}\left(x + 3\alpha_i \frac{\sqrt{2\lambda_3}}{2} t\right)\right]} \quad (4.76)$$

because the FN equation (in contrast to the generalized FN equation (3.245)) is invariant under discrete transformations $x \to -x$. Solutions of the form (4.76) were constructed for the first time in [146] using an ad hoc ansatz. In the 1990s they were independently found using Q-conditional symmetry of the FN equation in [17, 88, 191] (see also [213] where the known solutions of the FN equation are summarized).

It is worth noting that formula (4.76) presents real solutions under condition $\lambda_3 > 0$ only. In the case of the generalized FN equation (3.245), the corresponding solutions are real also for negative λ_3 provided $\lambda^2 + 8\lambda_3 \geq 0$.

Now we turn to the generalized Huxley equation (3.246)

$$u_t = u_{xx} + \lambda u u_x + \lambda_3 u^2(1-u), \quad \lambda_3 > 0.$$

Two families of exact solutions of this equation can be identified using formulae (4.67) and (4.69). In fact, formulae (3.248) and (3.249) produce

$$u = v + \frac{1}{3}, \quad y = x + \frac{\lambda}{3} t, \quad \alpha_1 = \frac{2}{3}, \quad \alpha_2 = -\frac{1}{3}, \quad (4.77)$$

hence the formulae generate the two families of exact solutions

$$u = \frac{1 + C_1 \exp\left[\frac{\lambda+\kappa}{4}(x + v_1 t)\right]}{\frac{\lambda+\kappa}{4}(x + 2v_2 t) + C_1 \exp\left[\frac{\lambda+\kappa}{4}(x + v_1 t)\right]} \quad (4.78)$$

and

$$u = \frac{1 + C_1 \exp\left[\frac{\lambda-\kappa}{4}(x + v_2 t)\right]}{\frac{\lambda-\kappa}{4}(x + 2v_1 t) + C_1 \exp\left[\frac{\lambda-\kappa}{4}(x + v_2 t)\right]}. \quad (4.79)$$

Here the velocities are specified as follows

$$v_1 = \frac{\lambda + \kappa}{4}, \quad v_2 = \frac{\lambda - \kappa}{4}. \quad (4.80)$$

We note that the exact solutions (4.79)–(4.80) are not continuous in the domain $(t, x) \in \mathbb{R}^+ \times \mathbb{R}$, hence their applicability cannot be readily suggested. However, those can be useful if the generalized Huxley equation is studied in a bounded domain.

Setting $\lambda = 0$ in (4.79)–(4.80), the known solutions of the Huxley equation can be identified

$$u = \frac{1 + C_1 \exp\left[\frac{\sqrt{2\lambda_3}}{2}(\pm x + \frac{\sqrt{2\lambda_3}}{2}t)\right]}{\frac{\sqrt{2\lambda_3}}{2}(\pm x - \sqrt{2\lambda_3}t) + C_1 \exp\left[\frac{\sqrt{2\lambda_3}}{2}(\pm x + \frac{\sqrt{2\lambda_3}}{2}t)\right]}, \quad (4.81)$$

which were found in [17, 88].

Traveling wave solutions of the generalized Huxley equation (3.246) can be easily constructed using solutions (4.70)–(4.71) and (4.77). Formally speaking, four traveling fronts can be derived, however, one notes that two of them only are indeed different

$$u = \frac{1}{1 + C_1 \exp\left[-\frac{\lambda+\kappa}{4}(x + v_1 t)\right]} \quad (4.82)$$

and

$$u = \frac{1}{1 + C_1 \exp\left[\frac{\kappa-\lambda}{4}(x + v_2 t)\right]}. \quad (4.83)$$

Traveling waves (4.82) and (4.83) with $C_1 > 0$ tend to the steady-state points $u = 0$ and $u = 1$ of Eq. (3.246) if $x \to \pm\infty$, i.e., they have the same behavior as the traveling fronts obtained above for the Fisher, Murray and the (generalized) FN equations.

Finally, we present exact solutions of the equation

$$u_t = u_{xx} + \lambda u u_x - \lambda_3 u(1-u)^2, \quad (4.84)$$

which follows from Eq. (3.245) by setting $\delta = 1$. Eq. (4.84) with $\lambda_3 < 0$ is a generalization of the KPP equation (3.244). Two families of exact solutions of the generalized KPP equation are

$$u = \frac{1 + \frac{1}{4}(\lambda + \kappa)(x + 2v_1 t)}{\frac{1}{4}(\lambda + \kappa)(x + 2v_1 t) + C_1 \exp\left[-\frac{\kappa+\lambda}{4}(x + \frac{3\lambda-\kappa}{4}t)\right]} \quad (4.85)$$

and

$$u = \frac{1 + \frac{1}{4}(\lambda - \kappa)(x + 2v_2 t)}{\frac{1}{4}(\lambda - \kappa)(x + 2v_2 t) + C_1 \exp\left[\frac{\kappa-\lambda}{4}(x + \frac{\kappa+3\lambda}{4}t)\right]}. \quad (4.86)$$

These solutions are obtained from (4.67) and (4.69) at $\alpha_1 = -\frac{2}{3}, \alpha_2 = \frac{1}{3}$ using formulae (3.248) and (3.249) with $\delta = 1$. Moreover, the traveling wave solutions (4.74) and (4.75) with $u_0 = \frac{2}{3}, \alpha_1 = -\frac{2}{3}, \alpha_2 = \frac{1}{3}$ are also the solutions of Eq. (4.84). Because $\lambda_3 < 0$, the condition $\lambda^2 + 8\lambda_3 \geq 0$ is needed in order to obtain real solutions. In particular, it means that the solutions (4.85) and (4.86) are complex ones in the case of the KPP equation (3.244). One may say that the above problem follows from the structure of Q-conditional symmetry corresponding to the RD equations with a positive coefficient next to u^3. In fact, such equations admit Q-conditional symmetry with a complex coefficient (see Theorem 3.4).

4.4 Solutions of reaction-diffusion-convection equations with power-law diffusivity

In this section, several reaction-diffusion-convection (RDC) equations are studied, which are either natural generalizations of those from Section 4.3, or important from other reasons. A common peculiarity of these equations consist of the structure of the diffusion and convection coefficients. We assume that both coefficients have the same structure given by power-law functions.

4.4.1 Lie's solutions of an equation with power-law diffusion and convection

Let us consider the equation from case 12 of Table 2.7, that reads as

$$u_t = (u^k u_x)_x + q u^m u_x + r u^{2m-k+1}. \tag{4.87}$$

Its particular cases describe a wide range of processes. For example, setting $q = 0$, $k > 0$, $2m - k + 1 > 0$, one obtains the porous media equation with a source/sink term. Such type equations were extensively studied (including application to combustion processes) in [225] (see also references therein). Setting $r = 0$, $k > 0$, $m > 0$, the porous media equation with the power-law convection is identified. An important subcase arises when $k = 1/2$, $m = 1$ leading to the foam drainage equation [90, 123].

In the case $k = m > 0$, Eq. (4.87) is related with the equation

$$u_t = (u^k u_x)_x + q u^k u_x + \lambda_1 u - \lambda_2 u^{k+1} \tag{4.88}$$

by the form-preserving transformation (FPTs) listed in case 10 of Table 2.6. Depending on the exponent k, Eq. (4.88) can be sought as a generalization of some models arising in mathematical biology. In particular, Eq. (4.88) with $k = 1$ is nothing else but the Murray equation with the porous diffusion.

Eq. (4.87) is invariant under three-dimensional MAI

$$AG^3 = <\partial_t,\ \partial_x,\ (k-2m)t\partial_t + (k-m)x\partial_x + u\partial_u>. \tag{4.89}$$

The corresponding MGI is

$$\bar{t} = t e^{(k-2m)v_1} + t_0,\quad \bar{x} = x e^{(k-m)v_1} + x_0,\quad \bar{u} = u e^{v_1}. \tag{4.90}$$

To construct all nonequivalent Lie's ansätze we need to consider the linear combination of the operators (4.89)

$$X = [\alpha_2(k-2m)t + \alpha_0]\partial_t + [\alpha_2(k-m)x + \alpha_1]\partial_x + \alpha_2 u \partial_u.$$

The invariance surface equation (1.25), takes the form

$$[\alpha_2(k-2m)t + \alpha_0]u_t + [\alpha_2(k-m)x + \alpha_1]u_x = \alpha_2 u. \tag{4.91}$$

TABLE 4.2: Lie's ansätze and reduction equations for Eq. (4.87)

	Ansatz	Reduction equation
1	$u = \varphi(\omega)$, $\omega = x - v_1 t$,	$(\varphi^k \varphi_\omega)_\omega + q\varphi^m \varphi_\omega + r\varphi^{2m-k+1} = -v_1 \varphi_\omega$
2	$u = \varphi(\omega) x^{\frac{1}{m}}$, $\omega = \alpha \ln x + t$, $k = 2m$	$\alpha[\alpha(\varphi^{2m}\varphi_\omega)_\omega + q\varphi^m \varphi_\omega] + r\varphi =$ $= [1 - \alpha(3 + \frac{2}{m})\varphi^{2m}]\varphi_\omega - \frac{m+1}{m^2}\varphi^{2m+1} - \frac{q}{m}\varphi^{m+1}$
3	$u = \varphi(\omega) t^{-\frac{1}{m}}$, $\omega = \alpha \ln t + x$, $k = m$	$(\varphi^m \varphi_\omega)_\omega + q\varphi^m \varphi_\omega + r\varphi^{m+1} = \alpha\varphi_\omega - \frac{1}{m}\varphi$
4	$u = \varphi(\omega) t^{\frac{1}{k-2m}}$, $\omega = xt^{\frac{m-k}{k-2m}}$, $k \neq m, 2m$	$(\varphi^k \varphi_\omega)_\omega + q\varphi^m \varphi_\omega + r\varphi^{2m-k+1} =$ $= \frac{1}{k-2m}[(m-k)\omega\varphi_\omega + \varphi]$

Solutions of (4.91) depend essentially on the values of the parameters α_0, α_1, α_2, m and k.

Of course, setting $\alpha_2 = 0$, we again obtain the plane wave ansatz and the corresponding reduction equation, which are listed in case 1 of Table 4.2 (v_1 and α are arbitrary parameters therein). There is also the trivial ansatz $u = \varphi(t)$ generated by the operator ∂_x, however one is skipped in what follows because it is irrelevant for any applications.

Eq. (4.91) with $\alpha_2 \neq 0$ has the similar structure to Eq. (4.12), hence one needs to consider three different cases: I $k = 2m$, II $k = m$, III $k \neq m, 2m$.

Each case can be examined in a quite similar way as it was done in Section 4.2. In particular the parameters α_0 and/or α_1 can be skipped without loosing a generality (see transformation (4.90)). As a result, all the inequivalent Lie ansätze and reduction equations were constructed and they are presented in cases 2–4 of Table 4.2.

All the reduction equations in Table 4.2 are nonintegrable ODEs provided their coefficients are arbitrary. However, several particular solutions can be found under correctly-specified restrictions on the coefficients. In particular, several exact solutions of Eq. (4.87) with $q = 0$ (i.e., no convective term), which are known in literature [132, 213, 225], can be identified. We present here some examples of exact solutions of Eq. (4.87) with $q \neq 0$ in order to extend the list of such solutions.

Setting $\alpha = 0$ in case 2 of Table 4.2, we obtain the first-order ODE

$$\varphi_\omega = \frac{m+1}{m^2}\varphi^{2m+1} + \frac{q}{m}\varphi^{m+1} + r\varphi, \qquad (4.92)$$

which possesses the general solution in the implicit form

$$\ln(m^2 r \varphi^{-2m} + mq\varphi^{-m} + m + 1) + 2mr(t + t_0) +$$
$$+ \frac{2q}{\sqrt{4(m+1)r - q^2}} \arctan\left[\frac{2(m+1)\varphi^m + mq}{m\sqrt{4(m+1)r - q^2}}\right] = 0. \quad (4.93)$$

Thus, we obtain the solution of the RDC equation (4.87) with $k = 2m$ of the form $u = x^{\frac{1}{m}}\varphi(t)$, where φ is a solution of the transcendent equation (4.93)(the constant t_0 can be skipped without loosing a generality).

In particular, setting $m = -1$ into (4.92) the exact solution

$$u = \frac{\pm e^{rt} + q}{rx}$$

of the RDC equation

$$u_t = (u^{-2}u_x)_x + qu^{-1}u_x + ru, \quad r \neq 0$$

is obtained.

Let us consider case 3 of Table 4.2 with $\alpha = 0$. Using the substitution $\varphi = \psi^{m+1}$, the reduction equation is transformed to the form

$$\psi_{\omega\omega} + q\psi_\omega + (m+1)r\psi + \frac{m+1}{m}\psi^{\frac{1}{m+1}} = 0.$$

Now we apply the ansatz [180]

$$\psi = \left(1 + Ce^{bx}\right)^{\frac{2(m+1)}{m}},$$

where C and b are to-be-determined coefficients. The relevant calculations lead to the exact solution

$$\psi = \left[1 + C\exp\left(\pm \frac{mx}{\sqrt{-2m(m+2)}}\right)\right]^{\frac{2(m+1)}{m}},$$

where C is an arbitrary parameter. Finally, we obtain the exact solution

$$u = t^{-\frac{1}{m}}\left[1 + C\exp\left(\pm \frac{mx}{\sqrt{-2m(2+m)}}\right)\right]^{\frac{2}{m}} \quad (4.94)$$

of the RDC equation

$$u_t = (u^m u_x)_x \mp \frac{3m+4}{\sqrt{-2m(m+2)}}u^m u_x - \frac{1}{m}u^{m+1}.$$

It should be noted that solution (4.94) is real under the restriction $m \in (-2, 0)$.

Let us consider case 4 of Table 4.2. Setting $r = 0$, $m = k+1$ and integrating, we obtain the first-order ODE

$$(k+2)\varphi^k \varphi_\omega + q\varphi^{k+2} + \omega\varphi = C_1,$$

which is reducible to the form

$$\varphi_\omega = -\frac{q}{3}\varphi^2 - \frac{1}{3}\omega \qquad (4.95)$$

by setting $k = 1$, $C_1 = 0$. Solving (4.95), we obtain [212]:

$$\varphi = \frac{3}{q}\frac{d}{d\omega}\ln w, \quad w = \omega^{\frac{1}{2}}\left[C_1 J_{\frac{1}{3}}(y) + C_2 J_{-\frac{1}{3}}(y)\right], \quad y = \frac{2\sqrt{q}}{9}\omega^{\frac{3}{2}}$$

where $J_{\pm\frac{1}{3}}(y)$ is the Bessel function. Using the known properties of the Bessel functions and making straightforward calculations, we obtain

$$w_\omega = \frac{C_1}{2}\omega^{-\frac{3}{2}}(\omega + \frac{9}{q})J_{\frac{1}{3}}(y) + \frac{C_2}{2}\omega^{-\frac{3}{2}}(\omega - \frac{9}{q})J_{-\frac{1}{3}}(y) - \frac{3C_1}{\sqrt{q}}J_{\frac{4}{3}}(y) - \frac{3C_2}{\sqrt{q}}J_{\frac{2}{3}}(y),$$

$$\varphi = \frac{3[\frac{C_1}{2}\omega^{-\frac{3}{2}}(\omega + \frac{9}{q})J_{\frac{1}{3}}(y) + \frac{C_2}{2}\omega^{-\frac{3}{2}}(\omega - \frac{9}{q})J_{-\frac{1}{3}}(y) - \frac{3C_1}{\sqrt{q}}J_{\frac{4}{3}}(y) - \frac{3C_2}{\sqrt{q}}J_{\frac{2}{3}}(y)]}{q\omega^{\frac{1}{2}}\left[C_1 J_{\frac{1}{3}}(y) + C_2 J_{-\frac{1}{3}}(y)\right]}.$$

Finally, we derive the exact solution

$$u = \frac{3[\frac{C_1}{2}\omega^{-\frac{3}{2}}(\omega + \frac{9}{q})J_{\frac{1}{3}}(y) + \frac{C_2}{2}\omega^{-\frac{3}{2}}(\omega - \frac{9}{q})J_{-\frac{1}{3}}(y) - \frac{3C_1}{\sqrt{q}}J_{\frac{4}{3}}(y) - \frac{3C_2}{\sqrt{q}}J_{\frac{2}{3}}(y)]}{qt^{\frac{1}{3}}\omega^{\frac{1}{2}}\left[C_1 J_{\frac{1}{3}}(y) + C_2 J_{-\frac{1}{3}}(y)\right]},$$

where $\omega = xt^{-\frac{1}{3}}$ and $y = \frac{2\sqrt{q}}{9}x^{\frac{3}{2}}t^{-\frac{1}{2}}$, of the nonlinear equation

$$u_t = (uu_x)_x + qu^2 u_x, \quad q \neq 0.$$

4.4.2 Non-Lie solutions of some equations with power-law diffusion and convection

Here, we construct exact solutions of the RDC equations presented in Theorems 3.5 and 3.6 using the Q-conditional symmetry operators obtained, discuss the properties of the solutions and show that they are non-Lie solutions. As it follows from the proofs of both theorems, the Q-conditional symmetry operators have essentially simpler structure if one uses the substitution (3.113). So, we will firstly construct exact solutions of the corresponding equations with the unknown function v and afterwards use (3.113) for finding those of the RDC equations arising in Theorems 3.5 and 3.6.

We start from case 1 of Theorem 3.5. Eq. (3.106) and operator (3.107) are transformed by the substitution (3.113) to the forms

$$v_{xx} = v^n v_t - \lambda v_x + (\lambda_1^* v + \lambda_2^*)(\lambda_3 - v^n), \qquad (4.96)$$

and
$$Q = \partial_t + (\lambda_1^* v + \lambda_2^*)\partial_v,$$

where $\lambda_i^* = \lambda_i(m+1)$, $i = 1, 2$ and $m \neq -1$. The relevant ansatz is constructed using the standard procedure, i.e., we solve the invariance surface condition $Q(v) = 0$. Since its general solution depends on λ_1^* two ansätze are obtained:

$$v = \begin{cases} \lambda_2^* t + \varphi(x), & \lambda_1^* = 0, \\ \varphi(x) e^{\lambda_1^* t} - \frac{\lambda_2^*}{\lambda_1^*}, & \lambda_1^* \neq 0 \end{cases} \qquad (4.97)$$

with $\varphi(x)$ being an unknown function. Substituting (4.97) with $\lambda_1^* = 0$ into (4.96), one arrives at the linear ODE

$$\varphi_{xx} + \lambda \varphi_x - \lambda_2^* \lambda_3 = 0,$$

with the general solution

$$\varphi = C_1 + C_2 e^{-\lambda x} + \frac{\lambda_2^* \lambda_3}{\lambda} x.$$

Hereafter C_1 and C_2 are arbitrary constants. Hence, Eq. (4.96) with $\lambda_1^* = 0$ possesses the exact solution

$$v = \lambda_2^* t + C_1 + C_2 e^{-\lambda x} + \frac{\lambda_2^* \lambda_3}{\lambda} x.$$

Using the substitution (3.113), we obtain the exact solution

$$u = \left[\lambda_2(m+1)t + C_1 + C_2 e^{-\lambda x} + \frac{\lambda_2 \lambda_3 (m+1)}{\lambda} x \right]^{\frac{1}{m+1}} \qquad (4.98)$$

of the nonlinear RDC equation

$$u_t = [u^m u_x]_x + \lambda u^m u_x + \lambda_2 u^{-m} - \lambda_2 \lambda_3, \quad m \neq -1 \qquad (4.99)$$

because Eq. (4.99) is invariant under the time and space translations.

Because the maximal invariance algebra (MAI) of Eq. (4.99) with $\lambda_2 \neq 0$ is the trivial Lie algebra with the basic operators ∂_t and ∂_x (it follows from Table 2.5), plane wave solutions are only obtainable by the Lie method. Obviously, the exact solution (4.98) has different structure and cannot be reduced to the form (1.29) provided $C_2 \neq 0$, therefore it is a non-Lie solution. Notably, the parameter C_1 can be skipped in (4.98) if $\lambda_2 \neq 0$.

Substituting (4.97) with $\lambda_1^* \neq 0$ into Eq. (4.96), one again obtains a linear second-order ODE, which is integrable in terms of elementary functions. The form of the solution depends on $D = \lambda^2 + 4\lambda_1^*\lambda_3 = \lambda^2 + 4\lambda_1(m+1)\lambda_3$. Dealing in quite similar way to the case $\lambda_1^* = 0$, we finally obtain three exact solutions

$$u = \left[\exp\left(\lambda_1(m+1)t - \frac{\lambda}{2}x\right)\left(C_1 e^{\frac{\sqrt{D}}{2}x} + C_2 e^{-\frac{\sqrt{D}}{2}x}\right) - \frac{\lambda_2}{\lambda_1}\right]^{\frac{1}{m+1}}, \quad D > 0 \quad (4.100)$$

$$u = \left[\exp\left(\lambda_1(m+1)t - \frac{\lambda}{2}x\right)(C_1 + C_2 x) - \frac{\lambda_2}{\lambda_1}\right]^{\frac{1}{m+1}}, \quad D = 0 \quad (4.101)$$

$$u = \left[\exp\left(\lambda_1(m+1)t - \frac{\lambda}{2}x\right)\left(C_1 \cos\frac{\sqrt{-D}}{2}x + C_2 \sin\frac{\sqrt{-D}}{2}x\right) - \frac{\lambda_2}{\lambda_1}\right]^{\frac{1}{m+1}}, \quad (4.102)$$
$$D < 0$$

of the nonlinear RDC equation

$$u_t = (u^m u_x)_x + \lambda u^m u_x + (\lambda_1 u^{m+1} + \lambda_2)(u^{-m} - \lambda_3), \quad m \neq -1. \quad (4.103)$$

Similarly to the exact solution (4.98), it is easily seen that solutions (4.100)–(4.102) are also non-Lie solutions. Of course, these solutions produce some plane wave solutions as particular cases (for example, by setting $C_2 = 0$ in (4.100) and (4.101)).

In the case $m = 1$ and $\lambda_2 = 0$, Eq. (4.103) takes the form

$$u_t = (uu_x)_x + \lambda u u_x + \lambda_1 u(1 - \lambda_3 u), \quad (4.104)$$

which is the Murray equation with the porous diffusion mentioned in Section 2.6. One notes that $D = \lambda^2 + 8\lambda_1\lambda_3 > 0$ if $\lambda_1 > 0$, $\lambda_3 > 0$. Hence solution (4.100) takes the form

$$u = \sqrt{\exp\left(2\lambda_1 t - \frac{\lambda}{2}x\right)\left(C_1 e^{\frac{\sqrt{D}}{2}x} + C_2 e^{-\frac{\sqrt{D}}{2}x}\right)}. \quad (4.105)$$

This solution unboundedly grows if $t \to \infty$ or $x \to \pm\infty$. More interesting solutions occur in the case of Eq. (4.104) with the anti-logistic term, i.e., $\lambda_1 < 0$, $\lambda_3 > 0$. Depending on D one obtains three types of solutions. In the case $m = 1$, $\lambda = 4$, $\lambda_1 = -1$, $\lambda_2 = 0$, $\lambda_3 = \frac{3}{2}$, $C_1 = -C_2 = 2.6$, solution (4.100) is presented in Figure 4.3. This solution is bounded, tends to zero if $t \to \infty$ and satisfies the zero boundary conditions for $x = 0$ and $x = \infty$. It means that the exact solution solves Eq. (4.104) with the anti-logistic term at the infinite interval $[0, \infty)$ with the zero Dirichlet conditions.

If $\lambda = 2$ (other coefficients are still the same) then solution (4.102) with $m = 1$ is valid. In the case $\lambda_1 = -1$, $\lambda_2 = 0$, $C_1 = 0.65, C_2 = 0$, this solution is presented in Figure 4.4. We note that the solution again vanishes if $t \to \infty$, but one satisfies the zero Dirichlet conditions on the bounded interval $[-\frac{\pi}{2}, \frac{\pi}{2}]$.

Consider Eq. (4.103) with $m = -2, \lambda_1 = -\lambda_2 > 0$ and $\lambda_3 = 0$

$$u_t = [u^{-2} u_x]_x + \lambda u^{-2} u_x + \lambda_1 u(1 - u). \quad (4.106)$$

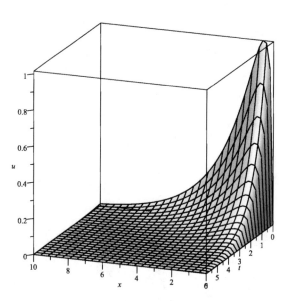

FIGURE 4.3: Exact solution (4.100) with $m = 1$, $\lambda = 4$, $\lambda_1 = -1$, $\lambda_2 = 0$, $\lambda_3 = \frac{3}{2}$, $C_1 = -C_2 = 2.6$

This equation can be called the generalized Murray equation with fast diffusion. Since $D = \lambda^2 > 0$ solution (4.100) takes the form

$$u = \frac{1}{1 + e^{-\lambda_1 t}(C_1 + C_2 e^{-\lambda x})} \qquad (4.107)$$

and possesses attractive properties. Assuming $C_1 > 0$ and $C_2 > 0$, one sees that this solution is positive and bounded for arbitrary $(t, x) \in \mathbb{R}^+ \times \mathbb{R}$. Moreover, the solution tends either to zero ($\lambda_1 < 0$) or to 1 ($\lambda_1 > 0$) if $t \to \infty$. Both values, $u = 0$ and $u = 1$, are steady-state points of (4.106). Solution (4.107) tends to the steady-state point $u = 0$ if $\lambda x \to -\infty$, while $u = [1 + C_1 \exp(-\lambda_1 t)]^{-1}$ if $\lambda x \to \infty$. An example of solution (4.107) is presented in Figure 4.5. It is worth noting that (4.107) with $C_1 = 0$ is a traveling wave solution and has the same structure as that for the Murray equation (see formula (4.33) in Section 4.3).

Now we turn to case 2 of Theorem 3.5. Eq. (3.108) and operator (3.109) are transformed by the substitution (3.113) to the forms

$$v_{xx} = e^v v_t - \lambda v_x + (\lambda_1 v + \lambda_2)(\lambda_3 - e^v), \qquad (4.108)$$

Solutions of RDC equations with power-law diffusivity   163

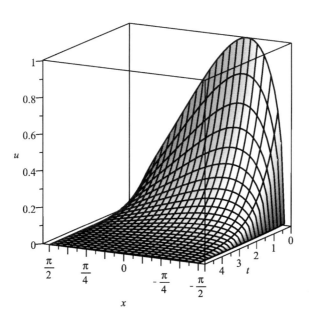

FIGURE 4.4: Exact solution (4.102) with $m = 1$, $\lambda = 2$, $\lambda_1 = -1$, $\lambda_2 = 0$, $\lambda_3 = 1$, $C_1 = 0.65$, $C_2 = 0$

and
$$Q = \partial_t + (\lambda_1 v + \lambda_2)\partial_v, \qquad (4.109)$$
respectively. Using operator (4.109) we obtain the ansatz
$$v = \begin{cases} \lambda_2 t + \varphi(x), & \lambda_1 = 0, \\ \varphi(x)e^{\lambda_1 t} - \frac{\lambda_2}{\lambda_1}, & \lambda_1 \neq 0, \end{cases} \qquad (4.110)$$
which has the same structure as (4.97). Substituting (4.110) into (4.108), one again obtains integrable second-order ODEs and easily constructs the relevant exact solutions of the RDC equation (3.108), i.e.,
$$u_t = (u^{-1}u_x)_x + \lambda u^{-1}u_x + (\lambda_1 \ln u + \lambda_2)(u - \lambda_3).$$
In the case $\lambda_1 = 0$, the solution is
$$u = \exp\left(\lambda_2 t + C_1 e^{-\lambda x} + \frac{\lambda_2 \lambda_3}{\lambda} x\right), \qquad (4.111)$$
while the case $\lambda_1 \neq 0$ produces three solutions depending on $D = \lambda^2 + 4\lambda_1\lambda_3$:
$$u = \exp\left[\exp\left(\lambda_1 t - \frac{\lambda}{2}x\right)\left(C_1 e^{\frac{\sqrt{D}}{2}x} + C_2 e^{-\frac{\sqrt{D}}{2}x}\right) - \frac{\lambda_2}{\lambda_1}\right], D > 0, \quad (4.112)$$

164 — Exact solutions of RDC equations

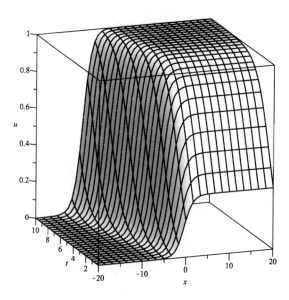

FIGURE 4.5: Exact solution (4.107) with $\lambda = 1$, $\lambda_1 = 1$, $C_1 = C_2 = 2$

$$u = \exp\left[\exp\left(\lambda_1 t - \frac{\lambda}{2}x\right)(C_1 + C_2 x) - \frac{\lambda_2}{\lambda_1}\right], \quad D = 0 \tag{4.113}$$

and

$$u = \exp\left[\exp\left(\lambda_1 t - \frac{\lambda}{2}x\right)\left(C_1 \cos \frac{\sqrt{-D}}{2}x + C_2 \sin \frac{\sqrt{-D}}{2}x\right) - \frac{\lambda_2}{\lambda_1}\right], \tag{4.114}$$
$$D < 0.$$

Because MAI of Eq. (3.108) with $\lambda_1 \neq 0$ is the trivial Lie algebra with the basic operators ∂_t and ∂_x (it follows from Table 2.5), formulae (4.112)–(4.114) present the non-Lie solutions of the RDC equation (3.108). Notably, Eq. (3.108) with $\lambda_1 = 0$ admits the three-dimensional Lie algebra (see case 24 of Table 2.5) and it can be shown that (4.111) is an invariant solution.

It turns out that properties of solutions (4.112)–(4.114) depend essentially on values of C_1 and C_2. For example, solution (4.112) with the negative C_1 and C_2 tends to zero or 1 if $x \to \pm \infty$ (see Figure 4.6 and Figure 4.7). Notably $u = 1$ is the steady-state point of Eq. (3.108) with $\lambda_2 = 0$. As it follows from Figure 4.8, solution (4.112) with $C_1 C_2 < 0$ has essentially different shape for small values of time but one is similar to one with the negative C_1 and C_2 for $t \gg 0$ (other parameters are the same).

Solutions of RDC equations with power-law diffusivity 165

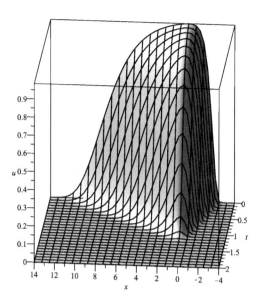

FIGURE 4.6: Exact solution (4.112) with $\lambda = 2$, $\lambda_1 = 5$, $\lambda_2 = 0$, $\lambda_3 = \frac{1}{4}$, $C_1 = C_2 = -0.01$

We note also that the time asymptotic of the solutions depends on the λ_1 sign (see Figure 4.6 and Figure 4.7). On the other hand, the parameter λ_2 is not so important, hence one is zero in the above figures. Finally, Figure 4.9 presents solution (4.112) with the nonvanish λ_2 and $\lambda_3 = 0$. As one may note, the shape of the picture differs from those presented in Figure 4.6–Figure 4.8.

Consider case 3 of Theorem 3.5. Since we were unable to solve the overdetermined system (3.112), we used the particular solution producing the Q-conditional operator (3.136). This operator generates the ansatz

$$u = [\beta(x)t + \varphi(x)]^2 \qquad (4.115)$$

to the nonlinear RDC equation

$$u_t = (u^{-\frac{1}{2}} u_x)_x + \lambda u^{-\frac{1}{2}} u_x + \lambda_2 u^{\frac{1}{2}} + \lambda_3. \qquad (4.116)$$

The functions β and φ arising in (4.115) satisfy the ODE system

$$\beta_{xx} + \lambda \beta_x - \beta^2 + \frac{\lambda_2}{2} \beta = 0,$$
$$\varphi_{xx} + \lambda \varphi_x - \beta \varphi + \frac{\lambda_2}{2} \varphi + \frac{\lambda_3}{2} = 0, \qquad (4.117)$$

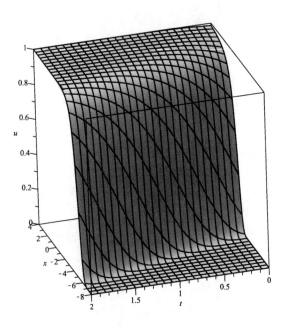

FIGURE 4.7: Exact solution (4.112) with $\lambda = 2$, $\lambda_1 = -3$, $\lambda_2 = 0$, $\lambda_3 = \frac{1}{4}$, $C_1 = C_2 = -0.1$

which is not integrable provided $\lambda\lambda_2 \neq 0$. Moreover, there are no particular solutions of this system in the known handbooks [142, 212]. The trivial solution of the first equation $\beta = \frac{\lambda_2}{2}$ leads to a particular case of solution (4.98).

We remind the reader that the special case $\lambda = \lambda_2 = \lambda_3 = 0$ leads to solution (4.21) of Eq. (4.18). It can be also noted that the first equation of system (4.117) with $\lambda = 0$ can be solved in terms of elliptic functions (see item 6.5 in [142]). Assuming additionally $\lambda_3 = 0$ (while $\lambda_2 \neq 0$), the solution of the second equation of system (4.117) is $\varphi = \beta(x)$. Thus, an exact solution of the nonlinear RDC equation (4.116) with $\lambda = \lambda_3 = 0$ is derived. However, this solution is invariant because one is obtainable via the Lie ansatz (with $m = -\frac{1}{2}$, $\alpha = 0$) arising in case 3 of Table 4.2.

Thus, we have constructed several exact solutions of the nonlinear RDC equations (4.99), (4.103), (4.104), (4.106), and (3.108), which were obtained by application of the Q-conditional symmetry operators listed in Theorem 3.5.

Now we apply the Q-conditional symmetries arising in Theorem 3.6 in order to construct exact solutions. Symmetry (3.140) arising in case 1 can be successfully applied to construct exact solutions in the explicit form. Omitting

Solutions of RDC equations with power-law diffusivity

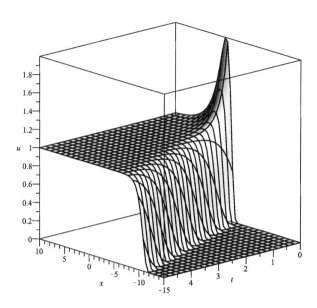

FIGURE 4.8: Exact solution (4.112) with $\lambda = 2$, $\lambda_1 = -3$, $\lambda_2 = 0$, $\lambda_3 = \frac{1}{4}$, $C_1 = 1$, $C_2 = -0.3$

rather trivial computations, we present the final result. The nonlinear equation

$$u_t = (u^m u_x)_x + \lambda u^{m+1} u_x + \lambda_2 u^{-m}, \quad m \neq -1;$$

possesses the solution

$$u = \left[\frac{\lambda_2 \lambda (m+1) t^2 - 2x}{2\lambda t} \right]^{\frac{1}{m+1}}, \qquad (4.118)$$

while Eq. (3.139) with $m \neq -1$, $\lambda_1 \neq 0$ has the solution

$$u = \left[\frac{m+1}{1 + Ce^{-\lambda_1(m+1)t}} \left(\lambda_2 t - \frac{\lambda_2 C}{\lambda_1(m+1)} e^{-\lambda_1(m+1)t} - \frac{\lambda_1}{\lambda} x \right) \right]^{\frac{1}{m+1}}. \qquad (4.119)$$

The Q-conditional symmetry operator occurring in case 2 of Theorem 3.6 has the most cumbersome structure. As a consequence, essential difficulties arise if one applies operator (3.142) for finding exact solutions. On the other hand, it will be shown that many of the exact solutions obtained possess remarkable properties.

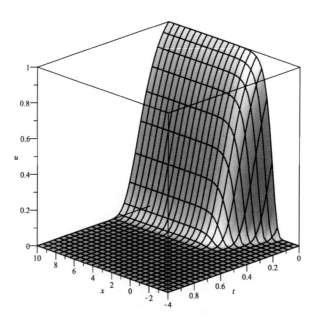

FIGURE 4.9: Exact solution (4.112) with $\lambda = 2$, $\lambda_1 = 10$, $\lambda_2 = -1$, $\lambda_3 = 0$, $C_1 = C_2 = -0.1$

Eq. (3.141) and operator (3.142) by the substitution (3.113) with $m = -\frac{1}{2}$ are transformed to the forms

$$v_{xx} = vv_t - \lambda v^2 v_x - \frac{\lambda_1^* + 3\lambda v}{3\lambda}\left(\frac{1}{3}\lambda_1^*\lambda v^3 + \lambda_2^* v + \lambda_3^*\right) \quad (4.120)$$

and

$$Q = \partial_t + (-\lambda v + \lambda_1^*)\partial_x + \left(\frac{1}{3}\lambda_1^*\lambda v^3 + \lambda_2^* v + \lambda_3^*\right)\partial_v, \quad (4.121)$$

respectively. Hereafter $\lambda_1^* = \frac{3\lambda_1}{2\lambda}$, $\lambda_2^* = \frac{\lambda_2}{2}$, $\lambda_3^* = \frac{\lambda_3}{2}$ and $v = u^{\frac{1}{2}} > 0$ is assumed since (3.141) contains terms $u^{\frac{1}{2}}$ and $u^{-\frac{1}{2}}$. Moreover, $\lambda_1^* \neq 0$ in what follows because case 2 with $\lambda_1 = 0$ reduces to the case 1 with $m = -\frac{1}{2}$.

Instead of construction of a non-Lie ansatz using operator (4.121), we use again the equation $Q(v) = 0$, i.e.,

$$v_t = (\lambda v - \lambda_1^*)v_x + \frac{1}{3}\lambda_1^*\lambda v^3 + \lambda_2^* v + \lambda_3^*, \quad (4.122)$$

to eliminate v_t from Eq. (4.120). In fact, substituting the right-hand side of

Eq. (4.122) into Eq. (4.120), one arrives at the equation

$$v_{xx} + \lambda_1^* v v_x + \frac{1}{9}\lambda_1^{*2}v^3 + \frac{1}{3\lambda}\lambda_1^*\lambda_2^* v + \frac{1}{3\lambda}\lambda_1^*\lambda_3^* = 0, \tag{4.123}$$

which is the nonlinear ODE containing the variable t as a parameter. Equation (4.123) is reduced to the form

$$v_{yy} + 3vv_y + v^3 + \frac{3\lambda_2^*}{\lambda_1^*\lambda}v + \frac{3\lambda_3^*}{\lambda_1^*\lambda} = 0 \tag{4.124}$$

by the simple substitution

$$y = \frac{\lambda_1^*}{3}x. \tag{4.125}$$

As we already know, Eq. (4.124) can be transformed into the linear third-order ODE

$$w_{yyy} + 3pw_y + 2qw = 0, \tag{4.126}$$

where $p = \frac{\lambda_2^*}{\lambda_1^*\lambda} = \frac{\lambda_2}{3\lambda_1}$, $2q = \frac{3\lambda_3^*}{\lambda_1^*\lambda} = \frac{\lambda_3}{\lambda_1}$, by the substitution [142](see item (6.38) therein)

$$v = \frac{w_y}{w}. \tag{4.127}$$

According to the classical ODE theory, one needs to solve the algebraic equation

$$\alpha^3 + 3p\alpha + 2q = 0. \tag{4.128}$$

Thus, four different subcases depending on the values of p and q should be separately considered.

Subcase 1. If $p = q = 0$ then $\alpha_1 = \alpha_2 = \alpha_3 = 0$. The general solution of (4.126) has the form $w = \phi_1 + \phi_2 y + \phi_3 y^2$ and we arrive at the expression

$$v = \frac{\phi_2 + 2\phi_3 y}{\phi_1 + \phi_2 y + \phi_3 y^2}, \tag{4.129}$$

giving the general solution of the nonlinear ODE (4.124). Here $\phi_1(t)$, $\phi_2(t)$ and $\phi_3(t)$ are arbitrary (at the moment) smooth function and at least one of them must be nonzero. So, (4.129) with (4.125) generates the general solution of (4.123) with $\lambda_1^* \neq 0$, $\lambda_2^* = \lambda_3^* = 0$. Finally, to obtain the general solution of system (4.120) and (4.122), we substitute (4.129) with $y = \frac{\lambda_1^*}{3}x$ into the second equation of this system. After the relevant calculations a cumbersome expression is obtained, however, one can be split into separate parts for x^n, $n = 0, 1, 2$. As a result, we arrive at the ODE system

$$\phi_2\dot{\phi}_3 - \dot{\phi}_2\phi_3 = \tfrac{2}{3}\lambda_1^{*2}\phi_3^2,$$
$$\phi_1\dot{\phi}_3 - \dot{\phi}_1\phi_3 = \tfrac{1}{3}\lambda_1^*\phi_3(2\lambda\phi_3 + \lambda_1^*\phi_2), \tag{4.130}$$
$$\phi_1\dot{\phi}_2 - \dot{\phi}_1\phi_2 = \tfrac{1}{3}\lambda_1^*(2\lambda\phi_2\phi_3 - 2\lambda_1^*\phi_1\phi_3 + \lambda_1^*\phi_2^2).$$

System (4.130) has the similar structure to one (4.57) and can be solved in a similar way. Substituting the general solution of (4.130) into (4.129) and using (4.125), we find the exact solutions

$$v = \frac{3}{\lambda_1^* x - \lambda_1^{*2} t} \tag{4.131}$$

and

$$v = \frac{6(x - \lambda_1^* t)}{\lambda_1^*(x - \lambda_1^* t)^2 - 18\lambda t} \tag{4.132}$$

(the integration constants were removed by the time and space invariance translations) of the equation

$$v_{xx} = v v_t - \lambda v^2 v_x - \frac{\lambda_1^*}{3}\lambda v^4 - \frac{\lambda_1^{*2}}{9} v^3. \tag{4.133}$$

Obviously, (4.131) is a primitive plane wave solution and one is neglected below. Applying substitution (3.113) with $m = -\frac{1}{2}$ to (4.132)–(4.133) and renaming parameters, we arrive at the exact solution

$$u = \left[\frac{2(x - \frac{3\lambda_1}{2\lambda} t)}{\frac{\lambda_1}{2\lambda}(x - \frac{3\lambda_1}{2\lambda} t)^2 - 6\lambda t} \right]^2$$

of the nonlinear RDC equation

$$u_t = (u^{-\frac{1}{2}} u_x)_x + \lambda u^{\frac{1}{2}} u_x + \lambda_1 u^2 + \frac{\lambda_1^2}{2\lambda} u^{\frac{3}{2}}.$$

This solution is a non-Lie solution provided $\lambda_1 \neq 0$.

Subcase 2. If $p^3 = -q^2 \neq 0$ then $\alpha_1 = -2\sqrt[3]{q}$ and $\alpha_2 = \alpha_3 = \sqrt[3]{q} = \sqrt[3]{\frac{\lambda_3}{2\lambda_1}}$. The general solution of Eq. (4.126) is

$$w = \phi_1 e^{\alpha_1 y} + (\phi_2 y + \phi_3) e^{\alpha_2 y}, \quad \alpha_1 = -2\alpha_2$$

so that the expression

$$v = \frac{\alpha_1 \phi_1 e^{\alpha_1 y} + [\alpha_2 \phi_3 + \phi_2(\alpha_2 y + 1)] e^{\alpha_2 y}}{\phi_1 e^{\alpha_1 y} + (\phi_2 y + \phi_3) e^{\alpha_2 y}} \tag{4.134}$$

presents the general solution of the nonlinear ODE (4.124). Dealing in quite similar way to Subcase 1, one obtains the ODE system

$$\phi_3 \dot{\phi}_2 - \dot{\phi}_3 \phi_2 = \frac{\lambda_1^*}{3}\phi_2^2(2\lambda\alpha_2 + \lambda_1^*),$$
$$\phi_1 \dot{\phi}_2 - \dot{\phi}_1 \phi_2 = \lambda_1^* \alpha_2 \phi_1 \phi_2(\lambda\alpha_2 - \lambda_1^*),$$
$$3\alpha_2(\phi_1 \dot{\phi}_3 - \dot{\phi}_1 \phi_3) + \phi_1 \dot{\phi}_2 - \dot{\phi}_1 \phi_2 =$$
$$= \lambda_1^* \alpha_2 \phi_1 \left[(\tfrac{2}{3}\lambda\alpha_2^2 - 3\alpha_2\lambda_1^* - \tfrac{5}{3})\phi_3 - (\lambda\alpha_2 + 2\lambda_1^*)\phi_2 \right]$$

(4.135)

to find the unknown functions $\phi_1(t)$, $\phi_2(t)$ and $\phi_3(t)$. It turns out that system (4.135) has the same structure as one (4.65), so that its general solution can be constructed. Finally, we find the exact solutions

$$u = \left[\frac{\alpha_2 - 2C\alpha_2 \exp\left(-\frac{3\lambda_1}{2\lambda}\alpha_2(x-v_1t)\right)}{1 + C\exp\left(-\frac{3\lambda_1}{2\lambda}\alpha_2(x-v_1t)\right)}\right]^2,$$

and

$$u = \left[\frac{1 + \frac{\lambda_1}{2\lambda}\alpha_2(x-v_2t) - 2C\alpha_2\exp\left(-\frac{3\lambda_1}{2\lambda}\alpha_2(x-v_1t)\right)}{\frac{\lambda_1}{2\lambda}(x-v_2t) + C\exp\left(-\frac{3\lambda_1}{2\lambda}\alpha_2(x-v_1t)\right)}\right]^2$$

of the nonlinear RDC equation

$$u_t = (u^{-\frac{1}{2}}u_x)_x + \lambda u^{\frac{1}{2}}u_x + (\lambda_1 u^{\frac{3}{2}} + \lambda_2 u^{\frac{1}{2}} + \lambda_3)\left(\frac{\lambda_1}{2\lambda^2} + u^{\frac{1}{2}}\right), \quad (4.136)$$

where

$$\lambda_2 = -3\sqrt[3]{\frac{\lambda_1\lambda_3^2}{4}}, \quad \alpha_2 = \sqrt[3]{\frac{\lambda_3}{2\lambda_1}} \neq 0, \quad v_1 = -\lambda\alpha_2 + \frac{3\lambda_1}{2\lambda}, \quad v_2 = 2\lambda\alpha_2 + \frac{3\lambda_1}{2\lambda}.$$

Subcase 3. If $p^3 + q^2 < 0$ then three roots of Eq. (4.128) are different and real. This case is the most cumbersome because the known formulae for solving the cubic equation must be used. Let us assume that $\alpha_i, i = 1, 2, 3$ are different real numbers, which are calculated by the formulae (see, e.g., [157])

$$\alpha_1 = 2\sqrt{-p}\cos\left(\frac{\beta}{3}\right),$$
$$\alpha_2 = -2\sqrt{-p}\cos\left(\frac{\beta}{3} + \frac{\pi}{3}\right), \quad (4.137)$$
$$\alpha_3 = -2\sqrt{-p}\cos\left(\frac{\beta}{3} - \frac{\pi}{3}\right),$$

where $\cos\beta = -\frac{q}{\sqrt{-p^3}}$.

The general solution of Eq. (4.126) is

$$w = \phi_1 e^{\alpha_1 y} + \phi_2 e^{\alpha_2 y} + \phi_3 e^{\alpha_3 y},$$

and it leads to the general solution

$$v = \frac{\alpha_1\phi_1 e^{\alpha_1 y} + \alpha_2\phi_2 e^{\alpha_2 y} + \alpha_3\phi_3 e^{\alpha_3 y}}{\phi_1 e^{\alpha_1 y} + \phi_2 e^{\alpha_2 y} + \phi_3 e^{\alpha_3 y}} \quad (4.138)$$

of the nonlinear ODE (4.124).

Substituting (4.138) with $y = \frac{\lambda_1^*}{3}x$ into Eq. (4.122) and conducting the relevant calculations, we again arrive at the ODE system

$$\dot{\phi}_1\phi_2 - \phi_1\dot{\phi}_2 = -\frac{\lambda_1^*}{3}\phi_1\phi_2[\lambda(\alpha_1^2 - \alpha_2^2) + \lambda_2(\alpha_1 - \alpha_2)],$$
$$\dot{\phi}_1\phi_3 - \phi_1\dot{\phi}_3 = -\frac{\lambda_1^*}{3}\phi_1\phi_3[\lambda(\alpha_1^2 - \alpha_3^2) + \lambda_2(\alpha_1 - \alpha_3)], \quad (4.139)$$
$$\dot{\phi}_2\phi_3 - \phi_2\dot{\phi}_3 = -\frac{\lambda_1^*}{3}\phi_2\phi_3[\lambda(\alpha_2^2 - \alpha_3^2) + \lambda_2(\alpha_2 - \alpha_3)]$$

to find the unknown functions $\phi_1(t)$, $\phi_2(t)$ and $\phi_3(t)$. Because system (4.139) has the same structure as (4.57), one is integrable. The general solution of the ODE system (4.139) leads to the exact solution

$$u = \left[\frac{\sum_{i=1}^{3} \alpha_i C_i \exp\left(\frac{\lambda_1}{2\lambda}\alpha_i(x - v_i t)\right)}{\sum_{i=1}^{3} C_i \exp\left(\frac{\lambda_1}{2\lambda}\alpha_i(x - v_i t)\right)} \right]^2 \quad (4.140)$$

of the nonlinear RDC equation (3.141) with $\lambda_1 \neq 0$. Here $v_i = \lambda\alpha_i + \frac{3\lambda_1}{2\lambda}$, $i = 1, 2, 3$ and the roots α_i, $i = 1, 2, 3$ are determined by the formulae (4.137). It is worth noting that a nonzero constant among C_i, $i = 1, 2, 3$ can be put equal 1 without losing a generality, hence (4.140) can be rewritten as a two-parameter family of solutions similarly to (4.59)–(4.60)).

In the case $C_i > 0, i = 1, 2, 3$, this type of exact solution is well-known in applications and often is called two-shock waves (see, e.g., [173]) and were already discussed in Section 4.3. However, formula (4.140) gives the exact solution of Eq. (3.141) with $\lambda_1 \neq 0$ only under condition that the expression in the right-hand side without the exponent 2 is nonnegative, i.e.,

$$\left[\sum_{i=1}^{3} \alpha_i C_i \exp\left(\frac{\lambda_1}{2\lambda}\alpha_i(x - v_i t)\right)\right]\left[\sum_{i=1}^{3} C_i \exp\left(\frac{\lambda_1}{2\lambda}\alpha_i(x - v_i t)\right)\right] \geq 0.$$

The same condition concerns all the solutions of Eq. (3.141) obtained here because from the very beginning we used the substitution $v = u^{\frac{1}{2}}$. So, the exact solution (4.140) with $C_i > 0, i = 1, 2, 3$ is not real for arbitrary $(t, x) \in \mathbb{R}^+ \times \mathbb{R}$ but in the domain Ω defined by the condition

$$\sum_{i=1}^{3} \alpha_i C_i \exp\left(\frac{\lambda_1}{2\lambda}\alpha_i(x - v_i t)\right) \geq 0.$$

An example of such solution is presented in Figure 4.10. One notes that this solution satisfies the zero flux condition $u_x = 0$ as $x = +\infty$. On the other hand, $u = 0$ on the line, which forms the finite boundary of Ω. Such properties are important for a physical interpretation if Eq. (3.141) is used for modeling processes involving moving boundaries (see, e.g., [4, 70, 71] and references cited therein).

Consider the particular case of Eq. (3.141) with $\lambda_3 = 0$ and $\lambda_1 = -\lambda_2$ of the form

$$u_t = (u^{-\frac{1}{2}}u_x)_x + \lambda u^{\frac{1}{2}} u_x + \lambda_2 u^{\frac{1}{2}}(u^{\frac{1}{2}} - \delta)(1 - u), \quad \delta = \frac{\lambda_2}{2\lambda^2} \neq 0, \quad (4.141)$$

which may be treated as a generalized Fitzhugh–Nagumo equation with the fast diffusion provided $0 < \delta < 1$.

Solutions of RDC equations with power-law diffusivity

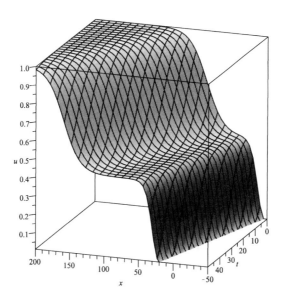

FIGURE 4.10: Exact solution (4.140) with $\lambda = 1$, $\lambda_1 = 0.5$, $C_1 = 0.1$, $C_2 = 2$, $C_3 = 0.001$, $\alpha_1 = -0.1$, $\alpha_2 = \frac{2}{3}$, $\alpha_3 = 1$

Since $\lambda_2 \neq 0$, we immediately obtain $p < 0$ and $q = 0$, therefore formulae (4.137) and (4.140) give the solution

$$u = \left[\frac{C_2 \exp\left(-\frac{\lambda_2}{2\lambda}(x - v_1 t)\right) - C_3 \exp\left(\frac{\lambda_2}{2\lambda}(x + v_2 t)\right)}{C_1 + C_2 \exp\left(-\frac{\lambda_2}{2\lambda}(x - v_1 t)\right) + C_3 \exp\left(\frac{\lambda_2}{2\lambda}(x + v_2 t)\right)} \right]^2, \quad (4.142)$$

where $v_1 = \lambda - \frac{3\lambda_2}{2\lambda}$, $v_2 = \lambda + \frac{3\lambda_2}{2\lambda}$.

Taking into account formula (3.113) with $m = -\frac{1}{2}$, we note that solution (4.142) with the positive C_1, C_2 and C_3 is real in the domain Ω defined by the condition

$$C_2 \exp\left(-\frac{\lambda_2}{2\lambda}(x - v_1 t)\right) \geq C_3 \exp\left(\frac{\lambda_2}{2\lambda}(x + v_2 t)\right).$$

An example of solution (4.142) is presented in Figure 4.11. One sees that this solution satisfies the same boundary conditions as the solution pictured in Figure 4.12. So, if the generalized FN equation (3.141) is used for modeling processes involving moving boundaries then solutions of the form (4.142) can be suitable for a biological interpretation.

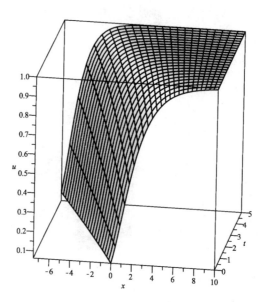

FIGURE 4.11: Exact solution (4.142) with $\lambda = \lambda_2 = 1, C_1 = C_2 = 1, C_3 = 2$

Setting in (4.142), for example, $C_3 = 0$ and positive constants C_1, C_2, we obtain the traveling front, which is valid in the domain $(t, x) \in \mathbb{R}^+ \times \mathbb{R}$. This traveling front is connecting two steady-state points $u = 0$ and $u = 1$ of Eq. (4.141) and it is in agreement with the standard requirement to such type solutions. An example of such solution is presented in Figure 4.12. It should be noted that the classical FN equation possesses similar traveling fronts firstly derived in [146] (see also handbook [213]).

Subcase 4. If $p^3 + q^2 > 0$ then three roots of Eq. (4.128) are different and two of them are complex conjugate. Setting $\alpha_1 = \alpha$, $\alpha_{2,3} = a \pm ib$, where

$$\alpha = -2a,$$
$$a = -\tfrac{1}{2}\left(\sqrt[3]{-q+\sqrt{p^3+q^2}} - \sqrt[3]{q+\sqrt{p^3+q^2}}\right),$$
$$b = \tfrac{\sqrt{3}}{2}\left(\sqrt[3]{-q+\sqrt{p^3+q^2}} + \sqrt[3]{q+\sqrt{p^3+q^2}}\right),$$
(4.143)

the general solution of (4.126) can be presented in the form

$$w = \phi_1 e^{-2ay} + \left[\phi_2 \cos(by) + \phi_3 \sin(by)\right] e^{ay}.$$

Solutions of RDC equations with power-law diffusivity   175

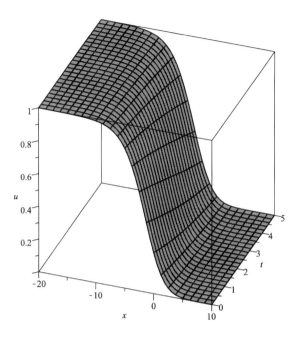

FIGURE 4.12: Exact solution (4.142) with $\lambda = \lambda_2 = 1, C_1 = C_2 = 1, C_3 = 0$

Using formulae (4.127), one arrives at the general solution

$$v = \frac{-2a\phi_1 e^{-2ay} + \left[(a\phi_2 + b\phi_3)\cos(by) + (a\phi_3 - b\phi_2)\sin(by)\right]e^{ay}}{\phi_1 e^{-2ay} + \left[\phi_2 \cos(by) + \phi_3 \sin(by)\right]e^{ay}} \quad (4.144)$$

of the nonlinear ODE (4.124). So, the analog of (4.139) takes the form

$$-3a(\dot{\phi}_1\phi_2 - \phi_1\dot{\phi}_2) + b(\phi_1\dot{\phi}_3 - \dot{\phi}_1\phi_3) = b\phi_1\phi_3(\lambda_2^* - 2\lambda_1^{*2}a) +$$
$$+ \tfrac{\lambda_1^*}{3}\lambda a\phi_1\phi_2(2a^2 + 2b^2 + 5a^3 + ab^2),$$

$$-3a(\dot{\phi}_1\phi_3 - \phi_1\dot{\phi}_3) + b(\dot{\phi}_1\phi_2 - \phi_1\dot{\phi}_2) = -b\phi_1\phi_2(\lambda_2^* - 2\lambda_1^{*2}a) + \quad (4.145)$$
$$+ \tfrac{\lambda_1^*}{3}\lambda a\phi_1\phi_3(2a^2 + 2b^2 + 5a^3 + ab^2),$$

$$\phi_2\dot{\phi}_3 - \dot{\phi}_2\phi_3 = \tfrac{\lambda_1^*}{3}b(2\lambda a + \lambda_2)(\phi_2^2 + \phi_3^2).$$

It should be stressed that the ODE system (4.145) has essentially different structure from those presented above and its solving takes a lot of efforts. We have done all necessary computations, which are omitting here, and to check the result using the program package Mathematica. As a result, the exact

solution
$$u = \left[\frac{a\cos\left(\frac{\lambda_1}{2\lambda}b(x-v_1t)\right) - b\sin\left(\frac{\lambda_1}{2\lambda}b(x-v_1t)\right) - 2aC\exp\left(-\frac{3\lambda_1}{2\lambda}a(x+v_2t)\right)}{\cos\left(\frac{\lambda_1}{2\lambda}b(x-v_1t)\right) + C\exp\left(-\frac{3\lambda_1}{2\lambda}a(x+v_2t)\right)}\right]^2$$

of the nonlinear RDC equation (3.141) with $\lambda_1 \neq 0$ has been found. Here a and b are determined by the formulae (4.143) and

$$v_1 = 2\lambda a + \frac{3\lambda_1}{2\lambda}, \quad v_2 = \frac{\lambda(b^2 + 3a^2)}{3a} - \frac{3\lambda_1}{2\lambda}.$$

Note that quasi-periodic solutions of the similar form were also obtained for the reaction-diffusion (RD) equation with a cubic nonlinearity [17, 88].

4.5 Solutions of reaction-diffusion-convection equations with exponential diffusivity

In this section, several nonlinear reaction-diffusion-convection (RDC) equations are studied in order to construct exact solutions and to provide their application and/or interpretation.

A common peculiarity of these equations consist of the structure of the diffusion and convection coefficients. We assume that both coefficients have the same structure given by exponential functions. Such nonlinearities are used as natural generalizations of those from Section 4.4. It should be noted that the RDC equations with exponential nonlinearities possess similar Lie symmetry properties to those with power-law nonlinearities. For example, both equations listed in cases 3 and 4 of Table 2.7 admit four-dimensional MAIs. Moreover, it can be easily shown that the corresponding Lie algebras are two different realizations of the same algebra. Similarly the RDC equations listed in cases 12 and 14 of Table 2.7 admit the same Lie algebra but with different realizations.

4.5.1 Lie's solutions of an equation with exponential diffusion and convection

Let us consider the equation from case 14 of Table 2.7, which reads as

$$u_t = (e^{ku}u_x)_x + qe^{mu}u_x + re^{(2m-k)u}, \quad (4.146)$$

where $|k| + |m| \neq 0$, $|q| + |2m - k| \neq 0$. Eq. (4.146) is invariant under three-dimensional MAI

$$AG^3 = <\partial_t,\ \partial_x,\ (k-2m)t\partial_t + (k-m)x\partial_x + \partial_u>. \quad (4.147)$$

TABLE 4.3: Lie's ansätze and reduction equations for Eq. (4.146)

	Ansatz	Reduction equation
1	$u = \varphi(\omega)$, $\omega = x - v_1 t$,	$(e^{k\varphi}\varphi_\omega)_\omega + qe^{m\varphi}\varphi_\omega + re^{(2m-k)\varphi} = -v_1\varphi_\omega$
2	$u = \varphi(\omega) + \frac{1}{m}\ln x$, $\omega = \alpha \ln x + t$, $k = 2m$	$\alpha^2\left(e^{2m\varphi}\varphi_\omega\right)_\omega + \left(3\alpha e^{2m\varphi} + q\alpha e^{m\varphi} - 1\right)\varphi_\omega +$ $+\frac{1}{m}(e^{2m\varphi} + qe^{m\varphi}) + r = 0$
3	$u = \varphi(\omega) - \frac{1}{m}\ln t$, $\omega = \alpha \ln t + x$, $k = m$	$(e^{m\varphi}\varphi_\omega)_\omega + qe^{m\varphi}\varphi_\omega + re^{m\varphi} = \alpha\varphi_\omega - \frac{1}{m}$
4	$u = \varphi(\omega) + \frac{1}{k-2m}\ln t$, $\omega = xt^{\frac{m-k}{k-2m}}$, $k \neq m; 2m$	$(e^{k\varphi}\varphi_\omega)_\omega + qe^{m\varphi}\varphi_\omega + re^{(2m-k)\varphi} =$ $= \frac{1}{2m-1}[(k-m)\omega\varphi_\omega - 1]$

The corresponding MGI is

$$\bar{t} = te^{(k-2m)u_0} + t_0, \quad \bar{x} = xe^{(k-m)u_0} + x_0, \quad \bar{u} = u + u_0. \tag{4.148}$$

To construct all nonequivalent Lie's ansätze we need to consider the linear combination of operators (4.147):

$$X = [\alpha_2(k - 2m)t + \alpha_0]\partial_t + [\alpha_2(k - m)x + \alpha_1]\partial_x + \alpha_2\partial_u.$$

The invariance surface condition (1.25), takes the form

$$[\alpha_2(k-2m)t + \alpha_0]u_t + [\alpha_2(k-m)x + \alpha_1]u_x = \alpha_2. \tag{4.149}$$

Solutions of (4.149) depend essentially on the values of the parameters α_0, α_1, α_2, m and k. Setting $\alpha_2 = 0$, we again obtain the plane wave ansatz and the corresponding reduction equation, which are listed in case 1 of Table 4.3 (v_1 and α are arbitrary parameters therein).

Since Eq. (4.149) has the same structure as Eq. (4.91), one needs again to consider three different cases: *I* $k = 2m$, *II* $k = m$, *III* $k \neq m, 2m$.

Each case can be examined in the same way as it was done in Section 4.2. In particular, the parameters α_0 and/or α_1 can be skipped without losing a generality (see transformation (4.148)). As a result, all the inequivalent Lie ansätze and reduction equations were constructed and they are presented in cases 2–4 of Table 4.3.

All the reduction equations in Table 4.3 are nonintegrable ODEs provided their coefficients are arbitrary. However, several particular solutions can be found under correctly-specified restrictions on the coefficients. In particular, some exact solutions of Eq. (4.146) with $q = 0$ (i.e., no convective term), which are known in literature [213], can be identified. We present here examples of exact solutions of Eq. (4.87) with $q \neq 0$ in order to extend the list of such solutions.

It can be noted that the parameter m can be reduced to a fixed number without losing a generality in cases 2–3 of Table 4.3. In fact, Eq. (4.146) admits the ET $ku \to u$ provided $k = 2m \neq 0$ and $k = m \neq 0$. Hence, we fix m for simplicity in what follows.

In case case 1 of Table 4.3, plane wave solutions can be obtained. Setting $r = 0$, one notes that the reduction equation can be immediately reduced to the first-order ODE

$$e^{k\varphi}\varphi_\omega + \frac{q}{m}e^{m\varphi} + v_1\varphi + C = 0,$$

which is integrable. As a result, the plane wave solution

$$\int \frac{e^{ku}du}{\frac{q}{m}e^{mu} + v_1 u + C} = v_1 t - x$$

of Eq. (4.146) with $r = 0$ is obtained.

The above integral can be expressed in terms of elementary functions only in particular cases. For example, setting $v_1 = 0$, $k = m$ and $C \neq 0$, the stationary solution

$$u = \frac{1}{m}\ln|e^{-qx} + C^*|, \quad C^* = -\frac{mC}{q}$$

of the diffusion-convection equation

$$u_t = (e^{mu}u_x)_x + qe^{mu}u_x$$

is obtained.

Let us consider case 2 of Table 4.3. Assuming $\alpha = 0$ and setting $m = \frac{1}{2}$, we obtain

$$\varphi_\omega = 2e^\varphi + 2qe^{\frac{\varphi}{2}} + r. \tag{4.150}$$

Solving Eq. (4.150) and using the ansatz from case 2 of Table 4.3, we arrive at a solution in the implicit form.

$$\int_0^{u-2\ln x} \frac{d\tau}{2e^\tau + 2qe^{\frac{\tau}{2}} + r} = t. \tag{4.151}$$

If $r = 0$ then we immediately obtain the solution

$$\ln|1 + qxe^{-\frac{u}{2}}| - qxe^{-\frac{u}{2}} = q^2 t$$

of Eq. (4.152) with $r = 0$.

If $r \neq 0$ then the integral in the left-hand side of (4.151) essentially depends on the value $D_1 = q^2 - 2r$. In the case $D_1 < 0$, the solution

$$u - \frac{2q}{\sqrt{-D_1}} \arctan\left(\frac{2e^{\frac{u}{2}} + qx}{\sqrt{-D_1} x}\right) - \ln\left|2e^u + 2qxe^{\frac{u}{2}} + rx^2\right| = rt,$$

of the equation

$$u_t = (e^u u_x)_x + qe^{\frac{u}{2}} u_x + r \qquad (4.152)$$

is obtained. In the case $D_1 = 0$ the solution

$$2u + \frac{4qx}{2e^{\frac{u}{2}} + qx} - 4\ln\left|qx + 2e^{\frac{u}{2}}\right| = q^2 t$$

of Eq. (4.152) with $r = \frac{q^2}{2}$ is derived. Finally, we obtain the solution

$$k_2 \ln\left|1 - k_1 x e^{-\frac{u}{2}}\right| - k_1 \ln\left|1 - k_2 x e^{-\frac{u}{2}}\right| = \pm\frac{r}{2}\sqrt{D_1} t$$

if $D_1 > 0$. Here $k_{1,2} = \frac{-q \pm \sqrt{D_1}}{2}$.

Let us consider case 3 of Table 4.3. Setting $m = 1$ and applying the substitution $\varphi = \ln \psi$ to the reduction equation we obtain

$$\psi_{\omega\omega} + q\psi_\omega + r\psi = \alpha \frac{\psi_\omega}{\psi} - 1. \qquad (4.153)$$

Obviously, ODE (4.153) reduces to the linear equation by setting $\alpha = 0$. Thus, we obtain three different exact solutions depending on $D_2 = q^2 - 4r$. As a result, three families of exact solutions of the nonlinear RDC equation

$$u_t = (e^u u_x)_x + qe^u u_x + re^u$$

were found, namely

$$u = \ln\left(\frac{C_1 e^{k_1 x} + C_2 e^{k_2 x} - 1}{rt}\right), \quad k_{1,2} = \frac{-q \pm \sqrt{D_2}}{2}, \quad D_2 > 0,$$

$$u = \ln\left[\frac{(C_1 + C_2 x) e^{-\frac{q}{2}x} - 4}{q^2 t}\right], \quad r = \frac{q^2}{4},$$

$$u = \ln\left[\frac{\left(C_1 \cos(\frac{\sqrt{-D_2}}{2} x) + C_2 \sin(\frac{\sqrt{-D_2}}{2} x)\right) e^{-\frac{q}{2}x} - 1}{rt}\right], \quad D_2 < 0.$$

Incidentally, the above solutions possess an interesting property. For example, the latter can be transformed by the time translation $t_0 > 0$ to the form

$$u = \ln\left[\frac{\left(C_1 \cos(\frac{\sqrt{-D_2}}{2} x) + C_2 \sin(\frac{\sqrt{-D_2}}{2} x)\right) e^{-\frac{q}{2}x} - 1}{r(t - t_0)}\right].$$

Obviously, it is a blow-up solution because one increases to infinity for the finite time t_0. Such solutions were extensively studied during the last decades because they are important in some real world applications (see, e.g., [225]).

4.5.2 Non-Lie solutions of an equation with exponential diffusion and convection

Here we use the results of Section 3.6 for construction of exact solutions in two essentially different cases. Firstly we find exact solutions of the RDC equation (3.164) using the Q-conditional symmetry (3.165). In what follows, any solution of the form $u(t+t_0, x+x_1)$ (t_0 and x_0 are some constants) was simplified to the form $u(t,x)$, since each RDC equation belonging to the class (3.34) is invariant w.r.t. the time and space translations (see (3.163)).

As is noted above, the Q-conditional symmetry operators have essentially simpler structure if one uses substitution (3.172). Taking this into account, we will map Eq. (3.164) and the corresponding operators to the simpler forms using (3.172) with $n=1$, construct exact solutions for the equations obtained and use inverse substitution at the final step [78].

The simplest calculations occur when the Q-conditional symmetry operator (3.195) corresponding case 1 of Table 3.1 is examined. So, we present the final list of exact solutions obtained. The RDC equation (3.164), i.e.,

$$u_t = (e^u u_x)_x + \lambda e^u u_x + \lambda_0 + \lambda_1 e^u + \lambda_2 e^{-u}$$

has the exact solutions

$$u = \ln\left[\frac{\lambda_2 t(\frac{t}{2}+C) - \frac{x}{\lambda}}{t+C}\right], \quad \lambda_0 = 0$$

and

$$u = \ln\left(\frac{\lambda_0 x - C\lambda_2 e^{-\lambda_0 t} - \lambda\lambda_2 t}{C\lambda_0 e^{-\lambda_0 t} - \lambda}\right), \quad \lambda_0 \neq 0,$$

provided $\lambda_1 = 0$. If $\lambda_1 \neq 0$ then the corresponding exact solutions are

$$u = \ln\left[\frac{C}{t}\exp\left(\frac{\lambda_0 t - (\lambda \pm \sqrt{P})x}{2}\right) - \frac{1}{\lambda_1 t} - \frac{\lambda_0}{2\lambda_1}\right], \quad D=0;$$

$$u = \ln\left[\frac{C\exp\left(\frac{(\lambda_0 \pm \sqrt{D})t - (\lambda \pm \sqrt{P})x}{2}\right)}{1 - e^{\pm\sqrt{D}t}} \mp \frac{\sqrt{D}}{2\lambda_1}\coth\left(\frac{\pm\sqrt{D}t}{2}\right) - \frac{\lambda_0}{2\lambda_1}\right], D>0;$$

$$u = \ln\left[\frac{C\exp\left(\frac{\lambda_0 t - (\lambda \pm \sqrt{P})x}{2}\right)}{\cos(\frac{\sqrt{-D}}{2}t)} + \frac{\sqrt{-D}}{2\lambda_1}\tan\left(\frac{\sqrt{-D}}{2}t\right) - \frac{\lambda_0}{2\lambda_1}\right], D<0,$$

where $D = \lambda_0^2 - 4\lambda_1\lambda_2$ and $P = \lambda^2 - 4\lambda_1$.

Now we consider the simplest form of operator (3.165) occurring in case 2 of Table 3.1. Substitution (3.172) reduces Eq. (3.164) and operator (3.165) with $\alpha = \frac{\lambda_0 \pm \sqrt{D}}{2}$ and $a = 0$ (see case 2 of Table 3.1) to the forms

$$v_{xx} = v^{-1}v_t - \lambda v_x - \lambda_0 - \lambda_1 v - \lambda_2 v^{-1} \qquad (4.154)$$

and
$$Q_{1,2} = \partial_t + \left(\frac{\lambda_0 \pm \sqrt{D}}{2} v + \lambda_2\right)\partial_v. \qquad (4.155)$$

To construct the corresponding solutions, one needs to solve the overdetermined system consisting of Eq. (4.154) and the invariance surface condition

$$Q_{1,2}(v) \equiv v_t - \frac{\lambda_0 \pm \sqrt{D}}{2} v - \lambda_2 = 0. \qquad (4.156)$$

So, extracting v_t from (4.156) and substituting into (4.154), we arrive at the linear ODE (with the time variable as a parameter)

$$v_{xx} + \lambda v_x + \lambda_1 v = -\frac{\lambda_0 \mp \sqrt{D}}{2}, \qquad (4.157)$$

which possesses the general solution

$$v = \begin{cases} \phi_1(t)e^{-\lambda x} + \phi_2(t) - \frac{\lambda_0 \mp |\lambda_0|}{2\lambda}x, & \lambda_1 = 0, \\ e^{-\frac{\lambda}{2}x}\left[\phi_1(t) + x\phi_2(t)\right] - \frac{2(\lambda_0 \mp \sqrt{D})}{\lambda^2}, & \lambda_1 \neq 0,\ P = 0, \\ e^{-\frac{\lambda}{2}x}\left[\phi_1(t)e^{\frac{\sqrt{P}}{2}x} + \phi_2(t)e^{-\frac{\sqrt{P}}{2}x}\right] - \frac{\lambda_0 \mp \sqrt{D}}{\lambda_1}, & \lambda_1 \neq 0,\ P > 0, \\ e^{-\frac{\lambda}{2}x}\left[\phi_1(t)\cos(\frac{\sqrt{-P}}{2}x) + \phi_2(t)\sin(\frac{\sqrt{-P}}{2}x)\right] - \frac{\lambda_0 \mp \sqrt{D}}{\lambda_1}, & \lambda_1 \neq 0,\ P < 0, \end{cases}$$

where $P = \lambda^2 - 4\lambda_1$, $\phi_i(t), i = 1, 2$ are arbitrary smooth functions at the moment. Substituting the expressions obtained above into (4.156), we obtain four linear systems of first-order ODEs to find the functions $\phi_i(t), i = 1, 2$ depending on λ_1 and P. These systems are integrable, therefore the solutions have been found in explicit forms. For example, the linear ODE system

$$\dot\phi_1 = \frac{\lambda_0 \pm |\lambda_0|}{2}\phi_1,\quad \dot\phi_2 = \frac{\lambda_0 \pm |\lambda_0|}{2}\phi_2 + \lambda_2$$

is obtained if $\lambda_1 = 0$. Finally, applying substitution (3.172), the following exact solutions of Eq. (3.164) have been constructed.

If $\lambda_1 = 0$ then the exact solutions are

$$u = \ln\left[e^{\lambda_0 t}(C_1 e^{-\lambda x} + C_2) - \frac{\lambda_2}{\lambda_0}\right],\ \lambda_0 \neq 0 \qquad (4.158)$$

and

$$u = \ln\left(C_1 e^{-\lambda x} - \frac{\lambda_0}{\lambda}x + \lambda_2 t + C_2\right). \qquad (4.159)$$

If $\lambda_1 \neq 0$ then the exact solutions have the form

$$u = \ln\left[b(x)\exp\left(\frac{\lambda_0 \pm \sqrt{D}}{2}t\right) + \frac{-\lambda_0 \pm \sqrt{D}}{2\lambda_1}\right], \qquad (4.160)$$

where

$$b(x) = \begin{cases} e^{-\frac{\lambda}{2}x}(C_1 + C_2 x), & P = 0, \\ e^{-\frac{\lambda}{2}x}\left(C_1 e^{\frac{\sqrt{P}}{2}x} + C_2 e^{-\frac{\sqrt{P}}{2}x}\right), & P > 0, \\ e^{-\frac{\lambda}{2}x}\left[C_1 \cos\left(\frac{\sqrt{-P}}{2}x\right) + C_2 \sin\left(\frac{\sqrt{-P}}{2}x\right)\right], & P < 0. \end{cases} \quad (4.161)$$

Now we examine the second case of Table 3.1 when the Q-conditional symmetry operator (3.165) contains the time dependent coefficient $\alpha \neq const$. Substitution (3.172) reduces Eq. (3.164) and operator (3.165) to the forms (4.154) and

$$Q = \partial_t + [\alpha(t) + \lambda_2]\partial_v. \qquad (4.162)$$

Operator (4.162) produces the linear equation

$$v_t = \alpha(t)v + \lambda_2. \qquad (4.163)$$

So, we again substitute v_t from Eq. (4.163) into (4.154) and obtain the linear ODE

$$v_{xx} + \lambda v_x + \lambda_1 v = \alpha(t) - \lambda_0, \qquad (4.164)$$

where t is a parameter. Depending on $P = \lambda^2 - 4\lambda_1$, Eq. (4.164) possesses the general solutions:

$$v = \begin{cases} \phi_1(t)e^{-\lambda x} + \phi_2(t) + \frac{\alpha(t)-\lambda_0}{\lambda}x, & \lambda_1 = 0, \\ e^{-\frac{\lambda}{2}x}\left[\phi_1(t) + x\phi_2(t)\right] + \frac{4[\alpha(t)-\lambda_0]}{\lambda^2}, & \lambda_1 \neq 0, \ P = 0, \\ e^{-\frac{\lambda}{2}x}\left[\phi_1(t)e^{\frac{\sqrt{P}}{2}x} + \phi_2(t)e^{-\frac{\sqrt{P}}{2}x}\right] + \frac{\alpha(t)-\lambda_0}{\lambda_1}, & \lambda_1 \neq 0, \ P > 0, \\ e^{-\frac{\lambda}{2}x}\left[\phi_1(t)\cos(\frac{\sqrt{-P}}{2}x) + \phi_2(t)\sin(\frac{\sqrt{-P}}{2}x)\right] + \frac{\alpha(t)-\lambda_0}{\lambda_1}, & \lambda_1 \neq 0, \ P < 0. \end{cases}$$

The next step is to substitute these solutions into (4.162) and to obtain the linear ODE systems to find the functions $\phi_1(t)$ and $\phi_2(t)$. We omit the relevant calculations because they are very similar to those we have done for operator (3.165) with $\alpha = \frac{\lambda \pm \sqrt{D}}{2}$ (see (4.155)). Thus, we present only the list of exact solutions of Eq. (3.164).

If $\lambda_1 = 0$ then its exact solutions are

$$u = \ln\left(\frac{C}{t}e^{-\lambda x} - \frac{x}{\lambda t} + \frac{\lambda_2}{2}t\right), \quad \lambda_0 = 0 \qquad (4.165)$$

and

$$u = \ln\left[\frac{e^{\lambda_0 t}(Ce^{-\lambda x} - \frac{\lambda_0}{\lambda}x + \lambda_2 t) \pm \frac{\lambda_2}{\lambda_0}}{e^{\lambda_0 t} \mp 1}\right], \quad \lambda_0 \neq 0. \qquad (4.166)$$

If $\lambda_1 \neq 0$ then its exact solutions are

$$u = \ln\left[e^{-\frac{\lambda}{2}x}\left(\phi_1(t) + \phi_2(t)x\right) + \frac{4(\alpha(t)-\lambda_0)}{\lambda^2}\right], \qquad (4.167)$$

if $P = 0$;

$$u = \ln\left[e^{-\frac{\lambda}{2}x}\left(\phi_1(t)e^{\frac{\sqrt{P}}{2}x} + \phi_2(t)e^{-\frac{\sqrt{P}}{2}x}\right) + \frac{\alpha(t) - \lambda_0}{\lambda_1}\right], \qquad (4.168)$$

if $P > 0$;

$$u = \ln\left[e^{-\frac{\lambda}{2}x}\left(\phi_1(t)\cos(\frac{\sqrt{-P}}{2}x) + \phi_2(t)\sin(\frac{\sqrt{-P}}{2}x)\right) + \frac{\alpha(t) - \lambda_0}{\lambda_1}\right], \qquad (4.169)$$

if $P < 0$. Here the function α is given in (3.182) and

$$\phi_{1,2}(t) = \begin{cases} \frac{C_{1,2}e^{\frac{\lambda_0}{2}t}}{t}, & D \equiv \lambda_0^2 - 4\lambda_1\lambda_2 = 0, \\ \frac{C_{1,2}e^{\frac{\lambda_0}{2}t}}{\cos(\frac{\sqrt{-D}}{2}t)}, & D < 0, \\ \frac{C_{1,2}e^{\frac{\lambda_0}{2}t}}{\sinh(\frac{\sqrt{D}}{2}t)}, & (2\alpha - \lambda_0)^2 > D > 0, \\ \frac{C_{1,2}e^{\frac{\lambda_0}{2}t}}{\cosh(\frac{\sqrt{D}}{2}t)}, & D > (2\alpha - \lambda_0)^2 > 0. \end{cases}$$

Thus, examination of case 2 is now completed.

Cases 3–6 of Table 3.1 can be treated in a similar way to case 2. In all the cases, the corresponding Q-conditional symmetry operators for unknown function v take the same structure

$$Q = \partial_t + av\partial_x + \left[(a_x - a^2 - \lambda a)v^2 + \alpha v + \lambda_2\right]\partial_v, \qquad (4.170)$$

hence, we obtain the invariance surface condition

$$v_t = -avv_x + (a_x - a^2 - \lambda a)v^2 + \alpha v + \lambda_2, \qquad (4.171)$$

where the functions a and α are listed in cases 3–6 of Table 3.1. Inserting v_t from (4.171) into (4.154), one arrives at the linear ODE

$$v_{xx} + (a + \lambda)v_x + (-a_x + a^2 + \lambda a + \lambda_1)v = \alpha - \lambda_0. \qquad (4.172)$$

Unfortunately, this equation takes a very cumbersome form when one substitutes a and α therein.

The simplest form of (4.172) occurs when the functions a and α are taken from case 3 of Table 3.1. In this case, Eq. (4.172) takes the form

$$\left(C_0 e^{\lambda x} - C_1 x + \theta\right) v_{xx} + (-C_1\lambda x + \lambda\theta + C_1)v_x + C_1\lambda v + \theta_t = 0, \qquad (4.173)$$

where $\theta(t)$ is an arbitrary smooth function. The general solution of (4.173) can be easily constructed because one notes that

$$v = x - \frac{1}{\lambda} - \frac{\theta}{C_1}, \quad C_1 \neq 0$$

is the particular solution of the corresponding homogenous ODE. Thus, applying the standard procedure the general solution of (4.173) in the form

$$v = \frac{1}{\lambda}\left[\left(x - \frac{\theta(t)}{C_1} + \frac{1}{\lambda}\right)\phi_1(t) + \left(e^{-\lambda x} - \frac{C_0\lambda}{C_1}\right)\phi_2(t) - \frac{\theta_t}{C_1}\right] \qquad (4.174)$$

was found. Inserting v from (4.174) into Eq. (4.154) and making the relevant calculations, we obtain the ODE system

$$\dot{\phi}_1 = \phi_1^2 + \lambda_0\phi_1, \quad \dot{\phi}_2 = \phi_1\phi_2 + \lambda_0\phi_2 \qquad (4.175)$$

for the functions ϕ_1 and ϕ_2. This system is integrable and produces two different solutions depending on the value of λ_0. Making straightforward calculations several two-parameter families of exact solutions of the nonlinear RDC

$$u_t = (e^u u_x)_x + \lambda e^u u_x + \lambda_0 + \lambda_2 e^{-u}$$

were constructed. However, we noted that they are nothing else but exact solutions (up to equivalence transformations) presented above in formulae (4.165) and (4.166).

Finally, setting $C_1 = 0$ into Eq. (4.173), we have shown that new solutions do not appear.

Thus, notwithstanding operator (3.165) with the coefficients a and α taken from the third case of Table 3.1 has essentially different structure than the operator with those taken from the second case of Table 3.1, this operator does not produce any new exact solutions.

It turns out that the same result is obtained if one examines cases 4–6 from Table 3.1 [78]. Now we show why these cases do not lead to new solutions. In fact, denoting

$$E(a) := -(-a_x + a^2 + \lambda a + \lambda_1) \qquad (4.176)$$

and using the second equation from (3.166) one easily derives the equation $\frac{\partial E(a)}{\partial x} = aE(a)$, i.e., $E(a) = c(t)\exp\int a(t,x)dx$, where $c(t)$ is an arbitrary smooth function. Because the function $a(t,x)$ has the form $a(t,x) = -\partial_x[\ln w(t,x)]$ (see cases 3–6 of Table 3.1 for the explicit form of $w(t,x)$), the formula for $E(a)$ can be rewritten as

$$E(a) = \frac{c(t)}{w(t,x)}. \qquad (4.177)$$

Now substituting (4.177) into (4.176) and taking into account the expression for $a(t,x)$, one obtains

$$c(t) = -w_{xx} + \lambda w_x - \lambda_1 w. \qquad (4.178)$$

Having formulae (4.176)–(4.178), one realizes that Eq. (4.172) takes the form

$$wv_{xx} + (\lambda w - w_x)v_x + (w_{xx} - \lambda w_x + \lambda_1 w)v = -w_t. \qquad (4.179)$$

Differentiating this equation with respect to (w.r.t.) the variable x and taking into account (4.178) and the identity $w_{tx} = 0$ (see cases 3–6 of Table 3.1), we immediately obtain the linear third-order equation

$$v_{xxx} + \lambda v_{xx} + \lambda_1 v_x = 0.$$

It means that (4.172) is reducible to the second order equation with constant coefficient

$$v_{xx} + \lambda v_x + \lambda_1 v = e(t), \qquad (4.180)$$

where the function $e(t)$ will be specified below. Moreover, subtracting (4.180) from (4.172) one obtains

$$a v_x + (-a_x + a^2 + \lambda a) v + e(t) = \alpha(t) - \lambda_0, \qquad (4.181)$$

therefore Eq. (4.171) can be simplified as follows

$$v_t = [e(t) + \lambda_0] v + \lambda_2. \qquad (4.182)$$

Thus, equations (4.171) and (4.172) with function $a(t,x)$ from Table 3.1 (see cases 3–6) are equivalent to the linear equations (4.180) and (4.182). Finally, setting $e(t) = \alpha(t) - \lambda_0$ we derive exactly Eqs. (4.163)–(4.164), which were solved above.

Thus, all the Q-conditional symmetries arising in case 1 of Theorem 3.8 and Table 3.1 were applied for search exact solutions of the nonlinear RDC (3.164).

The most cumbersome structure of the Q-conditional symmetries occurs in case 2 of Theorem 3.8. As consequence, essential difficulties arise if one applies operator (3.168) with the coefficients from Table 3.2. Here we examine only the simplest case, operator (3.168) with coefficients (3.227), for finding exact solutions of Eq. (3.167), i.e.

$$u_t = (e^u u_x)_x + \lambda e^{2u} u_x + \frac{1}{9} \lambda^2 e^{3u} + \lambda_0 + \lambda_1 e^u + \lambda_2 e^{-u}.$$

We again use substitution (3.172) to simplify calculations. The corresponding overdetermined system takes the form

$$v_{xx} = v^{-1} v_t - \lambda v v_x - \frac{1}{9} \lambda^2 v^3 - \lambda_0 - \lambda_1 v - \lambda_2 v^{-1}, \qquad (4.183)$$

$$Q(v) \equiv v_t + a v v_x + \frac{\lambda}{3} a v^3 + a^2 v^2 - \left(\frac{\lambda \lambda_2}{3a} + \lambda_0 \right) v - \lambda_2 = 0, \qquad (4.184)$$

where a is a root of the fourth-order polynomial

$$9 a^4 + 9 \lambda_1 a^2 + 3 \lambda \lambda_0 a + \lambda^2 \lambda_2 = 0. \qquad (4.185)$$

Extracting v_t from (4.184) and substituting into (4.183), we arrive at a nonlinear ODE (with the variable t as a parameter), which reduces to the form

$$v^*_{yy} + 3 v^* v^*_y + (v^*)^3 + 3 p v^* + 2 q = 0 \qquad (4.186)$$

by the substitution $v = v^* - \frac{a}{\lambda}$, $x = \frac{3}{\lambda}y$. Hereafter

$$p = \frac{1}{\lambda^2}(2a^2 + 3\lambda_1), \quad q = -\frac{1}{2\lambda^3 a}(7a^4 + 9\lambda_1 a^2 + 3\lambda_2 \lambda^2).$$

ODE (4.186) has the same structure as (4.48), hence, we use now the known substitution $v^* = \frac{w_y}{w}$ to linearize Eq. (4.186):

$$w_{yyy} + 3pw_y + 2qw = 0. \tag{4.187}$$

Eq. (4.187) is the linear third order ODE, hence its general solution can be easily constructed. Four different cases occur depending on p and q. Corresponding calculations are rather cumbersome but very similar to those presented above in Subsection 4.4.2. Here we present only the final results.

Case 1: $p = q = 0$. The exact solutions of the nonlinear RDC equation (3.167) are

$$u = \ln\left(\frac{3}{a^2 t + \lambda x} - \frac{a}{\lambda}\right) \tag{4.188}$$

and

$$u = \ln\left[\frac{6(a^2 t + \lambda x)}{(a^2 t + \lambda x)^2 + 6a\lambda t} - \frac{a}{\lambda}\right],$$

where the coefficient restrictions

$$\lambda_0 = -\frac{8a^3}{9\lambda}, \quad \lambda_1 = -\frac{2a^2}{3}, \quad \lambda_2 = -\frac{a^4}{3\lambda^2}$$

are assumed.

Case 2: $p^3 = -q^2 \neq 0$. The exact solutions of Eq. (3.167) are

$$u = \ln\left[\frac{a(1 - 2Ce^{-\alpha(x-vt)})}{\lambda(1 + Ce^{-\alpha(x-vt)})} - \frac{a}{\lambda}\right] \tag{4.189}$$

and

$$u = \ln\left[\frac{\alpha(x + v_2 t) + 3 - 2\alpha Ce^{-\alpha(x+v_1 t)}}{\lambda(x + v_2 t + Ce^{-\alpha(x+v_1 t)})} - \frac{a}{\lambda}\right],$$

where $\alpha = \pm\sqrt{-2a^2 - 3\lambda_1}$, $2a^2 + 3\lambda_1 < 0$, $v_1 = \frac{a}{\lambda}(a - \alpha)$, $v_2 = \frac{a}{\lambda}(a + 2\alpha)$.
In this case, the coefficients of Eq. (3.167) must satisfy the restriction

$$\Delta \equiv 4(2a^2 + 3\lambda_1)^3 + (20a^3 + 9\lambda\lambda_0 + 18\lambda_1 a)^2 = 0,$$

where a is a root of (4.185).

Case 3: $p^3 + q^2 < 0$. The corresponding exact solution involves three different exponents and has the form:

$$u = \ln\left[\frac{\sum_{i=1}^{3} \alpha_i C_i \exp\left(\frac{\lambda \alpha_i}{3}(x + v_i t)\right)}{\sum_{i=1}^{3} C_i \exp\left(\frac{\lambda \alpha_i}{3}(x + v_i t)\right)} - \frac{a}{\lambda}\right], \tag{4.190}$$

where $v_i = a(\frac{a}{\lambda} + \alpha_i)$, and the parameters $\alpha_i, i = 1, 2, 3$ are calculated by the Cardano formulae (4.137). In this case, the coefficients of Eq. (3.167) must satisfy the restriction $\Delta < 0$, where a is a root of (4.185).

Case 4: $p^3 + q^2 > 0$. The corresponding exact solution involves periodic functions and has the form

$$u = \ln\left[\frac{(\alpha_1 - \frac{a}{\lambda})\cos[\gamma(x+v_1t)] - \beta_1 \sin[\gamma(x+v_1t)] - C\left(2\alpha_1 + \frac{a}{\lambda}\right)e^{-\lambda\alpha_1(x+v_2t)}}{Ce^{-\lambda\alpha_1(x+v_2t)} + \cos[\gamma(x+v_1t)]}\right],$$

where

$$\gamma = \frac{\lambda\beta_1}{3}, \quad v_2 = a\left(\alpha_1 + \frac{\beta_1^2}{3\alpha_1} - \frac{a}{\lambda}\right), \quad v_1 = a\left(2\alpha_1 + \frac{a}{\lambda}\right),$$

$$\alpha_1 = -\frac{1}{2}\left(\sqrt[3]{-q + \sqrt{p^3 + q^2}} - \sqrt[3]{q + \sqrt{p^3 + q^2}}\right),$$

$$\beta_1 = \frac{\sqrt{3}}{2}\left(\sqrt[3]{-q + \sqrt{p^3 + q^2}} + \sqrt[3]{q + \sqrt{p^3 + q^2}}\right),$$

here $-2\alpha_1$ and $\alpha_1 \pm i\beta_1$ are the roots of the characteristic equation (4.187). In this case, the coefficients of Eq. (3.167) must satisfy the restriction $\Delta > 0$, where a is a root of (4.185).

Now we present a brief analysis of the exact solutions constructed above. First of all it should be stressed that all the solution obtained in this subsection have the structure $u = \ln f(t, x)$, where $f(t, x)$ is the correctly-specified function, hence each solution is real only under condition $f(t, x) > 0$.

One notes that the nonlinear RDC equation (3.164) with $\lambda_2 \neq 0$ admits only two-dimensional group of invariance consisting of the time and space translations. Thus, all the solutions with $\lambda_2 \neq 0$ are not obtainable by the classical Lie method, excepting the special cases when some of them are reduced to plane wave solution by vanishing some constants (e.g., $C_1 = 0$ in (4.158)).

If $\lambda_2 = 0$, then this equation (depending on λ, λ_0 and λ_1) admits three or four-dimensional group of invariance (see cases 17, 18, 27 and 28 in Table 2.5). It means that some solutions found above can be also obtainable via Lie symmetries. Let us present a very nontrivial example for Eq. (3.164) with $\lambda_2 = 0$ and $\lambda_1 = \frac{2\lambda^2}{9}$, which admits the four-dimensional MAI (see case 18 of Table 2.5). This equation possesses the invariant solution [48]

$$u(t, x) = \ln\left[\frac{9}{2\lambda_1^2} \frac{C_1 \exp(-\frac{1}{3}\lambda_1 x) + C_2 \exp(-\frac{2}{3}\lambda_1 x) - \lambda_0}{1 + a_0 \exp(-\lambda_0 t)}\right]. \quad (4.191)$$

Now one may check that the exact solution (4.168) with $\lambda_2 = 0$ and $\lambda_1 = \frac{2\lambda^2}{9}$ reduces exactly to (4.191)! So, we may guarantee that (4.168) is non-Lie solution only in the case $\lambda_2 \neq 0$.

Now we compare the solutions obtained above with those found earlier. We have also checked that solutions (4.160)–(4.161) with $\lambda = 0$ produces the solutions obtained in [16] (see formulae (3.11a), (3.11b), (3.11c) therein). In the paper [220], exact solutions of Eq. (3.164) were constructed using the generalized conditional symmetries. The first, second and third solutions listed in

Table 4 [220] are nothing else but particular cases of the exact solution (4.168) of the nonlinear equation (3.164) with $\lambda_2 \neq 0$, while the fourth and fifth solutions from Table 4 [220] are also obtainable by the Q-conditional symmetries (see solutions (4.158) and (4.159)). In Section 5.2, exact solutions of Eq. (3.164) are also constructed using another method and the corresponding comparison of the solutions obtained is presented in Section 5.3.

Finally, we note that the solutions found for the RDC equation (3.167) are not obtainable by using Lie symmetries provided $|\lambda_0|+|\lambda_1|+|\lambda_2| \neq 0$. In fact, Eq. (3.167) with $|\lambda_0| + |\lambda_1| + |\lambda_2| \neq 0$ admits the trivial Lie algebra, hence plane wave solutions can be only derived (see (4.188) and (4.189)). Obviously, all the other solutions derived above for Eq. (3.167) are non-Lie solutions. To the best of our knowledge, exact solutions of Eq. (3.167) were studied only in paper [77].

4.5.3 Application of the solutions obtained for population dynamics

Here we show how to apply the exact solutions obtained for solving boundary-value problems related to population dynamics.

Let us consider the following RD equation with exponential nonlinearities:

$$u_t = (e^{mu}u_x)_x + \lambda e^{mu}u_x + \lambda_0 + \lambda_1 e^{mu} + \lambda_2 e^{-mu}, \quad m \neq 0 \qquad (4.192)$$

This equation can be applied for mathematical modeling processes in population dynamics, when density (concentration) depends on the diffusivity and convection velocity by exponential law. Exponential law of growth for coefficients may be treated as the further generalization of the known models with power density-dependent coefficients [160, 180].

In the particular case, using the known series expansion

$$\begin{cases} e^{mu} = 1 + mu + \frac{m^2 u^2}{2} + \ldots \\ e^{-mu} = 1 - mu + \frac{m^2 u^2}{2} + \ldots, \end{cases} \qquad (4.193)$$

one can construct the generalized equation

$$u_t = [(1+mu)u_x]_x + \lambda(1+mu)u_x + \lambda_1^* u - \lambda_2^* u^2 \qquad (4.194)$$

from Eq. (4.192) as some approximation under the restriction

$$\lambda_0 + \lambda_1 + \lambda_2 = 0. \qquad (4.195)$$

This equation is simplified to the form

$$u_t = [(1+mu)u_x]_x + \lambda^* u u_x + \lambda_1^* u - \lambda_2^* u^2 \qquad (4.196)$$

by the Galilei boost $x \to x + \lambda t$ and renaming $\lambda = \frac{\lambda^*}{m}$. Equation (4.196) with $m = \lambda^* = 0$ is the classical Fisher equation [98] while one with $\lambda^* = 0$ is the Fisher equation with the simplest nonconstant diffusivity, which was investigated in [48, 61] (see Subsection 5.2.3 below). Equation (4.196) with $m = 0$ coincides with the Murray equation [179], which was studied in Section 4.3. It should be stressed that all these equations play an important role in mathematical biology and ecology [180, 181, 194].

Thus, it is important to look for the solutions of Eq. (4.192) and to study their properties. One notes that this equation reduces to the form

$$u_t = (e^u u_x)_x + \lambda e^u u_x + \lambda_0 + \lambda_1 e^u + \lambda_2 e^{-u} \qquad (4.197)$$

by the renaming $mu \to u, \lambda_i \to \frac{\lambda_i}{m}, i = 0, 1, 2$. Since the Fisher and Murray equations possess two nonnegative steady-state points and without a loss of generality they can be set $u_0 = 0, u_1 = 1$, we apply the same requirement for (4.197) and this leads to the condition (4.195) and $\lambda_2 = e\lambda_1$. Hence we consider the nonlinear RDC equation

$$u_t = (e^u u_x)_x + \lambda e^u u_x + \lambda_1 [e^u + e^{1-u} - (1+e)], \quad \lambda_1 > 0, \qquad (4.198)$$

possessing the steady-state points $u_0 = 0, u_1 = 1$.

Now we specify among the solutions constructed above those satisfying Eq. (4.198) and the zero flux boundary conditions, which are typical biologically motivated requirements. Because the coefficients in (4.198) satisfy the inequality $D = \lambda_1^2 (e-1)^2 > 0$, this restricts essentially our choice.

One notes that exact solutions (4.167)–(4.169) with the restriction $D > 0$ possess the structure

$$u = \ln\left[\frac{1+e}{2} - \frac{\sqrt{D}}{2\lambda_1}\tanh(\frac{\sqrt{D}}{2}t) + \frac{C_1 f_1(x) + C_2 f_2(x)}{\exp(\frac{\lambda}{2}x + \frac{\lambda_1(e+1)}{2}t)\cosh(\frac{\sqrt{D}}{2}t)}\right], \qquad (4.199)$$

where C_1 and C_2 are arbitrary constants while the functions f_1 and f_2 can be easily derived from (4.167)–(4.169). Notably, the second family of solutions can be obtained replacing in (4.199) the functions tanh and cosh by those coth and sinh, respectively.

Let us assume that Eq. (4.198) gives the population density in the unbounded domain $\Omega = \{(t,x) \in (0,+\infty) \times (0,+\infty)\}$ and zero flux conditions (zero Neumann conditions) take place for $x = 0$ and $x = \infty$. Using the solution (4.199) with $f_1 = 1$ and $f_2 = x$ (see formula (4.167)) and the correctly-specified constants C_1 and C_2, the following statement was proved by direct calculations.

Theorem 4.1 [78] *The exact solution of the boundary-value problem consisting of the nonlinear equation (4.198) (with $\lambda = 2\sqrt{\lambda_1}, \lambda_1 > 0$), the zero flux conditions*

$$u_x(t,0) = 0, \quad u_x(t,+\infty) = 0, \qquad (4.200)$$

and the initial condition

$$u(0,x) = \ln\left[\frac{1+e}{2} + C(\sqrt{\lambda_1}x + 1)e^{-\sqrt{\lambda_1}x}\right] \qquad (4.201)$$

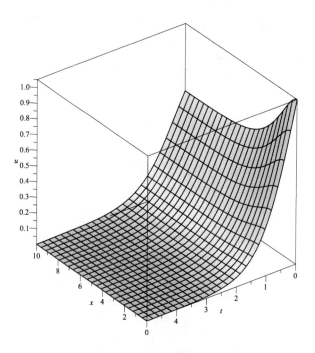

FIGURE 4.13: Exact solution (4.202) with $\lambda_1 = 1$, $C = 0.5$

is given in the domain Ω by the formula

$$u = \ln\left[\frac{e+1}{2} - \frac{e-1}{2}\tanh\left(\frac{\lambda_1(e-1)}{2}t\right) + \frac{2C(\sqrt{\lambda_1}x+1)}{e^{\sqrt{\lambda_1}x}(e^{\lambda_1 t} + e^{\lambda_1 et})}\right]. \quad (4.202)$$

Moreover, solution (4.202) is bounded and positive in the domain Ω provided $C > \frac{1-e}{2}$.

We remind the reader that Eq. (4.198) possesses two steady-state points $u_0 = 0$ and $u_1 = 1$. It can be easily established that one of them is stable while another is unstable, and this is the common peculiarity for many equations arising in population dynamics. Theorem 4.1 gives the space-time distribution of a population for the situation when the steady-state point $u_0 = 0$ is stable. One notes that solution (4.202) is vanishing if $t \to +\infty$, therefore the population die. This scenario is pictured in Figure 4.13.

To predict another scenario for the population, exact solutions of the RDC equation (4.198) with negative λ_1 should be examined. In this case, one obtains $P > 0$, hence f_1 and f_2 are the exponential functions (see formulae (4.168) and (4.199)). An interesting case occurs when Eq. (4.198) with $P < 0$ possesses quasi-periodic solutions of the form (4.168). As a result, the similar theorem can be formulated for a bounded domain Ω.

Chapter 5

The method of additional generating conditions for constructing exact solutions

5.1	Description of the method and the general scheme of implementation	191
5.2	Application of the method for solving nonlinear reaction-diffusion-convection equations	195
	5.2.1 Reduction of the nonlinear equations (5.10) and (5.11) to ODE systems	196
	5.2.2 Exact solutions of the nonlinear equations (5.10) and (5.11)	201
	5.2.3 Application of the solutions obtained for solving boundary-value problems	211
5.3	Analysis of the solutions obtained and comparison with the known results	216

5.1 Description of the method and the general scheme of implementation

As it was mentioned in Section 1.2, there are several methods and techniques allowing to construct exact solutions of nonlinear PDEs without knowledge Lie method and without using conditional (potential, higher-order etc.) symmetry. Probably the most general approach, called the method of differential constraints, was formulated in the 1960s [249] and further developed in monograph [232] (actually, the method have roots in the Monge, Ampère and Darboux works [119]). The main idea of the method of differential constraints is very simple: to define suitable constraint(s) for a given PDE in such a way that the overdetermined system obtained will be compatible and can be (partly) solved using the existing methods. However, the main problem of the method is how to define the suitable constraint(s). During the last decades several methods (techniques) were developed, which use the correctly-specified differential constraints in order to find exact solutions for some classes of nonlinear PDEs.

Here a method for the construction of exact solutions is presented, which is based on the consideration of a given nonlinear PDE together with an additional condition (constraint) in the form of a higher-order ordinary differential equation (ODE). The method was formulated and applied for solving some nonlinear equations and systems in [46, 48](the key idea of the method was firstly applied in [45]). Notwithstanding its efficiency was demonstrated on PDEs and systems involving only quadratic nonlinearities [45, 46, 47, 48, 51, 61, 62, 66, 72, 183], the method can be adopted also for solving nonlinear equations with a more complicated structure.

Consider the most general second-order evolution PDE involving quadratic nonlinearities

$$u_t = (d_0 + d_1 u + d_2 u_x + d_3 u_{xx}) u_{xx} + (\lambda u + r u_x) u_x + \lambda_0 + \lambda_1 u + \lambda_2 u^2, \quad (5.1)$$

where coefficients λ, λ_i and d_i ($i = 0, \ldots, 3$) are arbitrary constants. It turns out that several well-known PDEs can be reduced to evolution equations of the form (5.1) (see subsequent sections), hence we deal with an important class of PDEs. Notably, the linear convection term $v u_x$ was removed from Eq. (5.1) by the Galilei boost $x \to x + vt$.

Hereinafter we consider Eq. (5.1) together with the *additional generating conditions* (i.e., specified differential constraints) in the form of the linear mth-order homogeneous ODE:

$$\alpha_1(t,x) \frac{du}{dx} + \ldots + \alpha_{m-1}(t,x) \frac{d^{m-1} u}{dx^{m-1}} + \frac{d^m u}{dx^m} = 0, \quad (5.2)$$

where $\alpha_1(t,x), \ldots, \alpha_{m-1}(t,x)$ are arbitrary smooth functions and the variable t is considered as a parameter. Generally speaking, one may consider the general mth-order ODE and the method presented below works in the same way provided the general solution of the additional generating condition is known. Unfortunately, the list of nonlinear ODEs of high order with known general solutions is very short, hence we restrict ourselves to linear ODEs.

It is well-known that the general solution of (5.2) has the form

$$u = \varphi_0(t) g_0(t,x) + \ldots + \varphi_{m-1}(t) g_{m-1}(t,x), \quad (5.3)$$

where $\varphi_0(t), \varphi_1(t), \ldots, \varphi_{m-1}(t)$ are arbitrary functions and $g_0(t,x) = 1, g_1(t,x) \ldots, g_{m-1}(t,x)$ are fixed functions that form a fundamental system of solutions of (5.2). It is worth noting that the functions $g_1(t,x), \ldots, g_{m-1}(t,x)$ can be often expressed in an explicit form in terms of elementary ones. For example, this occurs when the functions α_i in (5.2) do not depend on the variable x. We also note that formula (5.2) with $m = 2$ reduces to the form $u = \varphi_0(t) + \varphi_1(t) g_1(t,x)$ and the latter can be thought as the ansatz corresponding to a Lie (or Q-conditional) symmetry (see ansatz (1.26) for comparison).

Let us consider formula (5.3) as an ansatz for PDEs of the form (5.1). It is important to note that this ansatz contains m yet-to-be determined functions

Description of the method and the general scheme of implementation   193

$\varphi_i, i = 1, ...m$ (we remind the reader that a single unknown function occurs in ansätze derived by using Lie and Q-conditional symmetries). This enables us to reduce the given PDE of the form (5.1) to a system of the first-order ODEs for the functions φ_i provided some conditions are fulfilled. It is well-known that such systems (usually called dynamical systems) have been extensively investigated because they arise as mathematical models in a wide range of applications. In particular, there is a vast literature devoted to construction of exact solutions of ODE systems (see, e.g., handbooks [142, 212] and papers cited therein).

Now we want to identify some sufficient conditions allowing to derive the ODE system mentioned above by using ansatz (5.3). Having this in mind, we insert ansatz (5.3) into Eq. (5.1) as follows. Calculating the derivatives u_t, u_x, u_{xx} using ansatz (5.3) and substituting those into PDE (5.1), one obtains the following cumbersome expression:

$$\varphi_{0,t} g_0 + \varphi_{1,t} g_1 + .. + \varphi_{m-1,t} g_{m-1} = \lambda_0$$

$$+\varphi_0 \Big(\lambda_1 g_0 + d_0 g_{0,xx} - g_{0,t} \Big) + ...$$

$$+\varphi_{m-1} \Big(\lambda_1 g_{m-1} + d_0 g_{m-1,xx} - g_{m-1,t} \Big)$$

$$+\varphi_0^2 \Big(d_1 g_0 g_{0,xx} + d_2 g_{0,x} g_{0,xx} + d_3 g_{0,xx}^2$$

$$+ r g_{0,x}^2 + \lambda g_0 g_{0,x} + \lambda_2 g_0^2 \Big) + ...$$

$$+\varphi_{m-1}^2 \Big(d_1 g_{m-1} g_{m-1,xx} + d_2 g_{m-1,x} g_{m-1,xx} + d_3 g_{m-1,xx}^2$$

$$+ r g_{m-1,x}^2 + \lambda g_{m-1} g_{m-1,x} + \lambda_2 g_{m-1}^2 \Big) \qquad (5.4)$$

$$+\varphi_0 \varphi_1 \Big(d_1 g_0 g_{1,xx} + d_1 g_1 g_{0,xx} + d_2 g_{0,x} g_{1,xx} + d_2 g_{1,x} g_{0,xx}$$

$$+2 d_3 g_{0,xx} g_{1,xx} + \lambda g_0 g_{1,x} + \lambda g_1 g_{0,x} + 2 r g_{0,x} g_{1,x} + 2 \lambda_2 g_0 g_1 \Big) +$$

$$\varphi_0 \varphi_2 \Big(d_1 g_0 g_{2,xx} + d_1 g_2 g_{0,xx} + d_2 g_{0,x} g_{2,xx} + d_2 g_{2,x} g_{0,xx}$$

$$+2 d_3 g_{0,xx} g_{2,xx} + \lambda g_0 g_{2,x} + \lambda g_2 g_{0,x} + 2 r g_{0,x} g_{2,x} + 2 \lambda_2 g_0 g_2 \Big) + ...$$

$$+\varphi_{m-2} \varphi_{m-1} \Big(d_1 g_{m-2} g_{m-1,xx} + d_1 g_{m-1} g_{m-2,xx} + d_2 g_{m-2,x} g_{m-1,xx}$$

$$+ d_2 g_{m-1,x} g_{m-2,xx} + 2 d_3 g_{m-2,xx} g_{m-1,xx} + \lambda g_{m-2} g_{m-1,x}$$

$$+ \lambda g_{m-1} g_{m-2,x} + 2 r g_{m-2,x} g_{m-1,x} + 2 \lambda_2 g_{m-2} g_{m-1} \Big),$$

where the indices t and x next to the functions $\varphi_i(t)$ and $g_i(t,x)$, $i = 0, 1, ...m-1$ denote differentiation with respect to (w.r.t.) t and x. Now the sufficient conditions for the reduction of expression (5.4) to a system of ODEs can be found. They have the following form:

$$\lambda_1 g_i + d_0 g_{i,xx} - g_{i,t} + \delta_{i,0}\lambda_0 = g_{i_1} Q_{ii_1}(t), \ i = 0,\ldots, m-1, \quad (5.5)$$

$$d_1 g_i g_{i,xx} + d_2 g_{i,x} g_{i,xx} + d_3 g_{i,xx}^2 + r g_{i,x}^2 + \lambda g_i g_{i,x} + \lambda_2 g_i^2 \\ = g_{i_1} R_{ii_1}(t), \ i = 0,\ldots, m-1, \quad (5.6)$$

$$d_1(g_i g_{j,xx} + g_j g_{i,xx}) + d_2(g_{i,x} g_{j,xx} + g_{j,x} g_{i,xx}) + 2d_3 g_{i,xx} g_{j,xx} + 2r g_{i,x} g_{j,x} \\ + \lambda(g_i g_{j,x} + g_j g_{i,x}) + 2\lambda_2 g_i g_j = g_{i_1} T_{ij}^{i_1}(t), \ i < j = 1,\ldots, m-1, \quad (5.7)$$

where functions Q_{ii_1}, R_{ii_1} and $T_{ij}^{i_1}$ on the right-hand side are defined by the expressions on the left-hand side. Notably, such functions may exist only under some restrictions on the coefficients arising in left-hand side. Hereafter a summation is assumed from 0 to m-1 over the repeated indices i_1, i_2 and $\delta_{i,0} = 0, 1$ is the Kronecker symbol.

Assuming that relations (5.5)–(5.7) take place, expression (5.5) simplifies to the form

$$\varphi_{0,t} g_0 + \varphi_{1,t} g_1 + \ldots + \varphi_{m-1,t} g_{m-1} = \\ = \lambda_0 + \varphi_0 g_{i_1} Q_{0i_1}(t) + \ldots + \varphi_{m-1} g_{i_1} Q_{m-1 i_1}(t) \\ + \varphi_0^2 g_{i_1} R_{0i_1}(t) + \ldots + \varphi_{m-1}^2 g_{i_1} R_{m-1 i_1}(t) \\ + \varphi_0 \varphi_1 g_{i_1} T_{01}^{i_1}(t) + \varphi_0 \varphi_2 g_{i_1} T_{02}^{i_1}(t) + \ldots + \varphi_{m-2} \varphi_{m-1} g_{i_1} T_{m-2 m-1}^{i_1}(t). \quad (5.8)$$

Because the functions $g_0, \ldots, g_{m-1}$ are linearly independent (they form the fundamental system of solutions of ODE (5.2)), the above expression can be split into m ODEs. As a result, the system of ODEs

$$\dot{\varphi}_i = Q_{i_1 i}(t)\varphi_{i_1} + R_{i_1 i}(t)\varphi_{i_1}^2 + T_{i_1 i_2}^i(t)\varphi_{i_1}\varphi_{i_2} + \delta_{i,0}\lambda_0, \ i = 0, \ldots, m-1 \quad (5.9)$$

is obtained for finding the functions $\varphi_i, i = 0, \ldots, m-1$.

Thus, we have proved the following statement.

Theorem 5.1 *[48] Any solution of the ODE system (5.9) generates the exact solution in the form (5.3) for nonlinear PDE (5.1) if the functions $g_i, i = 0, \ldots, m-1$ satisfy conditions (5.5)–(5.7).*

Remark 5.1 *The coefficients λ, λ_i and d_i $(i = 0, \ldots, 3)$ in Eq. (5.1) can be smooth functions of the variables t and x as well. In such cases, the theorem also works provided the sufficient conditions (5.5)–(5.7) take place.*

Remark 5.2 *Theorem 5.1 gives only sufficient conditions for the reduction of Eq. (5.1) to system (5.9). In some cases, noting additional relations between the fundamental system elements $g_0(t,x) = 1, g_1(t,x)\ldots, g_{m-1}(t,x)$, it is possible to find simpler sufficient conditions for the reduction (see an example in the next section).*

The above proof of Theorem 5.1 gives an algorithm for finding exact solutions of any PDE. The main steps of the algorithm are following.

1. To transform a nonlinear PDE in question to the form (5.1).

2. To identify the appropriate additional generating condition (5.3).

3. To construct the general solutions of ODE (5.3), taking into account all possible values of its coefficients.

4. To use the solutions of ODE (5.3) as ansätze for the nonlinear PDE in question.

5. To find sufficient conditions for reduction of the given equation to an ODE system of the form (5.9).

6. To construct solutions of the ODE system obtained.

7. To construct exact solutions of the nonlinear PDE in question.

So, we have the constructive method, called *the method of additional generating conditions* (MAGC), for finding new non-Lie ansätze and exact solutions and its efficiency will be shown in the subsequent sections.

The method was generalized on systems of PDEs [46] and several nonlinear systems arising in various applications were examined in order to construct exact solutions [46, 47, 62, 66, 72, 183]. In the case of PDEs with derivatives of second or higher orders with respect to (w.r.t.) t and x, MAGC is also applicable provided the equation in question involves quadratic nonlinearities only. In this case, the corresponding ODEs are those of the second and higher orders.

Formally speaking, MAGC can be adopted for solving PDEs with more complicated nonlinearities. For example, if one applies the method in a straightforward way to an equation involving cubic nonlinearities then a system of ODEs with cubic nonlinearities can be derived instead of system (5.9). However, we assume that the corresponding sufficient conditions are much more complicated than (5.5)–(5.7), hence it is not clear whether they will be useful. Clarification of this problem lies beyond this monograph.

5.2 Application of the method for solving nonlinear reaction-diffusion-convection equations

It can be easily noted that Eq. (5.1) contains as particular cases some well-known nonlinear equations. In fact, vanishing corresponding coefficients in Eq. (5.1), one obtains the Fisher, Murray and Burgers equations, which were studied in the previous chapters.

Here we apply MAGC in order to construct exact solutions of the reaction-diffusion-convection (RDC) equations with power-law and exponential nonlinearities [48]:

$$T_t = (T^\alpha T_x)_x + \lambda T^\alpha T_x + \frac{1}{\alpha}(\lambda_0 T^{1-\alpha} + \lambda_1 T + \lambda_2 T^{1+\alpha}) \qquad (5.10)$$

and

$$T_t = (e^{\beta T} T_x)_x + \lambda e^{\beta T} T_x + \frac{1}{\beta}(\lambda_0 e^{-\beta T} + \lambda_1 + \lambda_2 e^{\beta T}), \qquad (5.11)$$

where $T = T(t,x)$ is an unknown function and $\alpha\beta \neq 0$, $\lambda, \lambda_0, \lambda_1, \lambda_2$ are arbitrary constants.

Firstly, note that Eq. (5.10) is reduced by the substitution $u = T^\alpha$, $\alpha \neq 0$ to the equation

$$u_t = uu_{xx} + \frac{1}{\alpha}u_x^2 + \lambda uu_x + \lambda_0 + \lambda_1 u + \lambda_2 u^2. \qquad (5.12)$$

Analogously, the nonlinear evolution equation with exponential nonlinearities (5.11) is reduced by the substitution $u = e^{\beta T}$, $\beta \neq 0$ to the equation

$$u_t = uu_{xx} + \lambda uu_x + \lambda_0 + \lambda_1 u + \lambda_2 u^2. \qquad (5.13)$$

Thus, hereinafter we consider the nonlinear equation

$$u_t = uu_{xx} + ru_x^2 + \lambda uu_x + \lambda_0 + \lambda_1 u + \lambda_2 u^2, \qquad (5.14)$$

which is locally equivalent to either Eqs. (5.10) (if $r = \frac{1}{\alpha} \neq 0$) or (5.11) (if $r = 0$). In particular, any solution $u_0(t,x)$ of Eq. (5.14) generates a solution of the form

$$T_0(t,x) = u_0^{\frac{1}{\alpha}}, \quad r \neq 0 \qquad (5.15)$$

and

$$T_0(t,x) = \frac{1}{\beta}\ln u_0, \quad r = 0, \qquad (5.16)$$

for Eqs. (5.10) and (5.11), respectively.

5.2.1 Reduction of the nonlinear equations (5.10) and (5.11) to ODE systems

Consider an additional generating third-order condition of the form [48]

$$\alpha_1(t)\frac{du}{dx} + \alpha_2(t)\frac{d^2u}{dx^2} + \frac{d^3u}{dx^3} = 0, \qquad (5.17)$$

which is a particular case of (5.2) with $m = 3$. Condition (5.17) generates the following chain of the ansätze:

$$u = \varphi_0(t) + \varphi_1(t)e^{\gamma_1(t)x} + \varphi_2(t)e^{\gamma_2(t)x} \tag{5.18}$$

if $\gamma_{1,2}(t) = \frac{1}{2}(\pm\sqrt{D} - \alpha_2) \neq 0$ and $D = \alpha_2^2 - 4\alpha_1 > 0$;

$$u = \varphi_0(t) + \varphi_1(t)e^{\gamma(t)x} + x\varphi_2(t)e^{\gamma(t)x} \tag{5.19}$$

if $\gamma_1 = \gamma_2 = \gamma \neq 0$, i.e., $D = 0$;

$$u = \varphi_0(t) + \varphi_1(t)x + \varphi_2(t)e^{\gamma(t)x} \tag{5.20}$$

if $\alpha_1 = 0$;

$$u = \varphi_0(t) + \varphi_1(t)x + \varphi_2(t)x^2 \tag{5.21}$$

if $\alpha_1 = \alpha_2 = 0$.

Remark 5.3 *In the case $D = \alpha_2^2 - 4\alpha_1 < 0$, one obtains complex functions $\gamma_1 = \gamma_2^* = \frac{1}{2}(\pm i\sqrt{-D} - \alpha_2), i^2 = -1$ and then ansatz (5.18) reduces to the form*

$$u = \varphi_0(t) + \left[\psi_1(t)\cos\left(\frac{1}{2}\sqrt{-D}x\right) + \psi_2(t)\sin\left(\frac{1}{2}\sqrt{-D}x\right)\right]e^{-\frac{\alpha_2}{2}x}, \tag{5.22}$$

where $\varphi_0(t), \psi_1(t), \psi_2(t)$ are yet-to-be determined functions.

Substituting the functions $g_0 = 1, g_1 = e^{\gamma_1(t)x}, g_2 = e^{\gamma_2(t)x}$ from ansatz (5.18) into relations (5.5)–(5.7), one obtains

$$\begin{gathered} Q_{00} = Q_{11} = Q_{22} = \lambda_1, \quad R_{00} = \lambda_2, \\ T_{12}^0 = \frac{4\lambda_2 r}{1+r}, \quad T_{01}^1 = T_{02}^2 = \lambda_2\left(1 + \frac{r}{1+r}\right) \end{gathered} \tag{5.23}$$

and

$$R_{ii_1} = Q_{ii_1} = T_{ij}^{i_1} = 0 \tag{5.24}$$

for all combinations of the indices $i, i_1, j = 0, 1, 2$ and $k = 1, 2$ not listed in (5.23). Simultaneously the following constraints spring up: either

$$r = 0, \quad \gamma_{1,2}(t) = \frac{1}{2}[\pm(\lambda^2 - 4\lambda_2)^{\frac{1}{2}} - \lambda], \quad \lambda^2 - 4\lambda_2 \neq 0 \tag{5.25}$$

or

$$r \neq 0, \quad \lambda = 0, \quad \gamma_{1,2}(t) = \pm\left(\frac{-\lambda_2}{1+r}\right)^{\frac{1}{2}}, \quad \lambda_2(r+1) \neq 0. \tag{5.26}$$

Having the determined coefficients (5.23)–(5.24), system (5.9) reduces to the form

$$\begin{gathered} \dot{\varphi}_0 = \lambda_2\varphi_0^2 + \lambda_1\varphi_0 + \lambda_0, \\ \dot{\varphi}_1 = \lambda_1\varphi_1 + \lambda_2\varphi_0\varphi_1, \\ \dot{\varphi}_2 = \lambda_1\varphi_2 + \lambda_2\varphi_0\varphi_2 \end{gathered} \tag{5.27}$$

if constraint (5.25) takes place, and to the form

$$\dot\varphi_0 = \lambda_2\varphi_0^2 + \lambda_1\varphi_0 + \lambda_0 + \tfrac{4r\lambda_2}{1+r}\varphi_1\varphi_2,$$
$$\dot\varphi_1 = \lambda_1\varphi_1 + \lambda_2(1 + \tfrac{r}{1+r})\varphi_0\varphi_1, \qquad (5.28)$$
$$\dot\varphi_2 = \lambda_1\varphi_2 + \lambda_2(1 + \tfrac{r}{1+r})\varphi_0\varphi_2$$

in the case of constraint (5.26).

Substituting the functions $g_0 = 1, g_1 = e^{\gamma(t)x}, g_2 = xe^{\gamma(t)x}$ from ansatz (5.19) into relations (5.5)–(5.7), such corresponding functions R_{ii_1}, Q_{ii_1} and $T_{ii_1}^j$ are obtained, for which system (5.9) coincides with the system of ODEs (5.27). However, the constraints have the form

$$r = 0, \quad \lambda^2 - 4\lambda_2 = 0, \quad \gamma = -\frac{\lambda}{2}$$

instead of (5.25).

Similarly, ansatz (5.20) leads to the system of ODEs

$$\dot\varphi_0 = \lambda\varphi_0\varphi_1 + \lambda_1\varphi_0 + \lambda_0,$$
$$\dot\varphi_1 = \lambda\varphi_1^2 + \lambda_1\varphi_1, \qquad (5.29)$$
$$\dot\varphi_2 = \lambda\varphi_1\varphi_2 + \lambda_1\varphi_2$$

for finding the unknown functions $\varphi_i, i = 0, 1, 2$. The corresponding constraints are

$$r = q = 0, \quad \gamma = -\lambda.$$

Finally, ansatz (5.21) gives the system of ODEs

$$\dot\varphi_0 = 2\varphi_0\varphi_2 + \lambda_1\varphi_0 + \lambda_0 + r\varphi_1^2,$$
$$\dot\varphi_1 = (2 + 4r)\varphi_1\varphi_2 + \lambda_1\varphi_1, \qquad (5.30)$$
$$\dot\varphi_2 = (2 + 4r)\varphi_2^2 + \lambda_1\varphi_2$$

provided $\lambda = \lambda_2 = 0$ in Eq. (5.14).

Thus, each ansatz from set (5.18)–(5.21) can be applied for the reduction of the nonlinear PDE (5.14) to a system consisting of three first-order ODEs, i.e., the three-dimensional dynamical systems are obtained. However, additional constraints on the coefficients $r, \lambda, \lambda_0, \lambda_1,$ and λ_2 spring up in all the cases. These constraints follow from relations (5.5)–(5.7) and they guarantee the reduction.

On the other hand, Theorem 5.1 gives *sufficient conditions* for such reduction. In some cases, noting additional relations between the functions $g_0(t, x) = 1, g_1(t, x)..., g_{m-1}(t, x)$, it is possible to weak these conditions in order to find another reduction. Let us give an example using ansatz (5.20). Noting that $g_{2,t} = \gamma^{-2}\frac{d\gamma}{dt}g_1 g_{2,xx}$, one can transfer this term from relations (5.5),

Application of the method for solving nonlinear RDC equations

$i = 2$ into (5.7), $i = 1, j = 2$, so that two relations in (5.5)–(5.7) take new form. Taking into account this circumstance, one can identify at once the case $r = -1$ and $\lambda = \lambda_2 = 0$ when the nonlinear PDE (5.14) is reduced to the system of first-order ODEs

$$\begin{aligned}\dot{\gamma} &= \gamma^2 \varphi_1, \\ \dot{\varphi}_0 &= -\varphi_1^2 + \lambda_1 \varphi_0 + \lambda_0, \\ \dot{\varphi}_1 &= \lambda_1 \varphi_1, \\ \dot{\varphi}_2 &= -2\gamma \varphi_1 \varphi_2 + \lambda_1 \varphi_2 + \gamma^2 \varphi_0 \varphi_2.\end{aligned} \tag{5.31}$$

It is easily seen that in this case the function $\gamma(t) \neq const$ (otherwise $\varphi_1 = 0$ and the system reduces to two ODEs) and the first equation in (5.31) is additional for obtaining the function γ. Thus, the four-dimensional dynamical system is obtained for finding exact solutions of the nonlinear PDE (5.14) with $r = -1$ and $\lambda = \lambda_2 = 0$.

A natural question arises: Can other reductions be derived, leading to four-dimensional dynamical systems? The answer is positive. By analogy with the additional generating condition (5.17), let us consider the fourth-order condition of the form

$$\alpha_1(t)\frac{du}{dx} + \alpha_2(t)\frac{d^2 u}{dx^2} + \alpha_3(t)\frac{d^3 u}{dx^3} + \frac{d^4 u}{dx^4} = 0. \tag{5.32}$$

This condition generates a chain of the ansätze. Although there are other interesting cases, the most nontrivial one occurs when the ansatz

$$u = \varphi_0(t) + \varphi_1(t)e^{\gamma_1 x} + \varphi_2(t)e^{\gamma_2 x} + \varphi_3(t)e^{\gamma_3 x} \tag{5.33}$$

is examined. Here $\varphi_0(t), ..., \varphi_3(t)$ are yet-to-be determined functions and $\gamma_1, \gamma_2, \gamma_3$ are yet-to-be determined constants. We also assume that all the exponents $\gamma_1, \gamma_2, \gamma_3$ are nonzero and different.

It turns out that it is possible to reduce Eq. (5.14) using ansatz (5.33) to a system of ODEs of the form (5.9) only in a special case. Indeed, by substituting the functions $g_0 = 1, g_1 = e^{\gamma_1 x}, g_2 = e^{\gamma_2 x}$ and $g_3 = e^{\gamma_3 x}$ from ansatz (5.33) into relations (5.5) with $i = 0, \ldots, 3$, one obtains

$$Q_{00} = Q_{11} = Q_{22} = Q_{33} = \lambda_1, \quad Q_{ii_1} = 0, i \neq i_1. \tag{5.34}$$

Substitution of the functions $g_i, i = 0, \ldots, 3$ into relation (5.6) with $i = 0$ leads to

$$R_{00} = \lambda_2, \quad R_{0a} = 0, a = 1, 2, 3. \tag{5.35}$$

In the cases $i = 1, 2, 3$ relations (5.6) take the following form:

$$\left[(1+r)\gamma_a^2 + \lambda \gamma_a + \lambda_2\right] e^{2\gamma_a x} = g_{i_1} R_{ai_1}(t), \quad a = 1, 2, 3, \tag{5.36}$$

where a summation is assumed from 0 to 3 over the repeated indices i_1. It turns out that these relations can be fulfilled only in two cases:

(i)

$$\gamma_{1,2} = \frac{\pm[\lambda^2 - 4\lambda_2(1+r)]^{\frac{1}{2}} - \lambda}{2(1+r)}, \quad \gamma_3 = \frac{1}{2}\gamma_2, \quad [\lambda^2 - 4\lambda_2(1+r)](1+r) \neq 0 \quad (5.37)$$

and *(ii)*

$$\gamma_1 = \frac{\pm[\lambda^2 - 4\lambda_2(1+r)]^{\frac{1}{2}} - \lambda}{2(1+r)}, \quad \gamma_2 = \frac{1}{2}\gamma_1, \quad \gamma_3 = \frac{1}{4}\gamma_1. \quad (5.38)$$

It is easily checked that constraint (5.38) is too restrictive because it is impossible to satisfy relations (5.7). Hence case *(i)* only is considered in what follows.

Using relations (5.36) under constraint (5.37), one obtains

$$R_{32} = \frac{\lambda\gamma_2 + 3\lambda_2}{4}, \quad R_{ai_1} = 0, a = 1, 2, 3, \quad (5.39)$$

where the combinations of the indices $(a, i_1) \neq (3, 2)$.

Similarly, relations (5.7) at $(i,j) = (1,2)$ and $(i,j) = (2,3)$ can be fulfilled only in the case $\gamma_1 + \gamma_2 = \gamma_3$ (i.e., $\lambda_2 = -6\lambda^2 \neq 0$) and $r = -\frac{2}{3}$ (i.e., $\gamma_1 = 3\lambda$). Having the above constraints, the coefficients $T_{ij}^{i_1}$ are easily derived:

$$T_{02}^2 = 18\lambda^2, \quad T_{03}^3 = -6\lambda^2, \quad T_{13}^0 = 18\lambda^2, \quad T_{12}^3 = 54\lambda^2, \quad (5.40)$$

and

$$T_{ij}^{i_1} = 0 \quad (5.41)$$

for all combinations of the indices $i, i_1, j = 0, 1, 2, 3$ not listed in (5.40).

Finally, substituting coefficients (5.34)–(5.35) and (5.39)–(5.41) into system (5.9), we obtain the four-dimensional system of ODEs

$$\begin{aligned}
\dot{\varphi}_0 &= \lambda_1\varphi_0 + 18\lambda^2\varphi_1\varphi_3 - 6\lambda^2\varphi_0^2 + \lambda_0, \\
\dot{\varphi}_1 &= \lambda_1\varphi_1, \\
\dot{\varphi}_2 &= \lambda_1\varphi_2 + 18\lambda^2\varphi_0\varphi_2 - 6\lambda^2\varphi_3^2, \\
\dot{\varphi}_3 &= \lambda_1\varphi_3 - 6\lambda^2\varphi_0\varphi_3 + 54\lambda^2\varphi_1\varphi_2.
\end{aligned} \quad (5.42)$$

Thus, according to Theorem 5.1 each solution of the ODE system (5.42) generates an exact solution of the form

$$u = \varphi_0(t) + \varphi_1(t)e^{3\lambda x} + \varphi_2(t)e^{-6\lambda x} + \varphi_3(t)e^{-3\lambda x} \quad (5.43)$$

for the nonlinear equation

$$u_t = uu_{xx} - \frac{2}{3}u_x^2 + \lambda uu_x + \lambda_0 + \lambda_1 u - 6\lambda^2 u^2. \quad (5.44)$$

Because the nonlinear equations (5.14) and (5.10) are related via the substitution $u = T^\alpha$, we easily identify that Eq. (5.44) is equivalent to the nonlinear RDC equation

$$T_t = (T^{-\frac{3}{2}}T_x)_x + \lambda T^{-\frac{3}{2}}T_x - \frac{2}{3}\lambda_0 T^{\frac{5}{2}} - \frac{2}{3}\lambda_1 T + 4\lambda^2 T^{-\frac{1}{2}}. \quad (5.45)$$

Thus, any solution of Eq. (5.44) can be transformed into a solution of the nonlinear heat equation (5.45), using (5.15) with $\alpha = -\frac{3}{2}$.

We also point out that the additional generating condition (5.32) with $\alpha_1(t) = \alpha_2(t) = \alpha_3(t) = 0$ leads to the ansatz

$$u = \varphi_0(t) + \varphi_1(t)x + \varphi_2(t)x^2 + \varphi_3(t)x^3. \quad (5.46)$$

It turns out that application of ansatz (5.46) to Eq. (5.14) leads exactly to Eq. (5.44) with $\lambda = 0$. Thus, the corresponding equation for the function T is Eq. (5.45) with $\lambda = 0$, i.e.,

$$T_t = (T^{-\frac{3}{2}}T_x)_x - \frac{2}{3}\lambda_0 T^{\frac{5}{2}} - \frac{2}{3}\lambda_1 T. \quad (5.47)$$

The dynamical system for finding the functions $\varphi_0, \ldots, \varphi_3$ has the form

$$\begin{aligned}
\dot{\varphi}_0 &= \lambda_1\varphi_0 + 2\varphi_0\varphi_2 - \tfrac{2}{3}\varphi_1^2 + \lambda_0, \\
\dot{\varphi}_1 &= \lambda_1\varphi_1 + 6\varphi_0\varphi_3 - \tfrac{2}{3}\varphi_1\varphi_2, \\
\dot{\varphi}_2 &= \lambda_1\varphi_2 + 3\varphi_0\varphi_3 - \varphi_1\varphi_2 - \tfrac{8}{3}\varphi_2^2, \\
\dot{\varphi}_3 &= \lambda_1\varphi_3 + 4\varphi_1\varphi_3 - 8\varphi_2\varphi_3 + \varphi_2^2.
\end{aligned} \quad (5.48)$$

5.2.2 Exact solutions of the nonlinear equations (5.10) and (5.11)

The systems of ODEs, which were obtained in the previous section, enable us to construct multiparametric families of exact solutions of the nonlinear equation (5.14). Having the exact solutions of Eq. (5.14), we can easily construct solutions of the nonlinear RDC equations (5.10) and (5.11). It will be shown (see Section 5.3) that the exact solutions derived are not obtainable (excepting very special cases) vie Lie symmetries, i.e., they are the non-Lie solutions. Notably, Eq. (5.14) admits the time and space translations, hence we simplify the solutions in what follows using the transformations $t + t_0 \to t$ and/or $x + x_0 \to x$ (t_0 and x_0 are arbitrary parameters).

Let us start to examine the system of ODEs (5.27). The first equation in system (5.27) is autonomous and can be immediately integrated

$$\int \frac{d\varphi_0}{\lambda_2\varphi_0^2 + \lambda_1\varphi_0 + \lambda_0} = t + t_0. \quad (5.49)$$

So, the general solution of the first ODE essentially depends on the coefficients $\lambda_2, \lambda_1,$ and λ_0. Calculating the integral in (5.49), we obtain the following solutions

$$\varphi_0(t) = -\frac{1}{2\lambda_2} \begin{cases} \frac{2}{t} + \lambda_1, & D = 0; \\ -\sqrt{-D} \tan(\frac{\sqrt{-D}}{2}t) + \lambda_1, & D < 0; \\ \sqrt{D} \coth(\frac{\sqrt{D}}{2}t) + \lambda_1, & (2\lambda_2\varphi_0 + \lambda_1)^2 > D > 0; \\ \sqrt{D} \tanh(\frac{\sqrt{D}}{2}t) + \lambda_1, & D > (2\lambda_2\varphi_0 + \lambda_1)^2 > 0, \end{cases} \quad (5.50)$$

where $D = \lambda_1^2 - 4\lambda_0\lambda_2$ (here and in (5.52) the parameter t_0 was removed as it was noted above).

Having solution (5.50), it is easy to find the general solution for the system of ODEs (5.27) because the second and the third ODEs are linear w.r.t. φ_1 and φ_2, respectively. Thus, ansatz (5.18) generates the two-parameter family (one can be extended to the three-parameter family via time translation $t \to t - t_0$) of solutions of Eq. (5.14) with $r = 0$:

$$u = \varphi_0(t) + \frac{e^{\frac{\lambda_1}{2}t}}{\mu(t)} \left(C_1 e^{\gamma_1 x} + C_2 e^{\gamma_2 x} \right), \quad (5.51)$$

where $\gamma_{1,2} = \frac{1}{2}[\pm(\lambda^2 - 4\lambda_2)^{\frac{1}{2}} - \lambda]$ and

$$\mu(t) = \begin{cases} t, & D = 0; \\ |\cos(\frac{\sqrt{-D}}{2}t)|, & D < 0; \\ |\sinh(\frac{\sqrt{D}}{2}t)|, & (2\lambda_2\varphi_0 + \lambda_1)^2 > D > 0; \\ \cosh(\frac{\sqrt{D}}{2}t), & D > (2\lambda_2\varphi_0 + \lambda_1)^2 > 0. \end{cases} \quad (5.52)$$

Hereafter C_1 and C_2 are arbitrary constants. Notably, the family of solutions (5.51) in the case $D > 0$ can be rewritten in the form

$$u = \varphi_0(t) + \frac{\exp\left(\frac{\lambda_1 + \sqrt{D}}{2}t\right)\left(C_1 e^{\gamma_1 x} + C_2 e^{\gamma_2 x}\right)}{1 \pm e^{\sqrt{D}t}}. \quad (5.53)$$

Now we turn to the system of ODEs (5.28). Its solving is more complicated because the system does not involve an autonomous equation. As a result, the general solution can be not constructed in an explicit form, excepting some special cases, which are presented below. Notwithstanding the family of exact solutions

$$u = \varphi_0(t) + \exp\left(\lambda_1 t + \lambda_2 \kappa_r \int \varphi_0(t) dt\right) \left(C_1 e^{-\gamma_1 x} + C_2 e^{\gamma_1 x}\right) \quad (5.54)$$

of Eq. (5.14) with $\lambda = 0$, $\lambda_2(r+1) \neq 0$ has been found. Here $\gamma_1^2 = \frac{-\lambda_2}{1+r}$, $\kappa_r = \frac{2r+1}{r+1}$ and $\varphi_0(t)$ is an arbitrary solution of the integro-differential equation

$$\dot{\varphi}_0 = \lambda_2\varphi_0^2 + \lambda_1\varphi_0 + \lambda_0 + 4C_1C_2\lambda_2(\kappa_r - 1)\exp\left(2\lambda_1 t + 2\lambda_2\kappa_r\int\varphi_0(t)\mathrm{d}t\right). \quad (5.55)$$

Of course, Eq. (5.55) is reducible to a nonlinear second-order ODE, however, the latter cannot be integrated provided its coefficients are arbitrary. On the other hand, it is easily seen that the integro-differential equation (5.55) with $C_1C_2 = 0$, is reduced to the first-order ODE, which is already solved (see formulae (5.50)). So, the exact solutions

$$u = \varphi_0(t) + \frac{C_2}{[\mu(t)]^{\frac{2r+1}{r+1}}}\exp\left(\frac{\lambda_1}{2(r+1)}t + \sqrt{\frac{-\lambda_2}{r+1}}x\right) \quad (5.56)$$

and

$$u = \varphi_0(t) + \frac{C_1}{[\mu(t)]^{\frac{2r+1}{r+1}}}\exp\left(\frac{\lambda_1}{2(r+1)}t - \sqrt{\frac{-\lambda_2}{r+1}}x\right) \quad (5.57)$$

are obtained, where the functions $\varphi_0(t)$ and $\mu(t)$ are defined in (5.50) and (5.52).

Another special case when the integro-differential equation (5.55) reduces to a first-order ODE is $\kappa_r = 0$, i.e., $r = -\frac{1}{2}$. If additionally $\lambda_1 = 0$ then system (5.28) is rather trivial and its general solution can be immediately constructed. As a result, the two-parameter families (they can be extended to the three-parameter one via time translation $t \to t - t_0$) of solutions

$$u = \varphi_0(t) + C_1\exp(-\sqrt{-2\lambda_2}x) + C_2\exp(\sqrt{-2\lambda_2}x), \quad \lambda_2 < 0 \quad (5.58)$$

and

$$u = \varphi_0(t) + C_1\cos(\sqrt{2\lambda_2}x) + C_2\sin(\sqrt{2\lambda_2}x), \quad \lambda_2 > 0 \quad (5.59)$$

of the equation

$$u_t = uu_{xx} - \frac{1}{2}u_x^2 + \lambda_0 + \lambda_2 u^2, \quad \lambda_2 \neq 0 \quad (5.60)$$

has been found. Here the function $\varphi_0(t)$ has a similar structure to that in (5.50):

$$\varphi_0(t) = -\frac{1}{\lambda_2}\begin{cases} \frac{1}{t}, & \lambda_0 = 4\lambda_2 C_1 C_2; \\ -\sqrt{-D}\tan(\sqrt{-D}t), & D < 0; \\ \sqrt{D}\coth(\sqrt{D}t), & 4\lambda_2^2\varphi_0^2 > D > 0; \\ \sqrt{D}\tanh(\sqrt{D}t), & D > 4\lambda_2^2\varphi_0^2 > 0, \end{cases} \quad (5.61)$$

where $D = \lambda_2(4\lambda_2 C_1 C_2 - \lambda_0)$.

Assuming $\lambda_1 \neq 0$, Eq. (5.55) with $\kappa_r = 0$ reduces to the linear second-order ODE, which possesses the known general solution in terms of Bessel functions (see items 1.2.3.4 and 2.1.3.10 in [212]). So, the general solution

$$\varphi_0(t) = -\frac{\lambda_1}{\lambda_2} - \frac{1}{\lambda_2}\frac{d}{dt}\ln\left[C_3 J_\nu(\tau) + C_4 Y_\nu(\tau)\right], \quad (5.62)$$

of Eq. (5.55) with $\kappa_r = 0$ and $\lambda_1 \neq 0$ was derived. Here J_ν and Y_ν are the Bessel functions of the first and second kind and

$$\tau = 2\frac{|\lambda_2|}{|\lambda_1|}\sqrt{-C_1 C_2}e^{\lambda_1 t}, \quad \nu = \sqrt{\frac{1}{4} - \frac{\lambda_0 \lambda_2}{\lambda_1^2}}, \quad C_1 C_2(|C_3| + |C_4|) \neq 0.$$

Notably, J_ν and Y_ν reduces to the corresponding modified Bessel functions if ν takes complex values.

Thus, the family of solutions

$$u = \varphi_0(t) + C_1 \exp(\lambda_1 t - \sqrt{-2\lambda_2}x) + C_2 \exp(\lambda_1 t + \sqrt{-2\lambda_2}x) \quad (5.63)$$

(here $\varphi_0(t)$ is defined in (5.62)) of the equation

$$u_t = u u_{xx} - \frac{1}{2}u_x^2 + \lambda_0 + \lambda_1 u + \lambda_2 u^2, \quad \lambda_2 \lambda_1 \neq 0 \quad (5.64)$$

has been found.

It is worth noting that periodic solutions of Eq. (5.14) can be constructed in the case $\lambda = 0$, $\gamma_1^2 = -\frac{\lambda_2}{1+r} < 0$ and $r \neq -1$. In fact, it is easily seen that setting the complex constants $2C_1 = C_{10} + iC_{11}$, $2C_2 = C_{10} - iC_{11}$ in (5.54), we obtain the two-parameter family of solutions

$$u = \varphi_0(t) + \exp\left(\lambda_1 t + \lambda_2 \kappa_r \int \varphi_0(t)dt\right)\left(C_{10}\cos|\gamma_1|x + C_{11}\sin|\gamma_1|x\right), \quad (5.65)$$

where C_{10} and C_{11} are arbitrary real constants and $\varphi_0(t)$ is an arbitrary solution of the integro-differential equation

$$\dot\varphi_0 = \lambda_2\varphi_0^2 + \lambda_1\varphi_0 + \lambda_0 + (C_{10}^2 + C_{11}^2)\lambda_2(\kappa_r - 1)\exp\left(2\lambda_1 t + 2\lambda_2\kappa_r\int\varphi_0(t)dt\right). \quad (5.66)$$

As we already know, ansatz (5.19) reduces Eq. (5.14) to the system of ODEs (5.27) with $r = \lambda^2 - 4\lambda_2 = 0$. Thus, the two-parameter family of solutions

$$u = \varphi_0(t) + \frac{1}{\mu(t)}(C_1 + C_2 x)\exp\left[\frac{1}{2}(\lambda_1 t - \lambda x)\right] \quad (5.67)$$

of Eq. (5.14) with $r = \lambda^2 - 4\lambda_2 = 0$ was found. Here the functions φ_0 and μ are presented in formulae (5.50) and (5.52).

The ODE system (5.29) has an autonomous ODE (see the second equation therein), hence one has been integrated in the same way as system (5.27). So, the general solution of system (5.29) was found:

$$\varphi_0(t) = \begin{cases} \frac{\lambda_0 t}{2} + \frac{C_1}{t}, & \lambda_1 = 0; \\ \frac{\frac{\lambda_0}{\lambda_1} + e^{\lambda_1 t}(C_1 + \lambda_0 t)}{e^{\lambda_1 t} - 1}, & (2\lambda\varphi_1 + \lambda_1) > |\lambda_1| > 0; \\ \frac{-\frac{\lambda_0}{\lambda_1} + e^{\lambda_1 t}(C_1 + \lambda_0 t)}{e^{\lambda_1 t} + 1}, & |\lambda_1| > (2\lambda\varphi_1 + \lambda_1) > 0. \end{cases} \quad (5.68)$$

$$\varphi_1(t) = -\frac{1}{2\lambda} \begin{cases} \frac{2}{t}, & \lambda_1 = 0; \\ \lambda_1 \coth(\frac{\lambda_1}{2}t) + \lambda_1, & (2\lambda\varphi_1 + \lambda_1) > |\lambda_1| > 0; \\ \lambda_1 \tanh(\frac{\lambda_1}{2}t) + \lambda_1, & |\lambda_1| > (2\lambda\varphi_1 + \lambda_1) > 0, \end{cases} \quad (5.69)$$

$$\varphi_2(t) = C_2 \varphi_1(t) \quad (5.70)$$

Thus, ansatz (5.20) generates the two-parameter family of solutions of Eq. (5.14) with $r = \lambda_2 = 0, \lambda \neq 0$

$$u = \varphi_0(t) + \varphi_1(t)x + C_2\varphi_1(t)e^{-\lambda x} \quad (5.71)$$

where the functions φ_0 and φ_1 are presented in formulae (5.68)–(5.69).

Now we examine the ODE system (5.30) with $r \neq -\frac{1}{2}$. It can be noted that the third equations of the system coincide (up to notations) with the second in system (5.29). Making the quite similar calculations, one can make sure that ansatz (5.21) generates the following family of solutions of Eq. (5.14) with $\lambda = q = 0$

$$u = \varphi_0(t) + C_1\varphi_2(t)x + \varphi_2(t)x^2, \quad (5.72)$$

where

$$\varphi_0(t) = \exp(\lambda_1 t + I)\left[C_0 + \int \left(\frac{\lambda_0}{\exp(\lambda_1 t + I)} + rC_1^2 \exp(\lambda_1 t + (1+4r)I)\right)dt\right], \quad (5.73)$$

$$\varphi_2(t) = \frac{-1}{4(1+2r)} \begin{cases} \frac{2}{t}, & \lambda_1 = 0; \\ \lambda_1 \coth(\frac{\lambda_1}{2}t) + \lambda_1, & 4(1+2r)\varphi_1 + \lambda_1 > |\lambda_1| > 0; \\ \lambda_1 \tanh(\frac{\lambda_1}{2}t) + \lambda_1, & |\lambda_1| > 4(1+2r)\varphi_1 + \lambda_1 > 0, \end{cases} \quad (5.74)$$

and $I = 2\int \varphi_2(t) dt$. It seams to be that the expression in the right-hand side of (5.73) cannot be expressed in terms of elementary functions and/or the well-known special functions provided $\lambda_1 \neq 0$.

In the case $\lambda_1 = 0$, formula (5.72) can be simplified essentially, hence we obtain the exact solution (the linear term w.r.t. x can be removed by the appropriate space translation)

$$u = \frac{1}{t}\left[Ct^{\frac{2r}{1+2r}} + \lambda_0 \frac{1+2r}{2+2r}t^2 - \frac{1}{2(1+2r)}x^2\right], \quad (5.75)$$

of the equation
$$u_t = uu_{xx} + ru_x^2 + \lambda_0, \quad r \neq -1. \tag{5.76}$$

In particular, solution (5.75) with $r = 0$ takes the form
$$u = \frac{C}{t} + \frac{\lambda_0}{2}t - \frac{x^2}{2t}. \tag{5.77}$$

In the case $r = -1$, the solution of Eq. (5.75) takes the form
$$u = Ct + \lambda_0 t \ln t + \frac{x^2}{2t}. \tag{5.78}$$

Integration of the ODE system (5.30) with $r = -\frac{1}{2}$ is simpler because the second and the third equations are rather trivial. As a result, the exact solution
$$u = \pm e^{2Ct} + Cx^2 - \frac{\lambda_0}{2C}, \quad C \neq 0 \tag{5.79}$$

of Eq. (5.76) with $r = -\frac{1}{2}$ and $\lambda_1 = 0$ was obtained. In the case $\lambda_1 \neq 0$, the following family of exact solutions
$$u = \varphi_0(t) + e^{\lambda_1 t}\left(C_1 x + \frac{2C_2}{\lambda_1} x^2\right) \tag{5.80}$$

of the equation
$$u_t = uu_{xx} - \frac{1}{2}u_x^2 + \lambda_1 u + \lambda_0 \tag{5.81}$$

was found. Here the function
$$\varphi_0(t) = \begin{cases} C_0 e^{\lambda_1 t} - \frac{C_1^2}{2\lambda_1} e^{2\lambda_1 t} - \frac{\lambda_0}{\lambda_1}, & C_2 = 0 \\ \exp(\lambda_1 t + C_2 e^{\lambda_1 t})\left[C_0 + \frac{\lambda_0}{\lambda_1} Ei(1, C_2 e^{\lambda_1 t})\right] - \frac{C_1^2 e^{\lambda_1 t}}{2\lambda_1 C_2} - 1, & C_2 \neq 0 \end{cases} \tag{5.82}$$

and Ei is the exponential integral function.

Let us consider the four-dimensional dynamical system (5.31), which contains the subsystem
$$\dot{\gamma} = \gamma^2 \varphi_1,$$
$$\dot{\varphi}_0 = -\varphi_1^2 + \lambda_1 \varphi_0 + \lambda_0, \tag{5.83}$$
$$\dot{\varphi}_1 = \lambda_1 \varphi_1.$$

System (5.83) can be easily integrated and has the general solution
$$\gamma = \frac{\lambda_1}{C_1(t_0 - e^{\lambda_1 t})},$$
$$\varphi_0 = \frac{1}{\lambda_1}(C_0 \lambda_1 e^{\lambda_1 t} - C_1^2 e^{2\lambda_1 t} - \lambda_0), \tag{5.84}$$
$$\varphi_1 = C_1 e^{\lambda_1 t},$$

if $\lambda_1 \neq 0$ and
$$\gamma = \frac{1}{C_1(t_0-t)},$$
$$\varphi_0 = (\lambda_0 - C_1^2)t + C_0, \qquad (5.85)$$
$$\varphi_1 = C_1,$$

if $\lambda_1 = 0$. Here and t_0, C_0 and $C_1 \neq 0$ are arbitrary constants. So, the last ODE of (5.31) takes the forms

$$\dot{\varphi}_2 = \left[\lambda_1 - 2C_1\gamma e^{\lambda_1 t} + \frac{\gamma^2}{\lambda_1}(C_0\lambda_1 e^{\lambda_1 t} - C_1^2 e^{2\lambda_1 t} - \lambda_0)\right]\varphi_2 \qquad (5.86)$$

and

$$\dot{\varphi}_2 = \left[\gamma^2\left(C_0 + (\lambda_0 - C_1^2)t\right) - 2\gamma C_1\right]\varphi_2 \qquad (5.87)$$

for $\lambda_1 \neq 0$ and $\lambda_1 = 0$, respectively. Obviously the above ODEs are integrable.

Thus, ansatz (5.20) leads to the two essentially different families of exact solutions of Eq. (5.14) with $\lambda = \lambda_2 = 0, r = -1$, i.e.,

$$u_t = u u_{xx} - u_x^2 + \lambda_1 u + \lambda_0. \qquad (5.88)$$

If $\lambda_1 \neq 0$ then the corresponding calculations lead to the two-parameter family of solutions

$$u = C_1 x e^{\lambda_1 t} - \frac{C_1^2}{\lambda_1}e^{2\lambda_1 t} - \frac{\lambda_0}{\lambda_1} + C_2 \frac{(e^{\lambda_1 t} - 1)^{\kappa+1}}{e^{\lambda_1(\kappa-1)t}}\exp\left(\frac{1 + \kappa - \frac{\lambda_1}{C_1}x}{e^{\lambda_1 t} - 1}\right), \qquad (5.89)$$

where $\kappa = \frac{\lambda_0}{C_1^2}$ and $C_1 \neq 0$.

If $\lambda_1 = 0$ then the corresponding solution is much simpler (the time and space translations were used for its simplification):

$$u = (\lambda_0 - C_1^2)t + C_1 x + C_2 t^{\frac{\lambda_0}{C_1^2}+1}\exp\left(-\frac{x}{C_1 t}\right). \qquad (5.90)$$

Thus, we have examined all the ODE systems obtained by application of the additional generating condition (5.17). As a result, several families of exact solutions of the nonlinear equation (5.14) have been constructed. Using substitutions (5.15) and (5.16), all the solutions found for Eq. (5.14) can be transformed into solutions of the nonlinear RDC equations (5.10) and (5.11), respectively. In Tables 5.1–5.3, we list the families of the exact solutions of equations (5.10) and (5.11) that were found in this section.

In Figures 5.1–5.4, examples of exact solutions listed in Tables 5.1–5.3 are plotted. As one sees from Figures 5.1 and 5.2, the shape of the solution listed in case 3 of Table 5.1 depends essentially on the sign of the exponent α arising in Eq. (5.10). Figure 5.3 shows that the solution listed in case 1 of

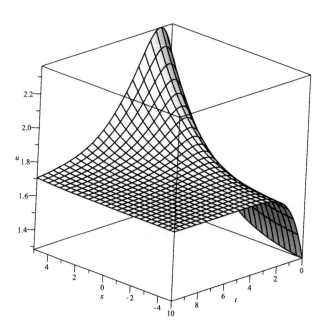

FIGURE 5.1: Exact solution from Table 5.1 case 3 with $\alpha = 3$, $\lambda_0 = 1$, $\lambda_1 = 1$, $\lambda_2 = -\frac{1}{4}$, $C_1 = 1$, $\gamma_1 = \frac{\sqrt{3}}{4}$

Table 5.2 has an interesting behavior in a vicinity of the line $x = 0$, while one sharply tends to zero when $|x| \gg 0$. The exact solution plotted in Figure 5.3 is essentially inhomogeneous for small values of time while one tends to the steady-state point of Eq. (5.11) (it is $T = \ln 3$ for the parameters indicated in Figure 5.4) provided $t \to \infty$.

Remark 5.4 *In Case 3 of Table 5.2, the solution can be rewritten in essentially different forms depending on the signs of λ_2 and ν^2 (see (5.62)) if one is looking for real solutions. For example, if $\lambda_2 > 0$ and $\nu^2 < 0$ then the corresponding solution has the form*

$$T = \left[\varphi_0(t) + e^{\lambda_1 t} \left(C_1 \cos(\sqrt{2\lambda_2}x) + C_2 \sin(\sqrt{2\lambda_2}x) \right) \right]^{-\frac{1}{2}},$$

where $\varphi_0(t)$ involves the modified Bessel functions instead of J_ν and Y_ν.

TABLE 5.1: Exact solutions of the nonlinear RDC equation (5.10) with $\alpha \neq -2$ and $\lambda = 0$

	Exact solutions	Restrictions
1	$\left[\varphi_0(t) + \exp\left(\lambda_1 t + \lambda_2 \frac{\alpha+2}{\alpha+1} \int \varphi_0(t)\mathrm{d}t\right)\right.$ $\left.(C_1 e^{-\gamma_1 x} + C_2 e^{\gamma_1 x})\right]^{\frac{1}{\alpha}},$ $\gamma_1^2 = -\frac{\alpha \lambda_2}{1+\alpha} > 0$	$\lambda_2 \neq 0, \alpha \neq -1$; $\varphi_0(t)$ is a solution of (5.55)
2	$\left[\varphi_0(t) + \exp\left(\lambda_1 t + \lambda_2 \frac{\alpha+2}{\alpha+1} \int \varphi_0(t)\mathrm{d}t\right)\right.$ $\left.(C_{10} \cos|\gamma_1|x + C_{11}\sin|\gamma_1|x)\right]^{\frac{1}{\alpha}},$ $\gamma_1^2 = -\frac{\alpha\lambda_2}{1+\alpha} > 0$	$\lambda_2 \neq 0, \alpha \neq -1$; $\varphi_0(t)$ is a solution of (5.66)
3	$\left[\varphi_0(t) + \frac{C_1}{\mu(t)^{\frac{\alpha+2}{\alpha+1}}}\exp\left(\frac{\lambda_1 \alpha}{2(1+\alpha)}t \pm \gamma_1 x\right)\right]^{\frac{1}{\alpha}},$ $\gamma_1^2 = -\frac{\alpha\lambda_2}{1+\alpha} > 0$	$\lambda_2 \neq 0, \alpha \neq -1$; $\varphi_0(t), \mu(t)$ see in (5.50), (5.52)
4	$\left[\varphi_0(t) + C_1\varphi_2(t)x + \varphi_2(t)x^2\right]^{\frac{1}{\alpha}}$	$\lambda_1 \neq 0, \lambda_2 = 0, \alpha \neq -1$; $\varphi_0(t), \mu(t)$ see in (5.73), (5.74)
5	$\left[Ct^{-\frac{\alpha}{\alpha+2}} + \frac{\lambda_0(\alpha+2)}{2(\alpha+1)}t - \frac{\alpha}{2(\alpha+2)}\frac{x^2}{t}\right]^{\frac{1}{\alpha}}$	$\lambda_1 = \lambda_2 = 0, \alpha \neq -1$
6	$\left[C_1 x e^{\lambda_1 t} - \frac{C_1^2}{\lambda_1}e^{2\lambda_1 t} - \frac{\lambda_0}{\lambda_1} + C_2 \frac{(e^{\lambda_1 t}-1)^{\kappa+1}}{e^{\lambda_1(\kappa-1)t}}\right.$ $\left.\exp\left(\frac{1+\kappa-\frac{\lambda_1}{C_1}x}{e^{\lambda_1 t}-1}\right)\right]^{-1}, \kappa = \frac{\lambda_0}{C_1^2}$	$\lambda_1 \neq 0, \lambda_2 = 0, \alpha = -1$, $C_1 \neq 0$
7	$\left[(\lambda_0 - C_1^2)t + C_1 x + C_2 t^{\frac{\lambda_0}{C_1^2}+1}\exp\left(-\frac{x}{C_1 t}\right)\right]^{-1}$	$\lambda_1 = \lambda_2 = 0, \alpha = -1$
8	$\left(Ct + \lambda_0 t \ln t + \frac{x^2}{2t}\right)^{-1}$	$\lambda_1 = \lambda_2 = 0, \alpha = -1$

TABLE 5.2: Exact solutions of the nonlinear RDC equation (5.10) with $\alpha = -2$ and $\lambda = 0$

	Exact solutions	Restrictions
1	$\left[\varphi_0(t) + C_1 e^{-\sqrt{-2\lambda_2}\,x} + C_2 e^{\sqrt{-2\lambda_2}\,x}\right]^{-\frac{1}{2}}$	$\lambda_1 = 0, \lambda_2 < 0$; $\varphi_0(t)$ see in (5.61)
2	$\left[\varphi_0(t) + C_1 \cos(\sqrt{2\lambda_2}\,x) + C_2 \sin(\sqrt{2\lambda_2}\,x)\right]^{-\frac{1}{2}}$	$\lambda_1 = 0, \lambda_2 > 0$; $\varphi_0(t)$ see in (5.61)
3	$\left[\varphi_0(t) + e^{\lambda_1 t}(C_1 e^{-\sqrt{-2\lambda_2}\,x} + C_2 e^{\sqrt{-2\lambda_2}\,x})\right]^{-\frac{1}{2}}$	$\lambda_1 \lambda_2 \neq 0$; $\varphi_0(t)$ see in (5.62)
4	$\left(\pm e^{2Ct} + Cx^2 - \frac{\lambda_0}{2C}\right)^{-\frac{1}{2}}$	$\lambda_1 = \lambda_2 = 0, C \neq 0$
5	$\left[\varphi_0(t) + C_1 e^{\lambda_1 t} x + \frac{2C_2}{\lambda_1} e^{\lambda_1 t} x^2\right]^{-\frac{1}{2}}$	$\lambda_1 \neq 0, \lambda_2 = 0$; $\varphi_0(t)$ see in (5.82)

TABLE 5.3: Exact solutions of the nonlinear RDC equation (5.11)

	Exact solutions	Restrictions
1	$\ln\left[\varphi_0(t) + \frac{e^{\frac{\lambda_1}{2}t}}{\mu(t)}\left(C_1 e^{\gamma_1 x} + C_2 e^{\gamma_2 x}\right)\right]^{\frac{1}{\beta}}$, $\gamma_{1,2} = \frac{1}{2}(\pm(\lambda^2 - 4\lambda_2)^{\frac{1}{2}} - \lambda)$	$\lambda^2 - 4\lambda_2 \neq 0$; $\varphi_0(t), \mu(t)$ see in (5.50), (5.52)
2	$\ln\left[\varphi_0(t) + \frac{C_1 + C_2 x}{\mu(t)} \exp\left(\frac{1}{2}(\lambda_1 t - \lambda x)\right)\right]^{\frac{1}{\beta}}$	$\lambda \neq 0, \lambda^2 - 4\lambda_2 = 0$; $\varphi_0(t), \mu(t)$ see in (5.50), (5.52)
3	$\ln\left[\varphi_0(t) + \varphi_1(t)x + C_2 \varphi_1(t)e^{-\lambda x}\right]^{\frac{1}{\beta}}$	$\lambda \neq 0, \lambda_2 = 0$; $\varphi_0(t), \varphi_1(t)$ see in (5.68), (5.69)
4	$\ln\left(\frac{C}{t} + \frac{\lambda_0}{2}t - \frac{x^2}{2t}\right)^{\frac{1}{\beta}}$	$\lambda = 0, \lambda_2 = 0$

Application of the method for solving nonlinear RDC equations   211

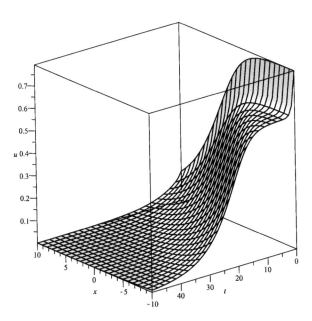

FIGURE 5.2: Exact solution from Table 5.1 case 3 with $\alpha = -3$, $\lambda_0 = 1$, $\lambda_1 = 1$, $\lambda_2 = -\frac{1}{4}$, $C_1 = 1$, $\gamma_1 = \frac{\sqrt{6}}{4}$

5.2.3 Application of the solutions obtained for solving boundary-value problems

Here we study the nonlinear reaction-diffusion (RD) equation of the form [51, 61]
$$u_t = [(d_0 + d_1 u)u_x]_x + \lambda u(1 - u), \tag{5.91}$$
where the function $u(t,x)$ means the concentration of cells (population, chemical etc.) and $d_1, d_0 > 0$ and $\lambda > 0$ are arbitrary parameters.

Eq. (5.91) with $d_1 = 0$ is equivalent the famous Fisher equation (4.25), which was discussed in Chapter 4, so that we assume $d_1 \neq 0$ in what follows. From the biological point of view, Eq. (5.91) can be thought as a natural generalization of the Fisher equation with the nonconstant diffusivity $d(u) = d_1 u + d_0$ and can be called the generalized Fisher equation. We remind the reader that nonconstant diffusion can play an important role in biologically motivated models (see, e.g., [180]–[182]). In particular, Eq. (5.91) with $d_0 = 0$ is the porous-Fisher equation, which was discussed (including the search for traveling fronts) in [180].

Here we restrict ourselves on the case $d_0 \neq 0$ in order to keep the Fisher

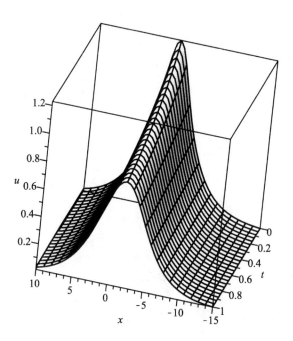

FIGURE 5.3: Exact solution from Table 5.2 case 1 with $\lambda_0 = 1$, $\lambda_2 = -2$, $C_1 = C_2 = \frac{1}{2}$

equation as a limiting case when $d_1 \to 0$. Thus, Eq. (5.91) without loosing a generality can be written as

$$u_t = [(1 + d_1 u)u_x]_x + \lambda u(1 - u). \qquad (5.92)$$

It should be also pointed out that the generalized Fisher equation (5.92) is a one-component analog of the Shigesada–Kawasaki–Teramoto cross-diffusion model [230]

$$\begin{aligned} u_t &= [(d_1 + \rho_1 v)u]_{xx} + u(a_1 - b_1 u - c_1 v) \\ v_t &= [(d_2 + \rho_2 u)v]_{xx} + v(a_2 - b_2 u - c_2 v), \end{aligned} \qquad (5.93)$$

which takes into account the pressures created by mutually competing species (here all the parameters are nonnegative constants). In particular, (5.93) reduces to (5.91) if the concentrations u and v are linearly depended. Notably, MAGC was used for construction of exact solutions of system (5.93) in [72] (see also the recent paper [60] for more solutions of this system).

It is easily seen that Eq. (5.92) is reducible to a particular case of Eq. (5.12) by the substitution $1 + d_1 u \to d_1 u$. Thus, the results obtained for Eq. (5.12) can be transferred onto Eq. (5.92) provided the coefficients satisfy the corresponding relations. As a result, it was established that solutions (5.56) and (5.57)

Application of the method for solving nonlinear RDC equations

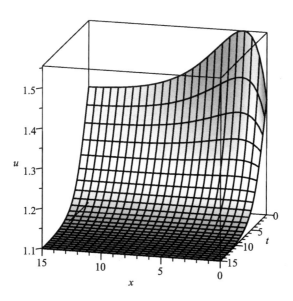

FIGURE 5.4: Exact solution from Table 5.3 case 1 with $\beta = 1$, $\lambda_0 = 1$, $\lambda_0 = 3.75$, $\lambda_1 = -2$, $\lambda_2 = \frac{1}{4}$, $C_1 = 0$, $C_2 = 1$

generate the following solutions of Eq. (5.92) with $d_1 \lambda > 0$

$$u = \frac{1}{2}\left[1 + \tanh \frac{\lambda(t-c_0)}{2}\right] + c_1 \frac{\exp \frac{\lambda(2+d_1)t}{4d_1}}{[\cosh \frac{\lambda(t-c_0)}{2}]^{\frac{3}{2}}} \exp\left(\sqrt{\frac{\lambda}{2d_1}}\, x\right) \qquad (5.94)$$

and

$$u = \frac{1}{2}\left[1 + \tanh \frac{\lambda(t-c_0)}{2}\right] + c_1 \frac{\exp \frac{\lambda(2+d_1)t}{4d_1}}{[\cosh \frac{\lambda(t-c_0)}{2}]^{\frac{3}{2}}} \exp\left(-\sqrt{\frac{\lambda}{2d_1}}\, x\right) \qquad (5.95)$$

with arbitrary parameters c_0 and c_1.

The solutions of the form (5.94)–(5.95) have attractive properties. In particular, any solution u^* of the form (5.95) with $d_1 > 1$ holds the conditions $u^* \to 1$ if $t \to \infty$ and $u^* \to \frac{1}{2}[1 + \tanh \frac{\lambda(t-c_0)}{2}] \leq 1$ if $x \to +\infty$. Taking into account these properties, the following statement can be formulated.

Theorem 5.2 *[52] The bounded exact solution of the boundary-value problem (BVP) for the generalized Fisher equation (5.92) with $d_1 > 1$, $\lambda > 0$, the initial*

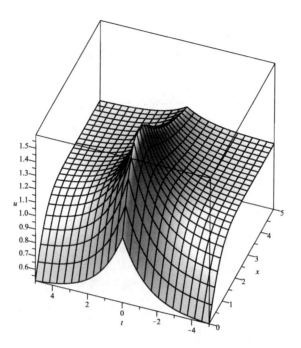

FIGURE 5.5: Exact solution (5.98) with $\lambda = 2$, $\lambda_0 = 2$, $c_0 = 0$, $c_1 = \frac{1}{2}$

condition

$$u(0, x) = C_0 + C_1 \exp\left(-\sqrt{\frac{\lambda}{2d_1}}|x|\right), \tag{5.96}$$

and the Neumann conditions

$$u_x(t, -\infty) = 0, \quad u_x(t, +\infty) = 0, \tag{5.97}$$

is given in the domain $\Omega = \{(t, x) \in (0, +\infty) \times (-\infty, +\infty)\}$ by the formula

$$u = \frac{1}{2}\left[1 + \tanh\frac{\lambda(t - c_0)}{2}\right] + c_1 \frac{\exp\frac{\lambda(2+d_1)t}{4d_1}}{[\cosh\frac{\lambda(t-c_0)}{2}]^{\frac{3}{2}}} \exp\left(-\sqrt{\frac{\lambda}{2d_1}}|x|\right), \tag{5.98}$$

where $C_0 = \frac{1}{2}(1 - \tanh\frac{\lambda c_0}{2})$, $C_1 = c_1(\cosh\frac{\lambda c_0}{2})^{-\frac{3}{2}}$.

An example of solution (5.98) is presented in Figure 5.5. Obviously, one is not analytic at $x = 0$, however the second derivative is well defined at this point since $u_{xx}\big|_{x \to 0+} = u_{xx}\big|_{x \to 0-}$ and $u_x^2\big|_{x \to 0+} = u_x^2\big|_{x \to 0-}$. Thus, this solution satisfies Eq. (5.92) at the point $x = 0$ as well. On the other hand, solution (5.98) of BVP (5.92), (5.96) and (5.97) is not unique (we remind the reader that examples of nonuniqueness for nonlinear BVP are well-known, see, e.g., [129]). Here we note that this issue for the given BVP was discussed in [61] using numerical methods.

If $d_1 \lambda < 0$ then solutions of the form (5.65) arise. In particular, setting $d_1 = -1$, the exact solution of Eq. (5.92)

$$u = 1 - \varphi_0(t) + C \exp\left(-\lambda t + \frac{3}{2}\lambda \int \varphi_0(t) dt\right) \sin\left(\sqrt{\frac{\lambda}{2}} x\right), \quad (5.99)$$

is obtained. Here $\varphi_0(t)$ is an arbitrary solution of the integro-differential equation

$$\dot{\varphi}_0 = \lambda \varphi_0^2 - \lambda \varphi_0 + \frac{1}{2} C^2 \lambda \exp\left(-2\lambda t + 3\lambda \int \varphi_0(t) dt\right). \quad (5.100)$$

Eq. (5.100) is reducible to the second-order ODE

$$\ddot{\varphi}_0 = 2\lambda \varphi_0 \dot{\varphi}_0 + (2 - \lambda)\dot{\varphi}_0 - \lambda \varphi_0^2 + \lambda \varphi_0, \quad (5.101)$$

which is not integrable. However, $\varphi_0 = 0$ is the steady-state point, therefore the solutions should exist, which tend to this point as $t \to \infty$. So, taking the solution with the property $\varphi_0(t) \to 0$ as $t \to \infty$, we note that $u \to 1$ as $t \to \infty$ for all values of C and $\lambda > 0$ (see formula (5.99)).

In contrast to the solutions of the forms (5.94) and (5.95), the periodic solution (5.99) satisfies the zero flux conditions (those are the typical requirements in the population dynamics) at the bounded interval $[A, B]$. In fact, setting $A = (\frac{1}{2} + k_1)\sqrt{\frac{2}{\lambda}}\pi$, $B = (\frac{1}{2} + k_2)\sqrt{\frac{2}{\lambda}}\pi$, and $k_1 < k_2 \in \mathbf{Z}$, one notes that the Neumann conditions

$$u_x(t, A) = 0, \quad u_x(t, B) = 0 \quad (5.102)$$

are fulfilled.

Assuming that the generalized Fisher equation (5.92) describes the time evolution of the population spread in the 1D space, the plot presented in Figure 5.5 represents the population density. The plot predicts the scenario when the population density increases in the initial time period $(0, t^*)$, but one decreases when time is larger than t^*. Finally, $u \approx 1$ for $t \gg t^*$.

The exact solution (5.99) with the suitable $\varphi_0(t)$ predicts the scenario when the initially periodic (w.r.t. the variable x) distribution of the population density is changing with time in such a way that the homogenous distribution $u \approx 1$ for $t \gg 0$ is obtained. Both scenarios are plausible and may occur in population dynamics.

5.3 Analysis of the solutions obtained and comparison with the known results

It can be easily established that a large majority of the solutions obtained above for the nonlinear RDC equations (5.10) and (5.11) are non-Lie solutions, i.e., they are not obtainable by the Lie method. Moreover, many of them cannot be obtained via Q-conditional symmetries, which have been studied in Chapter 3.

In fact, using LSC of the general class of the RDC equations derived in Chapter 2 (see Table 2.5 for instance), one notes that both equations are invariant with respect to (w.r.t.) the trivial algebra (2.85) provided (i) $\lambda_0 \lambda_1 \neq 0$ and/or (ii) $\lambda_0 \lambda_2 \neq 0$ (other coefficients are arbitrary). It means that the Lie method allows us to construct only the plane wave solutions of Eqs. (5.10) and (5.11) under the restrictions (i) and/or (ii). Obviously, *the exact solutions presented in Tables 5.1 (excepting Cases 5,7 and 8), 5.2 (excepting Case 4) and 5.3 are not plane wave solutions, hence they are non-Lie solutions.*

The four cases excluded above need the additional examination because the corresponding equations of the form (5.10) are invariant w.r.t. the three-dimensional maximal algebra of invariance (MAI) provided either $\lambda_0 \neq 0$, $\lambda_1 = \lambda_2 = 0$ (see Case 23 in Table 2.5), or $\lambda_0 = 0$, $\lambda_2 \neq 0$ (see Cases 23 and 24 in Table 2.5) and the four-dimensional MAI provided $\lambda_0 = \lambda_2 = 0$ (see Cases 3 and 11). Moreover, there is a special case $\alpha = -\frac{4}{3}$ when Eq. (5.10) with $\lambda_0 = \lambda = 0$ admits even the five-dimensional MAI (see Cases 2, 8 and 9). The corresponding exact solutions listed in Tables 5.1 and 5.2 may be reducible (not necessary!) to the Lie solutions provided the above coefficient restrictions take place. In particular, all the solutions listed in Cases 5,7 and 8 of Table 5.1 and Cases 4 of Table 5.2 are the invariant solutions if additionally $\lambda_0 = 0$. In fact, one notes that these solutions are obtainable via the corresponding ansätze from Table 4.1.

Let us consider an interesting example. The family of solutions (5.75) by the time shift t_0 takes the form

$$U = C(t_0 - t)^{\frac{-1}{1+2r}} - \lambda_0 \frac{1+2r}{2+2r}(t_0 - t) + \frac{1}{2+4r}\frac{x^2}{t_0 - t}. \tag{5.103}$$

On the other hand, (5.76) and (5.103) by the substitution

$$T(\tau, x) = U^{\frac{1}{\mu-1}}, \quad \tau = \frac{t}{\mu}, \quad r = \frac{1}{\mu - 1}, \quad \lambda_0 = \frac{\lambda_0^*(1-\mu)}{\mu} \tag{5.104}$$

are reduced to the equation

$$T_\tau = (T^\mu)_{xx} - \lambda_0^* T^{2-\mu}, \quad \mu \neq 0, 1 \tag{5.105}$$

Analysis of the solutions obtained and comparison with the known results 217

and to the solution

$$T^{\mu-1} = C(t_0 - \mu\tau)^{\frac{1-\mu}{1+\mu}} + \frac{\lambda_0^*(\mu^2 - 1)}{2\mu^2}(t_0 - \mu\tau) + \frac{\mu - 1}{2(1+\mu)} \frac{x^2}{t_0 - \mu\tau}, \quad (5.106)$$

respectively. Formula (5.106) with $\lambda_0^* = 0$ is nothing else but the well-known solution (4.17) (up to notations) of the nonlinear heat equation

$$T_\tau = (T^\mu)_{xx}, \quad \mu \neq 0, 1, \quad (5.107)$$

which was already discussed in Section 4.2. In particular, this solution is relevant to gas dynamics [19, 250] and can be derived using the Lie method.

To the best of our knowledge, solution (5.106) with $\lambda_0^* \neq 0$ was firstly obtained in [147]. In contrast to the case $\lambda_0^* = 0$, this solution is not obtainable by the Lie method because (5.105) with $\lambda_0^* \neq 0$ is invariant w.r.t. the three-dimensional MAI (see Case 23 in Table 2.5 setting $m = \lambda_2 = 0$), which does not produce a suitable ansatz in order to obtain solution (5.106). Thus, we conclude that the exact solution arising in Case 5 of Table 5.1 is a non-Lie solution provided $\lambda_0 \neq 0$.

As it was already established, the nonlinear RDC equation (5.10) with $\lambda = 0$ and $\lambda_0(|\lambda_1| + |\lambda_2|) \neq 0$ does not admit any Q-conditional symmetry of the form (3.11) (see Theorem 3.5). Thus, the exact solutions listed in Cases 1–4, 6 of Table 5.1 and Case 1–3, 5 of Table 5.2 are not obtainable by any Q-conditional symmetry of the form (3.11)(of course, it may happen that there exist suitable symmetries of the form (3.12)). Notably, Eq. (5.10) with $\lambda = \lambda_1 = \lambda_2 = 0$ admits Q-conditional symmetries, which coincide with Lie's and this case was discussed above.

Now we turn to the solutions listed in Table 5.3. All those are non-Lie provided the restrictions (i) and/or (ii) take place. The solutions listed in Cases 1–3 are highly nontrivial. They were independently derived for the first time in [220] using generalized conditional symmetries and in [48] via MAGC. It was established much later [78] that these solutions are also obtainable via the Q-conditional symmetries (see Subsection 4.5.2 for details).

Thus, all the three-dimensional ODE systems derived in Subsection 5.2.1 were examined and the corresponding solutions were used for constructing exact solutions of the nonlinear RDC equations (5.10) and (5.11). The solutions are summarized in Tables 5.1–5.3 and briefly discussed above.

We point out that there are two four-dimensional ODE systems (see (5.42) and (5.48). To the best of our knowledge they are still not solved and this task lies beyond this monograph. Here we only note that the reduction of Eq. (5.45) to (5.42) was firstly found in [48] (later rediscovered in [141]), while the reduction to (5.48) was noted in [117, 150, 151]. A generalization of the above reduction with the diffusion exponent $-\frac{3}{2}$ on the reaction-diffusion systems is presented in [66] (recently rediscovered in [140] using generalized conditional symmetries).

In conclusion, it is worth noting that the MAGC key idea is to reduce solving of the given nonlinear PDE to a system of ODEs, while application of the Lie method and the method of Q-conditional symmetries lead to a single ODE (excepting special cases when the symmetry operators in question involve arbitrary functions). It is a main reason why the exact solutions obtained via MAGC are usually non-Lie solutions, moreover, as it was shown above, they often cannot be derived via Q-conditional (nonclassical) symmetries of the form (3.11).

On the other hand, there are other methods allowing to reduce nonlinear PDEs (with a correctly-specified structure) to the ODE systems too. The most popular among them are the method of generalized conditional symmetries [101, 173, 251] and the method of linear invariant subspaces [119]. An analysis of exact solutions derived by these methods and MAGC shows that almost each solution can be independently derived by any of the three methods. For example, if one compares the reductions of the RDC equations derived here using MAGC (actually, they are taken from [48]) and those obtained in [141] via generalized conditional symmetries then one concludes that many of them are identical or very similar. In our opinion, MAGC is simpler for realization and can be easily implemented in a symbolic program (like it has been already done for finding Lie symmetries).

Finally, we point out that different methods (techniques, approaches) can be used for finding exact solutions of nonintegrable PDEs with "poor" Lie symmetry. However, efficiency of each method depends essentially on structure of the nonlinear equation in question. Thus, creation of any new method (technique, approach) allowing to solve (at least partly) a class of nonintegrable PDEs is of fundamental importance because there is no existing general theory for integrating nonlinear PDEs.

References

[1] M. J. Ablowitz and P. A. Clarkson. *Solitons, nonlinear evolution equations and inverse scattering*, volume 149 of *London Mathematical Society Lecture Note Series*. Cambridge University Press, Cambridge, 1991.

[2] M. J. Ablowitz and A. Zeppetella. Explicit solutions of Fisher's equation for a special wave speed. *Bulletin of Mathematical Biology*, 41(6):835–840, 1979.

[3] I. S. Akhatov, R. K. Gazizov, and N. K. Ibragimov. Nonlocal symmetries. Heuristic approach. *Journal of Soviet Mathematics*, 55(1):1401–1450, 1991.

[4] V. Alexiades and A. D. Solomon. *Mathematical Modeling of Melting and Freezing Processes*. Taylor & Francis, 1993.

[5] W. F. Ames. *Nonlinear partial differential equations in engineering. Vol. I*. Academic Press, New York-London, 1965.

[6] W. F. Ames. *Nonlinear partial differential equations in engineering. Vol. II*. Academic Press, New York-London, 1972. Mathematics in Science and Engineering, Vol. 18-II.

[7] W. F. Ames, R. L. Anderson, V. A. Dorodnitsyn, E. V. Ferapontov, R. K. Gazizov, N. H. Ibragimov, and S. R. Svirshchevskii. *CRC handbook of Lie group analysis of differential equations. Vol. 1*. CRC Press, Boca Raton, FL, 1994. Symmetries, exact solutions and conservation laws.

[8] R. L. Anderson, V. A. Baikov, R. K. Gazizov, W. Hereman, N. H. Ibragimov, F. M. Mahomed, S. V. Meleshko, M. C. Nucci, P. J. Olver, M. B. Sheftel, A. V. Turbiner, and E. M. Vorob'ev. *CRC handbook of Lie group analysis of differential equations. Vol. 3*. CRC Press, Boca Raton, FL, 1996. New trends in theoretical developments and computational methods.

[9] P. Appell. Sur l'équation $\frac{\partial^2 z}{\partial x^2} - \frac{\partial z}{\partial y} = 0$ et la théorie de la chaleur. *Journal de Mathématiques Pures et Appliquées. Paris*, 8:187–216, 1892 (in French).

[10] R. Aris. *The mathematical theory of diffusion and reaction in permeable catalysts: the theory of the steady state*. Oxford, Clarendon Press, 1975.

References

[11] R. Aris. *The mathematical theory of diffusion and reaction in permeable catalysts: Vol. 2: questions of uniqueness, stability, and transient behavior.* Oxford, Clarendon Press, 1975.

[12] D. J. Arrigo. *Symmetry analysis of differential equations.* John Wiley & Sons, Inc., Hoboken, NJ, 2015. An introduction.

[13] D. J. Arrigo, P. Broadbridge, and J. M. Hill. Nonclassical symmetry solutions and the methods of Bluman–Cole and Clarkson–Kruskal. *J. Math. Phys.*, 34(10):4692–4703, 1993.

[14] D. J. Arrigo, D. A. Ekrut, J. R. Fliss, and L. Le. Nonclassical symmetries of a class of Burgers' systems. *J. Math. Anal. Appl.*, 371(2):813–820, 2010.

[15] D. J. Arrigo and F. Hickling. On the determining equations for the nonclassical reductions of the heat and Burgers' equation. *J. Math. Anal. Appl.*, 270(2):582–589, 2002.

[16] D. J. Arrigo and J. M. Hill. Nonclassical symmetries for nonlinear diffusion and absorption. *Stud. Appl. Math.*, 94(1):21–39, 1995.

[17] D. J. Arrigo, J. M. Hill, and P. Broadbridge. Nonclassical symmetry reductions of the linear diffusion equation with a nonlinear source. *IMA J. Appl. Math.*, 52(1):1–24, 1994.

[18] V. A. Baikov, R. K. Gazizov, N. H. Ibragimov, and V. F. Kovalev. Water redistribution in irrigated soil profiles: invariant solutions of the governing equation. *Nonlinear Dynam.*, 13(4):395–409, 1997.

[19] G. I. Barenblatt. On some unsteady motions of a liquid and gas in a porous medium. *Akad. Nauk SSSR. Prikl. Mat. Meh.*, 16:67–78, 1952 (in Russian).

[20] G. I. Barenblatt and Y. B. Zeldovich. On the dipole type solution in the problem of unsteady gas filtration in the polytropic regime. *Prikl. Math. Mech.*, 21:718–720, 1957 (in Russian).

[21] J. Bear. *Dynamics of fluids in porous media.* Dover Civil and Mechanical Engineering Series. Dover, 1972.

[22] M. Bertsch, R. Kersner, and L. A. Peletier. Positivity versus localization in degenerate diffusion equations. *Nonlinear Anal.*, 9(9):987–1008, 1985.

[23] G. Birkhoff. *Hydrodynamics: a study in logic, fact, and similitude.* Dover books on science. Princeton University Press for University of Cincinnati, 1950.

[24] G. W. Bluman and S. C. Anco. *Symmetry and integration methods for differential equations*, volume 154 of *Applied Mathematical Sciences.* Springer, New York, 2002.

References

[25] G. W. Bluman, A. F. Cheviakov, and S. C. Anco. *Applications of symmetry methods to partial differential equations*, volume 168 of *Applied Mathematical Sciences*. Springer, New York, 2010.

[26] G. W. Bluman and J. D. Cole. The general similarity solution of the heat equation. *J. Math. Mech.*, 18:1025–1042, 1968/69.

[27] G. W. Bluman and J. D. Cole. *Similarity methods for differential equations*, volume 13 of *Applied Mathematical Sciences*. Springer, New York-Heidelberg, 1974.

[28] G. W. Bluman and S. Kumei. On the remarkable nonlinear diffusion equation $(\partial/\partial x)[a(u+b)^{-2}(\partial u/\partial x)] - (\partial u/\partial t) = 0$. *J. Math. Phys.*, 21(5):1019–1023, 1980.

[29] G. W. Bluman and S. Kumei. *Symmetries and differential equations*, volume 81 of *Applied Mathematical Sciences*. Springer, New York, 1989.

[30] G. W. Bluman, G. J. Reid, and S. Kumei. New classes of symmetries for partial differential equations. *J. Math. Phys.*, 29(4):806–811, 1988.

[31] L. Boltzmann. Zur Integration der Diffusionsgleichung bei variabeln Diffusionscoefficienten. *Annalen der Physik*, 289(13):959–964, 1894 (in German).

[32] J. Boussinesq. Recherches théoriques sur l'écoulement des nappes d'eau infiltrées dans le sol et sur débit de sources. *J. Math. Pures Appl.*, 10:5–78, 1904 (in French).

[33] R. H. Boyer. On some solutions of a non-linear diffusion equation. *Journal of Mathematics and Physics*, 40(1):41–45, 1961.

[34] B. H. Bradshaw-Hajek, M. P. Edwards, P. Broadbridge, and G. H. Williams. Nonclassical symmetry solutions for reaction-diffusion equations with explicit spatial dependence. *Nonlinear Anal.*, 67(9):2541–2552, 2007.

[35] N. F. Britton. *Essential mathematical biology*. Springer Undergraduate Mathematics Series. Springer, London, 2003.

[36] P. Broadbridge. Integrable forms of the one-dimensional flow equation for unsaturated heterogeneous porous media. *J. Math. Phys.*, 29(3):622–627, 1988.

[37] P. Broadbridge and I. White. Constant rate rainfall infiltration: a versatile nonlinear model 1. analytic solution. *Water Resour. Res.*, 24(1):145–154, 1988.

[38] W. Brutsaert. Some exact solutions for nonlinear desorptive diffusion. *Zeitschrift für angewandte Mathematik und Physik ZAMP*, 33(4):540–546, 1982.

[39] J. M. Burgers. *The nonlinear diffusion equation. Asymptotic solutions and statistical problems*. Dordrecht – Boston: D. Reidel Publishing Company, 1974.

[40] H. M. Byrne, J. R. King, D. L. S. McElwain, and L. Preziosi. A two-phase model of solid tumour growth. *Appl. Math. Lett.*, 16(4):567–573, 2003.

[41] M. Carini, D. Fusco, and N. Manganaro. Wave-like solutions for a class of parabolic models. *Nonlinear Dynam.*, 32(2):211–222, 2003.

[42] E. Cartan. Sur la théorie des systèmes en involution et ses applications à la relativité. *Bull. Soc. Math. France*, 59:88–118, 1931 (in French).

[43] Z.-X. Chen and B.-Y. Guo. Analytic solutions of the Nagumo equation. *IMA J. Appl. Math.*, 48(2):107, 1992.

[44] R. Cherniha. On exact solutions of a nonlinear diffusion-type system. *Symmetry Analysis and Exact Solutions of Equations of Mathematical Physics, Kyiv, Institute of Mathematics, Ukrainian Acad. Sci.*, pages 49–53, 1988 (in Russian).

[45] R. Cherniha. Symmetry and exact solutions of heat-and-mass transfer equations in tokamak plasma. *Dopovidi Akad. Nauk Ukrain.*, 4:17–21, 1995 (in Ukrainian).

[46] R. Cherniha. A constructive method for obtaining new exact solutions of nonlinear evolution equations. *Rep. Math. Phys.*, 38(3):301–312, 1996.

[47] R. Cherniha. Application of one constructive method for the construction of non-Lie solutions of nonlinear evolution equations. *Ukrainian Math. J.*, 49(6):910–924, 1997.

[48] R. Cherniha. New non-Lie ansätze and exact solutions of nonlinear reaction-diffusion-convection equations. *J. Phys. A*, 31(40):8179–8198, 1998.

[49] R. Cherniha. New symmetries and exact solutions of nonlinear reaction-diffusion-convection equations. In *1st International Workshop on Similarity Methods*, pages 323–336. Universität Stuttgart, Stuttgart, 1998.

[50] R. Cherniha. Lie symmetries of nonlinear two-dimensional reaction-diffusion systems. *Rep. Math. Phys.*, 46(1-2):63–76, 2000.

[51] R. Cherniha. New exact solutions of one nonlinear equation in mathematical biology and their properties. *Ukrainian Math. J.*, 53(10):1712–1727, 2001.

[52] R. Cherniha. Nonlinear Galilei-invariant PDEs with infinite-dimensional Lie symmetry. *J. Math. Anal. Appl.*, 253(1):126–141, 2001.

[53] R. Cherniha. New Q-conditional symmetries and exact solutions of some reaction-diffusion-convection equations arising in mathematical biology. *J. Math. Anal. Appl.*, 326(2):783–799, 2007.

[54] R. Cherniha. Conditional symmetries for systems of PDEs: new definitions and their application for reaction-diffusion systems. *J. Phys. A*, 43(40):405207, 13, 2010.

[55] R. Cherniha and N. Cherniha. Exact solutions of a class of nonlinear boundary value problems with moving boundaries. *J. Phys. A*, 26(18):L935–L940, 1993.

[56] R. Cherniha and V. Davydovych. Conditional symmetries and exact solutions of the diffusive Lotka–Volterra system. *Math. Comput. Modelling*, 54(5-6):1238–1251, 2011.

[57] R. Cherniha and V. Davydovych. Lie and conditional symmetries of the three-component diffusive Lotka–Volterra system. *J. Phys. A*, 46(18):185204, 14, 2013.

[58] R. Cherniha and V. Davydovych. Reaction-diffusion systems with constant diffusivities: conditional symmetries and form-preserving transformations. In *Algebra, geometry and mathematical physics*, volume 85 of *Springer Proc. Math. Stat.*, pages 533–553. Springer, Heidelberg, 2014.

[59] R. Cherniha and V. Davydovych. Nonlinear reaction-diffusion systems with a non-constant diffusivity: conditional symmetries in no-go case. *Appl. Math. Comput.*, 268:23–34, 2015.

[60] R. Cherniha, V. Davydovych, and L. Muzyka. Lie symmetries of the Shigesada–Kawasaki–Teramoto system. *Commun. Nonlinear Sci. Numer. Simul.*, 45:81–92, 2017.

[61] R. Cherniha and V. Dutka. Exact and numerical solutions of the generalized Fisher equation. *Rep. Math. Phys.*, 47(3):393–411, 2001.

[62] R. Cherniha and J. D. Fehribach. New exact solutions for a free boundary system. *J. Phys. A*, 31(16):3815–3829, 1998.

[63] R. Cherniha and M. Henkel. On non-linear partial differential equations with an infinite-dimensional conditional symmetry. *J. Math. Anal. Appl.*, 298(2):487–500, 2004.

[64] R. Cherniha and M. Henkel. The exotic conformal Galilei algebra and nonlinear partial differential equations. *J. Math. Anal. Appl.*, 369(1):120–132, 2010.

[65] R. Cherniha and J. R. King. Lie symmetries of nonlinear multidimensional reaction-diffusion systems. II. *J. Phys. A*, 36(2):405–425, 2003.

[66] R. Cherniha and J. R. King. Non-linear reaction-diffusion systems with variable diffusivities: Lie symmetries, ansätze and exact solutions. *J. Math. Anal. Appl.*, 308(1):11–35, 2005.

[67] R. Cherniha and J. R. King. Lie symmetries and conservation laws of non-linear multidimensional reaction-diffusion systems with variable diffusivities. *IMA J. Appl. Math.*, 71(3):391–408, 2006.

[68] R. Cherniha and J. R. King. Lie and conditional symmetries of a class of nonlinear (1 + 2)–dimensional boundary value problems. *Symmetry*, 7(3):1410–1435, 2015.

[69] R. Cherniha, J. R. King, and S. Kovalenko. Lie symmetry properties of nonlinear reaction-diffusion equations with gradient-dependent diffusivity. *Commun. Nonlinear Sci. Numer. Simul.*, 36:98–108, 2016.

[70] R. Cherniha and S. Kovalenko. Lie symmetries and reductions of multi-dimensional boundary value problems of the Stefan type. *J. Phys. A*, 44(48):485202, 25, 2011.

[71] R. Cherniha and S. Kovalenko. Lie symmetries of nonlinear boundary value problems. *Commun. Nonlinear Sci. Numer. Simul.*, 17(1):71–84, 2012.

[72] R. Cherniha and L. Myroniuk. New exact solutions of a nonlinear cross-diffusion system. *J. Phys. A*, 41(39):395204, 15, 2008.

[73] R. Cherniha and I. H. Odnorozhenko. Exact solutions of a nonlinear boundary problem of the metal melting and evaporation under a powerful flux of energy. *Dopovidi Akad. Nauk Ukrain. Ser. A*, 12:44–47, 1990 (in Ukrainian).

[74] R. Cherniha and I. H. Odnorozhenko. Studies of the processes of melting and evaporation of metals under the action of laser radiation pulses. *Promyshlennaya Teplotekhnika*, 13:51–59, 1991 (in Russian).

[75] R. Cherniha and O. Pliukhin. New conditional symmetries and exact solutions of nonlinear reaction-diffusion-convection equations. *J. Phys. A*, 40(33):10049–10070, 2007.

[76] R. Cherniha and O. Pliukhin. New conditional symmetries and exact solutions of nonlinear reaction-diffusion-convection equations. III. *arXiv:0902.2290*, 2009.

[77] R. Cherniha and O. Pliukhin. Nonlinear evolution equations with exponential nonlinearities: conditional symmetries and exact solutions. In *Algebra, geometry and mathematical physics*, volume 93 of *Banach Center Publ.*, pages 105–115. Polish Acad. Sci. Inst. Math., Warsaw, 2011.

References

[78] R. Cherniha and O. Pliukhin. New conditional symmetries and exact solutions of reaction-diffusion-convection equations with exponential nonlinearities. *J. Math. Anal. Appl.*, 403(1):23–37, 2013.

[79] R. Cherniha and M. Serov. Lie and non-Lie symmetries of nonlinear diffusion equations with convection term. In *Symmetry in nonlinear mathematical physics, Vol. 1, 2 (Kyiv, 1997)*, pages 444–449. Natl. Acad. Sci. Ukraine, Inst. Math., Kyiv, 1997.

[80] R. Cherniha and M. Serov. Symmetries, ansätze and exact solutions of nonlinear second-order evolution equations with convection terms. *European J. Appl. Math.*, 9(5):527–542, 1998.

[81] R. Cherniha and M. Serov. Nonlinear systems of the Burgers-type equations: Lie and Q-conditional symmetries, ansätze and solutions. *J. Math. Anal. Appl.*, 282(1):305–328, 2003.

[82] R. Cherniha and M. Serov. Symmetries, ansätze and exact solutions of nonlinear second-order evolution equations with convection terms. II. *European J. Appl. Math.*, 17(5):597–605, 2006.

[83] R. Cherniha, M. Serov, and I. Rassokha. Lie symmetries and form-preserving transformations of reaction-diffusion-convection equations. *J. Math. Anal. Appl.*, 342(2):1363–1379, 2008.

[84] R. Cherniha and J. Waniewski. Exact solutions of a mathematical model for fluid transport in peritoneal dialysis. *Ukrainian Math. J.*, 57(8):1316–1324, 2005.

[85] A. F. Cheviakov. Gem software package for computation of symmetries and conservation laws of differential equations. *Computer Physics Communications*, 176(1):48–61, 2007.

[86] P. A. Clarkson. New similarity reductions and Painlevé analysis for the symmetric regularised long wave and modified Benjamin-Bona-Mahoney equations. *J. Phys. A*, 22(18):3821–3848, 1989.

[87] P. A. Clarkson and M. D. Kruskal. New similarity reductions of the Boussinesq equation. *J. Math. Phys.*, 30(10):2201–2213, 1989.

[88] P. A. Clarkson and E. L. Mansfield. Symmetry reductions and exact solutions of a class of nonlinear heat equations. *Phys. D*, 70(3):250–288, 1994.

[89] J. D. Cole. On a quasi-linear parabolic equation occurring in aerodynamics. *Quart. Appl. Math.*, 9:225–236, 1951.

[90] T. Colin and P. Fabrie. A free boundary problem modeling a foam drainage. *Math. Models Methods Appl. Sci.*, 10(6):945–961, 2000.

[91] J. Crank. *The mathematics of diffusion.* Oxford University Press, 1979.

[92] V. G. Danilov, V. P. Maslov, and K. A. Volosov. *The Flow Around a Flat Plate*, pages 254–294. Springer, Dordrecht, 1995.

[93] R. K. Dodd, J. C. Eilbeck, J. D. Gibbon, and H. C. Morris. *Solitons and nonlinear wave equations.* Academic Press, Inc. [Harcourt Brace Jovanovich, Publishers], London-New York, 1982.

[94] V. A. Dorodnitsyn. Invariant solutions of the nonlinear heat equation with a source. *Zh. Vychisl. Mat. i Mat. Fiz.*, 22(6):1393–1400, 1533, 1982 (in Russian).

[95] V. A. Dorodnitsyn, I. V. Knyazeva, and S. R. Svirshchevskii. Group properties of the nonlinear heat equation with source in the two- and three-dimensional cases. *Differential'niye Uravneniya*, 19:1215–1223, 1983 (in Russian).

[96] L. Edelstein-Keshet. *Mathematical Models in Biology*, volume 46 of *Classics in Applied Mathematics*. Society for Industrial and Applied Mathematics, 2005.

[97] P. Fife. *Mathematical aspects of reacting and diffusing systems*, volume 28 of *Lecture notes in biomathematics*. Springer, Berlin-Heidelberg-New York, 1979.

[98] R. A. Fisher. The wave of advance of advantageous genes. *Ann. Eugenics*, 7:353–369, 1937.

[99] R. A. Fisher. *The genetical theory of natural selection.* Oxford University Press, Oxford, variorum edition, 1999. Revised reprint of the 1930 original, Edited, with a foreword and notes, by J. H. Bennett.

[100] R. Fitzhugh. Impulses and physiological states in theoretical models of nerve membrane. *Biophysical Journal*, 1:445–466, 1961.

[101] A. S. Fokas and Q. M. Liu. Generalized conditional symmetries and exact solutions of non-integrable equations. *Teoret. Mat. Fiz.*, 99(2):263–277, 1994.

[102] A. S. Fokas and Y. C. Yortsos. On the exactly solvable equation $S_t = [(\beta S + \gamma)^{-2} S_x]_x + \alpha(\beta S + \gamma)^{-2} S_x$ occurring in two-phase flow in porous media. *SIAM J. Appl. Math.*, 42(2):318–332, 1982.

[103] A. R. Forsyth. *Theory of differential equations. 5, 6. Partial differential equations.* Six volumes bound as three. Dover Publications, Inc., New York, 1959.

[104] H. Fujita. The exact pattern of a concentration-dependent diffusion in a semi-infinite medium, part I. *Textile Research Journal*, 22(11):757–760, 1952.

[105] H. Fujita. The exact pattern of a concentration-dependent diffusion in a semi-infinite medium, part II. *Textile Research Journal*, 22(12):823–827, 1952.

[106] H. Fujita. The exact pattern of a concentration-dependent diffusion in a semi-infinite medium, part III. *Textile Research Journal*, 24(3):234–240, 1954.

[107] W. I. Fushchych and R. M. Cherniha. The Galilean relativistic principle and nonlinear partial differential equations. *J. Phys. A*, 18(18):3491–3503, 1985.

[108] W. I. Fushchych and R. M. Cherniha. Galilei invariant non-linear equations of Schrödinger type and their exact solutions. I. *Ukrainian Math. J.*, 41(10):1161–1167, 1989.

[109] W. I. Fushchych and R. M. Cherniha. Galilei invariant nonlinear equations of Schrödinger type and their exact solutions. II. *Ukrainian Math. J.*, 41(12):1456–1463, 1989.

[110] W. I. Fushchych and R. M. Cherniha. Galilei-invariant nonlinear systems of evolution equations. *J. Phys. A*, 28(19):5569–5579, 1995.

[111] W. I. Fushchych, M. I. Serov, and T. K. Amerov. Conditional invariance of nonlinear heat equation. *Dopovidi Akad. Nauk Ukrain. Ser. A*, (11):16–18, 1990 (in Ukrainian).

[112] W. I. Fushchych, M. I. Serov, and V. I. Chopyk. Conditional invariance and nonlinear heat equations. *Dopovidi Akad. Nauk Ukrain. Ser. A*, (9):17–21, 86, 1988 (in Ukrainian).

[113] W. I. Fushchych, M. I. Serov, and L. A. Tulupova. The conditional invariance and exact solutions of the nonlinear diffusion equation. *Dopovidi Akad. Nauk Ukrain.*, (4):37–40, 1993.

[114] W. I. Fushchych, W. M. Shtelen, and M. I. Serov. *Symmetry analysis and exact solutions of equations of nonlinear mathematical physics*, volume 246 of *Mathematics and Its Applications*. Kluwer Academic Publishers Group, Dordrecht, 1993.

[115] W. I. Fushchych, W. M. Shtelen, M. I. Serov, and R. O. Popovych. Q-conditional symmetry of the linear heat equation. *Dopovidi Akad. Nauk Ukrain.*, (12):28–33, 170, 1992 (in Ukrainian).

[116] W. I. Fushchych and I. M. Tsyfra. On a reduction and solutions of nonlinear wave equations with broken symmetry. *J. Phys. A*, 20(2):L45–L48, 1987.

[117] W. I. Fushchych and R. Z. Zhdanov. Antireduction and exact solutions of nonlinear heat equations. *J. Nonlinear Math. Phys.*, 1(1):60–64, 1994.

[118] V. A. Galaktionov. Invariant subspaces and new explicit solutions to evolution equations with quadratic nonlinearities. *Proc. Roy. Soc. Edinburgh Sect. A*, 125(2):225–246, 1995.

[119] V. A. Galaktionov and S. R. Svirshchevskii. *Exact solutions and invariant subspaces of nonlinear partial differential equations in mechanics and physics*. Chapman & Hall/CRC Applied Mathematics and Nonlinear Science Series. Chapman & Hall/CRC, Boca Raton, FL, 2007.

[120] M. L. Gandarias. New symmetries for a model of fast diffusion. *Phys. Lett. A*, 286(2-3):153–160, 2001.

[121] M. L. Gandarias, J. L. Romero, and J. M. Díaz. Nonclassical symmetry reductions of a porous medium equation with convection. *J. Phys. A*, 32(8):1461–1473, 1999.

[122] J.-P. Gazeau and P. Winternitz. Symmetries of variable coefficient Korteweg–de Vries equations. *J. Math. Phys.*, 33(12):4087–4102, 1992.

[123] B. H. Gilding and R. Kersner. *Travelling waves in nonlinear diffusion-convection reaction*, volume 60 of *Progress in Nonlinear Differential Equations and Their Applications*. Birkhäuser, Basel, 2004.

[124] J. A. Goff. Transformations leaving invariant the heat equation of physics. *Amer. J. Math.*, 49(1):117–122, 1927.

[125] T. R. Goodman. Application of integral methods to transient nonlinear heat transfer. *Advances in Heat Transfer*, 1:51–122, 1964.

[126] M. S. Hashemi and M. C. Nucci. Nonclassical symmetries for a class of reaction-diffusion equations: the method of heir-equations. *J. Nonlinear Math. Phys.*, 20(1):44–60, 2013.

[127] M. Henkel. *Conformal invariance and critical phenomena*. Texts and Monographs in Physics. Springer, Berlin, 1999.

[128] M. Henkel and J. Unterberger. Schrödinger invariance and spacetime symmetries. *Nuclear Phys. B*, 660(3):407–435, 2003.

[129] D. Henry. *Geometric theory of semilinear parabolic equations*, volume 840 of *Lecture Notes in Mathematics*. Springer, Berlin-New York, 1981.

[130] W. Hereman. Review of symbolic software for Lie symmetry analysis. *Math. Comput. Modelling*, 25(8):115–132, 1997.

[131] J. M. Hill. Similarity solutions for nonlinear diffusion – a new integration procedure. *Journal of Engineering Mathematics*, 23(2):141–155, 1989.

[132] J. M. Hill, A. J. Avagliano, and M. P. Edwards. Some exact results for nonlinear diffusion with absorption. *IMA J. Appl. Math.*, 48(3):283–304, 1992.

[133] A. L. Hodgkin and A. F. Huxley. A quantitative description of membrane current and its application to conduction and excitation in nerve. *The Journal of Physiology*, 117(4):500–544, 1952.

[134] E. Hopf. The partial differential equation $u_t + uu_x = \mu u_{xx}$. *Comm. Pure Appl. Math.*, 3:201–230, 1950.

[135] P. E. Hydon. *Difference equations by differential equation methods*, volume 27 of *Cambridge Monographs on Applied and Computational Mathematics*. Cambridge University Press, Cambridge, 2014.

[136] N. H. Ibragimov, A. V. Aksenov, V. A. Baikov, V. A. Chugunov, R. K. Gazizov, and A. G. Meshkov. *CRC handbook of Lie group analysis of differential equations. Vol. 2*. CRC Press, Boca Raton, FL, 1995. Applications in engineering and physical sciences.

[137] N. H. Ibragimov and S. V. Meleshko. Linearization of third-order ordinary differential equations by point and contact transformations. *J. Math. Anal. Appl.*, 308(1):266–289, 2005.

[138] N. H. Ibragimov, M. Torrisi, and A. Valenti. Preliminary group classification of equations $v_{tt} = f(x, v_x)v_{xx} + g(x, v_x)$. *J. Math. Phys.*, 32(11):2988–2995, 1991.

[139] M. Janet. Sur les systèmes d'équations aux dérivées partielles. *C. R. Acad. Sci., Paris*, 170:1101–1103, 1920 (in French).

[140] L. Ji. The method of linear determining equations to evolution system and application for reaction-diffusion system with power diffusivities. *Symmetry*, 8(12):Art. 157, 21, 2016.

[141] L. Ji and C. Qu. Conditional Lie-Bäcklund symmetries and invariant subspaces to nonlinear diffusion equations with convection and source. *Stud. Appl. Math.*, 131(3):266–301, 2013.

[142] E. Kamke. *Differentialgleichungen*. B. G. Teubner, Stuttgart, 1977 (in German). Lösungsmethoden und Lösungen. I: Gewöhnliche Differentialgleichungen, Neunte Auflage, Mit einem Vorwort von Detlef Kamke.

[143] O. V. Kaptsov. Determining equations and differential constraints. *J. Nonlinear Math. Phys.*, 2(3-4):283–291, 1995.

[144] O. V. Kaptsov and I. V. Verevkin. Differential constraints and exact solutions of nonlinear diffusion equations. *J. Phys. A*, 36(5):1401, 2003.

[145] V. L. Katkov. A class of exact solutions of the equation for the forecast of the geopotential. *Izv. Acad. Sci. USSR Atmospher. Ocean. Phys.*, 1:630–631, 1965 (in Russian).

[146] T. Kawahara and M. Tanaka. Interactions of traveling fronts: an exact solution of a nonlinear diffusion equation. *Phys. Lett. A*, 97(8):311–314, 1983.

[147] R. Kersner. Some properties of generalized solutions of quasilinear degenerate parabolic equations. *Acta Math. Acad. Sci. Hungar.*, 32(3-4):301–330, 1978.

[148] J. R. King. Exact results for the nonlinear diffusion equations $\partial u/\partial t = (\partial/\partial x)(u^{-4/3}\partial u/\partial x)$ and $\partial u/\partial t = (\partial/\partial x)(u^{-2/3}\partial u/\partial x)$. *J. Phys. A*, 24(24):5721–5745, 1991.

[149] J. R. King. Some non-self-similar solutions to a nonlinear diffusion equation. *J. Phys. A*, 25(18):4861–4868, 1992.

[150] J. R. King. Exact multidimensional solutions to some nonlinear diffusion equations. *The Quarterly Journal of Mechanics and Applied Mathematics*, 46(3):419, 1993.

[151] J. R. King. Exact polynomial solutions to some nonlinear diffusion equations. *Phys. D*, 64(1-3):35–65, 1993.

[152] J. G. Kingston. On point transformations of evolution equations. *J. Phys. A*, 24(14):L769–L774, 1991.

[153] J. G. Kingston and C. Sophocleous. On form-preserving point transformations of partial differential equations. *J. Phys. A*, 31(6):1597–1619, 1998.

[154] G. Kirchhoff. *Vorlesungen über mathematische Physik. Bd. 4. Theorie der Wärme.* Teubner, Leipzig, 1894 (in German).

[155] I. V. Knyazeva and M. D. Popov. A system of two diffusion equations. *CRC handbook of Lie group analysis of differential equations,* Editor N.H. Ibragimov, Boca Raton, CRC Press, 1:171–176, 1994.

[156] A. Kolmogorov, I. Petrovskii, and N. Piscunov. Etude de l'èquations de la diffusion avec croissance de la quantité de matière et son application à un problème biologique. *Moscow University Bull. Math,* 1(6):1–25, 1937 (in French).

[157] G. A. Korn and T. M. Korn. *Mathematical handbook for scientists and engineers.* Second, enlarged and revised edition. McGraw-Hill Book Co., New York-Toronto, Ont.-London, 1968.

[158] L. Kozdoba. *Methods for solving nonlinear heat conduction problems.* Nauka, Moscow, 1975 (in Russian).

[159] I. S. Krasil'shchik and A. M. Vinogradov. Nonlocal trends in the geometry of differential equations: symmetries, conservation laws, and Bäcklund transformations. *Acta Appl. Math.*, 15(1-2):161–209, 1989.

[160] Y. Kuang, J. D. Nagy, and S. E. Eikenberry. *Introduction to mathematical oncology*. Chapman & Hall/CRC Mathematical and Computational Biology Series. CRC Press, Boca Raton, FL, 2016.

[161] N. A. Kudryashov. Comment on: "A novel approach for solving the Fisher equation using Exp-function method" [Phys. Lett. A 372 (2008) 3836] [mr2418599]. *Phys. Lett. A*, 373(12-13):1196–1197, 2009.

[162] A. A. Lacey, J. R. Ockendon, and A. B. Tayler. "Waiting-time" solutions of a nonlinear diffusion equation. *SIAM J. Appl. Math.*, 42(6):1252–1264, 1982.

[163] V. I. Lahno, S. V. Spichak, and V. I. Stognii. *Symmetry analysis of evolution type equations*. Institute of Mathematics of NAS of Ukraine, Kyiv, 2002 (in Ukrainian).

[164] D. Levi and P. Winternitz. Nonclassical symmetry reduction: example of the Boussinesq equation. *J. Phys. A*, 22(15):2915–2924, 1989.

[165] J.-M. Levi-Leblond. Galilei group and Galilei invariance. *Group theory and its applications. II, ed. E.M. Loebl*, (15):221–299, 1971.

[166] S. Lie. Über die Integration durch bestimmte Integrale von einer Klasse lineare partiellen Differentialgleichungen. *Arch. Math.*, 6(3):328–368, 1881 (in German).

[167] S. Lie. Algemeine Untersuchungen über Differentialgleichungen, die eine continuirliche endliche Gruppe gestatten. *Math. Annalen*, 25, 1885 (in German).

[168] S. Lie. *Gesammelte Abhandlungen. Band 3*. Johnson Reprint Corp., New York-London, German edition, 1973. Herausgegeben von Friedrich Engel und Poul Heegaard, Abhandlungen zur Theorie der Differentialgleichungen. Erste Abteilung, Herausgegeben von Friedrich Engel (in German).

[169] S. Lie. *Gesammelte Abhandlungen. Band 4*. Johnson Reprint Corp., New York-London, German edition, 1973. Herausgegeben von Friedrich Engel und Poul Heegaard, Abhandlungen zur Theorie der Differentialgleichungen. Abteilung 2, Herausgegeben von Friedrich Engel (in German).

[170] S. Lie. *Gesammelte Abhandlungen. Band 5*. Johnson Reprint Corp., New York-London, 1973 (in German). Herausgegeben von Friedrich Engel und Poul Heegaard, Abhandlungen über die Theorie der Transformationsgruppen. Abteilung 1, Herausgegeben von Friedrich Engel, Reprint of the 1924 edition.

[171] S. Lie. *Gesammelte Abhandlungen. Band 6.* Johnson Reprint Corp., New York-London, 1973 (in German). Herausgegeben von Friedrich Engel und Poul Heegaard, Abhandlungen über die Theorie der Transformationsgruppen. Abteilung 2, Herausgegeben von Friedrich Engel, Reprint of the 1927 edition.

[172] S. Lie. *Theory of transformation groups. I.* Springer, Heidelberg, 2015. General properties of continuous transformation groups, A contemporary approach and translation, With the collaboration of Friedrich Engel, Edited and translated from the German and with a foreword by Joël Merker.

[173] Q. M. Liu and A. S. Fokas. Exact interaction of solitary waves for certain nonintegrable equations. *J. Math. Phys.*, 37(1):324–345, 1996.

[174] W.-X. Ma, T. Huang, and Y. Zhang. A multiple exp-function method for nonlinear differential equations and its application. *Physica Scripta*, 82(6):065003, 2010.

[175] W. Malfliet. The tanh method: a tool for solving certain classes of nonlinear evolution and wave equations. *Journal of Computational and Applied Mathematics*, 164:529 – 541, 2004.

[176] E. L. Mansfield. The nonclassical group analysis of the heat equation. *J. Math. Anal. Appl.*, 231(2):526–542, 1999.

[177] A. V. Mikhailov, A. B. Shabat, and R. I. Yamilov. The symmetry approach to the classification of non-linear equations. Complete lists of integrable systems. *Russian Mathematical Surveys*, 42(4):1–64, 1987.

[178] S. Murata. Non-classical symmetry and Riemann invariants. *Internat. J. Non-Linear Mech.*, 41(2):242–246, 2006.

[179] J. D. Murray. *Nonlinear differential equation models in biology.* Clarendon Press, Oxford, 1977.

[180] J. D. Murray. *Mathematical biology*, volume 19 of *Biomathematics*. Springer, Berlin, 1989.

[181] J. D. Murray. *Mathematical biology. I*, volume 17 of *Interdisciplinary Applied Mathematics*. Springer, New York, third edition, 2002. An introduction.

[182] J. D. Murray. *Mathematical biology. II*, volume 18 of *Interdisciplinary Applied Mathematics*. Springer, New York, third edition, 2003. Spatial models and biomedical applications.

[183] L. Myroniuk and R. Cherniha. Reduction and solutions of a class of nonlinear reaction-diffusion systems with the power nonliniarities. *Proc. Inst. Mathematics of NAS of Ukraine*, 3:217–224, 2006 (in Ukrainian).

[184] T. Nagatani, H. Emmerich, and K. Nakanishi. Burgers equation for kinetic clustering in traffic flow. *Physica A: Statistical Mechanics and Its Applications*, 255:158–162, 1998.

[185] J. Nagumo, S. Arimoto, and S. Yoshizawa. An active pulse transmission line simulating nerve axon. *Proceeding IRE*, 50:2061–2070, 1962.

[186] G. Nariboli. Self-similar solutions of some nonlinear equations. *Appl. Scientific Res.*, 22:449–461, 1970.

[187] A. C. Newell and J. A. Whitehead. Finite bandwidth, finite amplitude convection. *Journal of Fluid Mechanics*, 38:279–303, 1969.

[188] U. Niederer. The maximal kinematical invariance group of the free Schrödinger equation. *Helv. Phys. Acta*, 45(5):802–810, 1972/73.

[189] U. Niederer. Schrödinger invariant generalized heat equations. *Helv. Phys. Acta*, 51(2):220–239, 1978.

[190] E. Noether. Invariante variationsprobleme. *Nachrichten von der Gesellschaft der Wissenschaften zu Göttingen, Mathematisch-Physikalische Klasse*, 1918:235–257, 1918 (in German).

[191] M. C. Nucci. Symmetries of linear, C-integrable, S-integrable, and non-integrable equations. In *Nonlinear evolution equations and dynamical systems (Baia Verde, 1991)*, pages 374–381. World Sci. Publ., River Edge, NJ, 1992.

[192] M. C. Nucci. Iterations of the non-classical symmetries method and conditional Lie-Bäcklund symmetries. *J. Phys. A*, 29(24):8117–8122, 1996.

[193] M. C. Nucci and P. A. Clarkson. The nonclassical method is more general than the direct method for symmetry reductions. An example of the FitzHugh-Nagumo equation. *Phys. Lett. A*, 164(1):49–56, 1992.

[194] A. Okubo and S. A. Levin. *Diffusion and ecological problems: modern perspectives*, volume 14 of *Interdisciplinary Applied Mathematics*. Springer, New York, second edition, 2001.

[195] F. Oliveira-Pinto and B. W. Conolly. *Applicable mathematics of nonphysical phenomena*. Ellis Horwood Ltd., Chichester; Halsted Press [John Wiley & Sons, Inc.], New York, 1982. Including reprints of papers by V. Volterra, A. Kolmogorov [A. N. Kolmogorov], W. Feller, G. W. Hardy, I. Petrovsky [I. G. Petrovskiĭ], N. Piskounov [N. S. Piskunov], A. G. McKendrick, W. O. Kermack and F. W. Lanchester, Ellis Horwood Series in Mathematics and its Applications.

[196] P. J. Olver. *Applications of Lie groups to differential equations*, volume 107 of *Graduate Texts in Mathematics*. Springer, New York, second edition, 1993.

[197] P. J. Olver. Direct reduction and differential constraints. *Proc. Roy. Soc. London Ser. A*, 444(1922):509–523, 1994.

[198] P. J. Olver and P. Rosenau. Group-invariant solutions of differential equations. *SIAM J. Appl. Math.*, 47(2):263–278, 1987.

[199] P. J. Olver and E. M. Vorob'ev. Nonclassical and conditional symmetries. *CRC handbook of Lie group analysis of differential equations*, Editor N.H. Ibragimov, Boca Raton, CRC Press, 3:291–328, 1996.

[200] A. Oron and P. Rosenau. Some symmetries of the nonlinear heat and wave equations. *Phys. Lett. A*, 118(4):172–176, 1986.

[201] L. V. Ovsiannikov. Group relations of the equation of non-linear heat conductivity. *Dokl. Akad. Nauk SSSR*, 125:492–495, 1959 (in Russian).

[202] L. V. Ovsiannikov. *The group properties of differential equations*. Siberian branch of AN SSSR, Novosibirsk, 1962 (in Russian).

[203] L. V. Ovsiannikov. *Group analysis of differential equations*. Academic Press, Inc. [Harcourt Brace Jovanovich, Publishers], New York-London, 1982. Translated from the Russian by Y. Chapovsky, Translation edited by William F. Ames.

[204] J. Patera and P. Winternitz. Subalgebras of real three- and four-dimensional Lie algebras. *J. Math. Phys.*, 18(7):1449–1455, 1977.

[205] J. Patera, P. Winternitz, and H. Zassenhaus. Continuous subgroups of the fundamental groups of physics. I. General method and the Poincaré group. *J. Math. Phys.*, 16:1597–1614, 1975.

[206] J. Patera, P. Winternitz, and H. Zassenhaus. Continuous subgroups of the fundamental groups of physics. II. The similitude group. *J. Math. Phys.*, 16:1615–1624, 1975.

[207] R. E. Pattle. Diffusion from an instantaneous point source with a concentration-dependent coefficient. *Quart. J. Mech. Appl. Math.*, 12:407–409, 1959.

[208] J. R. Philip. General method of exact solution of the concentration-dependent diffusion equation. *Austral. J. Phys.*, 13:1–12, 1960.

[209] J. R. Philip. Flow in porous media. *Annual Review of Fluid Mechanics*, 2(1):177–204, 1970.

[210] O. Pliukhin. Q-conditional symmetries and exact solutions of nonlinear reaction-diffusion systems. *Symmetry*, 7(4):1841–1855, 2015.

[211] O. A. Pocheketa and R. O. Popovych. Reduction operators and exact solutions of generalized Burgers equations. *Phys. Lett. A*, 376(45):2847–2850, 2012.

[212] A. D. Polyanin and V. F. Zaitsev. *Handbook of exact solutions for ordinary differential equations*. Chapman & Hall/CRC, Boca Raton, FL, second edition, 2003.

[213] A. D. Polyanin and V. F. Zaitsev. *Handbook of nonlinear partial differential equations*. Chapman & Hall/CRC, Boca Raton, FL, 2004.

[214] J.-F. Pommaret. *Systems of partial differential equations and Lie pseudogroups*, volume 14 of *Mathematics and Its Applications*. Gordon & Breach Science Publishers, New York, 1978.

[215] R. O. Popovych and N. M. Ivanova. New results on group classification of nonlinear diffusion-convection equations. *J. Phys. A*, 37(30):7547–7565, 2004.

[216] R. O. Popovych, O. O. Vaneeva, and N. M. Ivanova. Potential nonclassical symmetries and solutions of fast diffusion equation. *Phys. Lett. A*, 362(2-3):166–173, 2007.

[217] E. Pucci. Similarity reductions of partial differential equations. *J. Phys. A*, 25(9):2631–2640, 1992.

[218] E. Pucci and G. Saccomandi. On the weak symmetry groups of partial differential equations. *J. Math. Anal. Appl.*, 163(2):588–598, 1992.

[219] E. Pucci and G. Saccomandi. Evolution equations, invariant surface conditions and functional separation of variables. *Phys. D*, 139(1-2):28–47, 2000.

[220] C. Qu. Group classification and generalized conditional symmetry reduction of the nonlinear diffusion-convection equation with a nonlinear source. *Stud. Appl. Math.*, 99(2):107–136, 1997.

[221] C. Qu. Exact solutions to nonlinear diffusion equations obtained by a generalized conditional symmetry method. *IMA J. Appl. Math.*, 62(3):283–302, 1999.

[222] G. Rideau and P. Winternitz. Evolution equations invariant under two-dimensional space-time Schrödinger group. *J. Math. Phys.*, 34(2):558–570, 1993.

[223] G. Rosen. Nonlinear heat conduction in solid h_2. *Phys. Rev. B*, 19:2398–2399, 1979.

[224] G. Saccomandi. A personal overview on the reduction methods for partial differential equations. *Note Mat.*, 23(2):217–248, 2004/05.

[225] A. A. Samarskii, V. A. Galaktionov, S. P. Kurdyumov, and A. P. Mikhailov. *Blow-up in quasilinear parabolic equations*, volume 19 of *De Gruyter Expositions in Mathematics*. Walter de Gruyter & Co., Berlin, 1995.

[226] M. I. Serov. Conditional invariance and exact solutions of the nonlinear equation. *Ukrainian Math. J.*, 42(10):1216–1222, 1990.

[227] M. I. Serov. Conditional and nonlocal symmetry of nonlinear heat equation. *J. Nonlinear Math. Phys.*, 3(1-2):63–67, 1996.

[228] M. I. Serov and R. M. Cherniha. Lie symmetries and exact solutions of nonlinear equations of heat conductivity with convection term. *Ukrainian Math. J.*, 49(9):1423–1433, 1997.

[229] M. I. Serov, L. O. Tulupova, and N. V. Andreeva. Q-conditional symmetry of a nonlinear two-dimensional heat-conduction equation. *Ukrainian Math. J.*, 52(6):969–973, 2000.

[230] N. Shigesada, K. Kawasaki, and E. Teramoto. Spatial segregation of interacting species. *J. Theoret. Biol.*, 79(1):83–99, 1979.

[231] R. Shonkwiler and J. Herod. *Mathematical Biology: An Introduction with Maple and MATLAB®*. Undergraduate Texts in Mathematics. Springer, New York, 2009.

[232] A. F. Sidorov, V. P. Shapeev, and N. N. Yanenko. *Method of differential relations and its application to gas dynamics*. Nauka, Novosibirsk, 1984 (in Russian).

[233] J. D. Sokolov. Some particular solutions of the Boussinesq equation. *Ukrainian Math. J.*, 8:54–58, 1956 (in Russian).

[234] C. Sophocleous. Transformation properties of a variable-coefficient Burgers equation. *Chaos Solitons Fractals*, 20(5):1047–1057, 2004.

[235] T. L. Stepien, E. M. Rutter, and Y. Kuang. A data-motivated density-dependent diffusion model of in vitro glioblastoma growth. *Math. Biosci. Eng.*, 12(6):1157–1172, 2015.

[236] M. L. Storm. Heat conduction in simple metals. *Journal of Applied Physics*, 22(7):940–951, 1951.

[237] S. R. Svirshchevskii. Invariant linear spaces and exact solutions of nonlinear evolution equations. *J. Nonlinear Math. Phys.*, 3(1-2):164–169, 1996.

[238] S. Titov. Solutions of non-linear partial differential equations in the form of polynomials with respect to one of variables. *Chislennye Metody Mekhaniki Sploshnoy Sredy*, 8:144–149, 1977 (in Russian).

[239] B. Tuck. Some explicit solutions to the non-linear diffusion equation. *Journal of Physics D: Applied Physics*, 9(11):1559, 1976.

[240] A. M. Turing. The chemical basis of morphogenesis. *Philosophical Transactions of the Royal Society of London B: Biological Sciences*, 237(641):37–72, 1952.

[241] V. Tychynin. New nonlocal symmetries of diffusion-convection equations and their connection with generalized hodograph transformation. *Symmetry*, 7(4):1751–1767, 2015.

[242] V. Vanaja. Exact solutions of a nonlinear diffusion-convection equation. *Physica Scripta*, 80(4):045402, 2009.

[243] O. O. Vaneeva, R. O. Popovych, and C. Sophocleous. Enhanced group analysis and exact solutions of variable coefficient semilinear diffusion equations with a power source. *Acta Appl. Math.*, 106(1):1–46, 2009.

[244] A. I. Volpert, V. A. Volpert, and V. A. Volpert. *Traveling wave solutions of parabolic systems*, volume 140 of *Translations of Mathematical Monographs*. American Mathematical Society, Providence, RI, 1994.

[245] J. Waniewski. *Theoretical foundations for modeling of mebrane transport in medicine and biomedical engineering*. Institute of Computer Science, PAS: Warsaw, 2015.

[246] A.-M. Wazwaz. The extended tanh method for the Zakharov-Kuznetsov (ZK) equation, the modified ZK equation, and its generalized forms. *Commun. Nonlinear Sci. Numer. Simul.*, 13(6):1039–1047, 2008.

[247] G. M. Webb. Painlevé analysis of a coupled Burgers–heat-equation system, and nonclassical similarity solutions of the heat equation. *Phys. D*, 41(2):208–218, 1990.

[248] T. Wolf. Crack, LiePDE, ApplySym and ConLaw, section 4.3.5. *Foundations, Applications, Systems, Computer Algebra Handbook, J. Grabmeier, E. Kaltofen, V. Weispfenning (Eds.)*, pages 465–468, 2002.

[249] N. N. Yanenko. Compatibility theory and integration methods for systems of nonlinear partial differential equations. In *Proc. Fourth All-Union Math. Congr. (Leningrad, 1961), Vol. II*, pages 247–252. Nauka, Leningrad, 1964 (in Russian).

[250] Y. B. Zeldovich and G. I. Barenblatt. Asymptotic properties of self-preserving solutions of equations of unsteady motion of gas through porous media. *Dokl. Akad. Nauk SSSR*, 118(4):671–674, 1958 (in Russian).

[251] R. Z. Zhdanov. Conditional Lie–Bäcklund symmetry and reduction of evolution equations. *J. Phys. A*, 28(13):3841–3850, 1995.

[252] R. Z. Zhdanov and V. I. Lahno. Conditional symmetry of a porous medium equation. *Phys. D*, 122(1-4):178–186, 1998.

[253] R. Z. Zhdanov, I. M. Tsyfra, and R. O. Popovych. A precise definition of reduction of partial differential equations. *J. Math. Anal. Appl.*, 238(1):101–123, 1999.

Index

Q-conditional symmetry, 4, 78–85, 87–93, 95, 96, 99, 103–105, 107, 110, 114, 117–119, 129, 131–133, 137, 142–145, 147, 153–155, 159, 166, 167, 180, 182, 183, 185, 188, 193

Q-conditional symmetry classification (QSC), 92, 101, 110, 112, 118, 141

Burgers equation, 4, 30, 32, 33, 36, 59, 64, 71–74, 93, 95, 131, 195

conditional symmetry, 4, 5, 74, 78, 82, 83, 87, 91, 92, 101, 191

differential consequence, 35, 44, 45, 82, 89, 97–99, 125, 126, 129

equivalence transformation (ET), 22, 23, 26, 27, 29–33, 35, 47, 49, 52, 55, 57–60, 69, 74, 92, 93, 95, 101, 112, 131, 144, 178, 184

evolution equation, 7, 78, 81, 82, 137, 192, 196

exact solution, 3, 4, 6–9, 11, 15–17, 71, 72, 91, 124, 132, 135–144, 146, 147, 149–152, 154, 155, 157–161, 163, 166, 167, 170–172, 176–182, 184–190

fast diffusion, 87, 112, 131, 136, 142, 172

Fitzhugh–Nagumo (FN) equation, 132, 146, 154, 155, 174

form-preserving transformation (FPT), 29, 30, 32–37, 60, 64–68, 74, 92, 93, 95, 101, 112, 118, 156

general solution, 14, 25, 38, 41–47, 53, 56, 57, 65, 67, 83–86, 93–96, 100, 102, 105, 107, 109–111, 114, 119–123, 125, 127, 128, 138, 146–150, 158, 160, 169–172, 174, 175, 181–184, 186, 192, 195, 202, 205, 206

generalized conditional symmetry, 5, 78, 137

generalized Fitzhugh–Nagumo (FN) equation, 133, 148, 152, 154, 155, 172

heat equation, 4, 17, 20, 21, 23, 26, 30, 35, 36, 38, 49, 52, 58, 59, 64, 69, 70, 72–74, 77, 83, 84, 141, 201

Huxley equation, 132, 133, 154, 155

invariance criteria, 11, 16, 31, 79, 82, 88

Kolmogorov–Petrovskii–Piskunov (KPP) equation, 132, 146, 148, 155

Lie algebra, 10–13, 15, 17, 21–27, 29, 32, 39, 41, 45, 47, 49, 57–59, 64, 69, 72, 74, 75, 81, 131, 137, 138, 141, 147, 164, 176, 188

239

Lie ansatz, 14, 15, 136, 141, 151, 157, 166, 177
Lie solution, 15, 136, 137
Lie symmetry, 4, 7, 8, 11, 13–15, 20, 21, 26, 29, 30, 36, 37, 39, 49, 52, 60, 68–71, 73, 74, 77–79, 81–84, 86, 87, 90, 91, 95, 96, 100, 101, 104, 105, 109, 110, 114–116, 118, 131, 136–138, 141, 143, 144, 147, 176, 187, 188, 193, 201
Lie symmetry classification (LSC), 13, 21, 29, 60, 83, 92

manifold, 9, 12, 24, 37, 79, 80, 82, 88, 89
maximal algebra of invariance (MAI), 12–16, 20, 22, 26, 27, 39, 40, 49, 52–55, 57–60, 64, 66–68, 71, 72, 74, 139, 156, 160, 164, 176, 187
Murray equation, 74, 131, 144–146, 152, 155, 156, 161, 162, 189

non-Lie ansatz, 14, 136, 137, 141, 142, 145, 147, 168
non-Lie solution, 136, 137, 141–143, 146, 147, 151, 160, 161, 164, 170, 187, 188, 201
nonclassical symmetry, 4, 78

overdetermined system, 6, 13, 66, 68, 82, 94, 96, 99, 108, 112, 113, 117, 118, 124, 125, 148–151, 165, 181, 185, 191

plane wave solution, 15, 71, 136, 138, 144, 145, 147, 151, 152, 160, 161, 170, 178, 187, 188
population, 2, 3, 132, 143, 188–190
porous diffusion, 10, 156, 161
potential symmetry, 5, 78, 191
power-law diffusivity, 2, 13, 70, 131
principal algebra, 23, 25–27, 30, 55, 56, 138

reaction-diffusion (RD) equation, 7, 28, 30, 82, 99, 132, 176, 188, 211
reaction-diffusion-convection (RDC) equation, 19, 29, 30, 32, 33, 36–41, 43–47, 49, 55, 57, 58, 60, 64, 68, 69, 72–75, 78, 84, 87, 90–93, 95, 100–102, 105, 107, 110–112, 115, 116, 118, 119, 124, 125, 129, 132, 135–137, 143, 147, 149, 151, 156, 158–161, 163–166, 170–172, 176, 179, 180, 186–190, 196, 201
reduction, 7, 14–16, 30, 81, 140, 157, 177, 193–195, 198
reduction equation, 14–16, 138–142, 157, 158, 177–179

smooth function, 192, 194
system of determining equations (DEs), 25, 30, 36–39, 41, 78, 82, 84–86, 88–90, 92, 94, 95, 102, 103, 106, 108–113

traveling front, 15, 71, 136, 144, 145, 151, 152, 155, 174

weak symmetry, 5, 78